Molecular Marker-Assisted Improvement and Production of Medicinal Plants

This volume explores the fascinating world of plant-based bioactive compounds that have influenced healthcare systems across centuries. From ancient Ayurvedic and Chinese traditions to modern biotechnology, it reveals how nature's pharmacy continues to inspire the development of medicines, nutraceuticals, and cosmeceuticals. It highlights the incredible diversity of medicinal plants and their secondary metabolites, such as alkaloids, flavonoids, terpenoids, and glycosides, which have led to the discovery of many important therapeutic agents. The text emphasizes the growing importance of medicinal plants in the post-COVID era, reflecting the global shift towards sustainable and holistic health solutions. Blending traditional wisdom with advancements in genomics, metabolomics, molecular markers, and gene editing, it illustrates how biotechnology is reshaping plant-based drug discovery. This work serves as a valuable resource for researchers and professionals committed to uniting nature's legacy with modern scientific progress.

Key Features:

This book

- Provides a clear conceptual foundation of molecular markers and medicinal plants that smoothly transitions readers into advanced literature and technical discussions.
- Delivers expert-driven insights for industrial-scale production, standardization, and commercialization of plant-based bioactives.
- Explores innovative roles of molecular markers in tissue culture and in vitro propagation, unlocking scalable cultivation strategies.
- Maps future directions in molecular medicine, highlighting the transformative potential of medicinal plants in precision healthcare.

Exploring Medicinal Plants

Series Editor: Azamal Husen, Wolaita Sodo University, Ethiopia

Medicinal plants render a rich source of bioactive compounds used in drug formulation and development; they play a key role in traditional or indigenous health systems. As the demand for herbal medicines increases worldwide, supply is declining as most of the harvest is derived from naturally growing vegetation. Considering global interests and covering several important aspects associated with medicinal plants, the Exploring Medicinal Plants series comprises volumes valuable to academia, practitioners, and researchers interested in medicinal plants. Topics provide information on a range of subjects including diversity, conservation, propagation, cultivation, physiology, molecular biology, growth response under extreme environment, handling, storage, bioactive compounds, secondary metabolites, extraction, therapeutics, mode of action, and healthcare practices.

Led by Azamal Husen, PhD, this series is directed to a broad range of researchers and professionals consisting of topical books exploring information related to medicinal plants. It includes edited volumes, references, and textbooks available for individual print and electronic purchases.

Tissue Culture Techniques and Medicinal Plants: Enhancing Propagation and Production
Edited by Azamal Husen and Manu Pant

Genetic Improvement and Conservation Practices of Medicinal Plants
Edited by Manu Pant, Sayed Sartaj Sohrab and Azamal Husen

Sustainable Use of Plants and their Products in Neurodegenerative Diseases Management
Edited by Johnson O. Oladele, Ebenezer I. O. Ajayi, María L. Flores-López, and Oluwaseyi E. Okoro

Wild Edibles of Deserts: Food Security and Sustainability
Salman Majeed, Muhammad Zafar, Mushtaq Ahmad, Mohd Kafeel Ahmad Ansari, Shazia Sultana, and Andrea Pieroni

Impact of Climate Change on Medicinal and Herbal Plant micro-RNA
Kanchanlata Tungare, Parul Johri, Sachidanand Singh, and Surojeet Das

Exploring Medicinal Bamboos
Edited by Manu Pant and Azamal Husen

Exploring Herbal Synergies for Optimal Human Health
Edited by Art Devi, Nancy George, Zaved Ahmed Khan, and Vagish Dwibedi

Exploring Medicinal and Commercial Uses of Flowers
Edited by Rakesh Kumar Bachheti, Archana Bachheti, and Azamal Husen

Phenolic Compounds from Medicinal Plants: Pharmaceutical and Health Benefits
Edited by Rakesh Kumar Bachheti, Archana Bachheti, and Azamal Husen

Therapeutic Potential of Vegetables
Edited by Azamal Husen and Megha Barot

Molecular Marker-Assisted Improvement and Production of Medicinal Plants
Edited by Arpan Modi, Megha Barot, Sachidanand Singh, and Azamal Husen

For more information about this series, please visit: www.routledge.com/Exploring-Medicinal-Plants/book-series/CRCEMP

Molecular Marker-Assisted Improvement and Production of Medicinal Plants

Edited by
Arpan Modi, Megha Barot, Sachidanand Singh, and Azamal Husen

CRC Press
Taylor & Francis Group
Boca Raton London New York

CRC Press is an imprint of the
Taylor & Francis Group, an informa business

Designed cover image: Shutterstock

First edition published 2027
by CRC Press
2385 NW Executive Center Drive, Suite 320, Boca Raton FL 33431

and by CRC Press
4 Park Square, Milton Park, Abingdon, Oxon, OX14 4RN

CRC Press is an imprint of Taylor & Francis Group, LLC

ISBN: 978-1-032-89417-1 (hbk)
ISBN: 978-1-032-89419-5 (pbk)
ISBN: 978-1-003-54268-1 (ebk)

DOI: 10.1201/9781003542681

Typeset in Times
by Newgen Publishing UK

Contents

Preface vii
About the Editors ix
List of Contributors xiii

Chapter 1 Introduction to Molecular Markers 1
Ravi Kant Rahi, Nalinee Kumari, and Antra Kumari

Chapter 2 Advances in Biotechnological Interventions of Medicinal Plants and Herbal Formulations: Current Status and Regulatory Affairs 15
Pragati Khare, Noopur Khare, Tanvi Jain, Sachidanand Singh, Megha Barot, and Azamal Husen

Chapter 3 Scope of Molecular Markers against Other Conventional Markers in the Genetic Improvement of Medicinal Plants 33
Noopur Khare, Pragati Khare, Tanvi Jain, Sachidanand Singh, Megha Barot, and Azamal Husen

Chapter 4 Species Identification, Diversity, and Conservation of Medicinal Plants Using Molecular Markers 50
Ravi Shah

Chapter 5 Advances in Techniques Used in Molecular Marker Development 58
Susan Maria Binu, Rebecca Jacey Kurian, and Anu Jacob

Chapter 6 Quantitative Trait Loci (QTLs) of Secondary Metabolism 78
Manu Yadav, Anjali Pathak, and Ananya Sharma

Chapter 7 Molecular Markers Associated with Secondary Metabolism and Value Addition Using Biotechnological Interventions 91
Anshul Jaiswal, R. Sagar, Mohammad Sirazuddin Ansari, Aakansha Pandey, Deepankar Yadav, and Richa Rawat

Chapter 8 From Genotyping to Chemotyping of Medicinal Plants: Current Status and Perspective 116
Shreya Arora and Kalpana Katiyar

Chapter 9 Clonal Fidelity in Micropropagated Medicinal Plant 129

Kajal Kumari, Pritom Biswas, Ankita Kumari, and Nitish Kumar

Chapter 10 Role of Transcriptional Markers during *in vitro* Organogenesis 141

Gareema Pandey and Arpan Modi

Chapter 11 Genome-Editing Tools for Plant-Based Improved Medicines 151

Reut Peer, Arpan Modi, and Megha Barot

Chapter 12 Molecular Insights of Hairy Root Culture 161

Mahak Sachdev

Chapter 13 Scope of Molecular Markers in Cannabis Research 166

Anna Zilov and Arpan Modi

Chapter 14 Genetic Characterization of *Artemisia annua* L. Using EST-SSR Markers 172

Surojeet Das, Divyang Ghanshyambhai Savaliya, Raviraj Bharatbhai Barot, Hiral Sanghavi, and Sachidanand Singh

Chapter 15 Past, Present, and Future of Molecular Markers in Herbal Medicines 183

Amanullakhan Pathan, Tanvi Nayak, Dipika Pandya, and Sohilkhan Chauhan

Index 203

Preface

All the species of the plant kingdom contain important secondary chemicals apart from the primary ones like carbohydrates, proteins, lipids, nucleic acids, and vitamins. These secondary chemicals are not only important for the survival and growth of the plants but also useful for the other species living on the planet. Some of the plant species contain specialized chemicals that are widely used in ancient and modern medicines, and therefore, such species are categorized as medicinal plants. Medicinal plants have become an important part of the lifestyle in the post-COVID era. Apart from the routine medication, they are also being used in other products such as skincare, nutraceuticals, and recreational purposes. It has been noted that, to date, more than 72,000 medicinal plant species have been identified, and out of them, more than 50,000 have been used for various medicinal purposes worldwide.

Alkaloids, flavonoids, terpenoids, and glycosides are just a few of the many bioactive substances that medicinal plants produce and which have therapeutic effects on a variety of illnesses. Medicinal plants continue to provide important leads for drug discovery and holistic treatment, from traditional Chinese and Ayurvedic medicine to contemporary phytotherapy. In light of chronic illnesses, growing antibiotic resistance, and the global movement towards sustainable and natural solutions, their importance has only increased. Medicinal plants continue to be an essential link between traditional knowledge and modern biomedical innovation as scientific study expands our understanding of metabolites generated from plants and their modes of action. The foundation of the global healthcare sector is made up of medicinal plants, which provide a wealth of bioactive substances that are used in a variety of commercial goods. These comprise traditional formulations like Chinese and Ayurvedic medicines, as well as standardized herbal extracts, essential oils, phytopharmaceuticals, nutraceuticals, and cosmeceuticals. Numerous substances originating from plants, such as paclitaxel from *Taxus brevifolia*, artemisinin from *Artemisia annua*, and morphine from *Papaver somniferum*, have been turned into common medications used in cancer treatment, infectious disease treatment, and pain management. In addition to medications, nutritional supplements, functional foods, and personal care products frequently contain medicinal plant materials, which reflects customer demand for all-natural and holistic health solutions. The economic potential of medicinal plants keeps growing as biotechnological advancements allow for better extraction, standardization, and formulation, connecting conventional wisdom with cutting-edge therapeutic innovation. Conversely, molecular markers are tiny DNA fragments that aid researchers in identifying variations among plant species. These markers can reveal the relationships between plants, their responses to illnesses, or their growth patterns in various habitats. They are helpful for safeguarding rare species, identifying plants, and enhancing breeding techniques. They provide certain and trustworthy findings as they are founded on the genetic material of the plant. They are therefore a valuable instrument in agriculture and plant research.

Novel drug discovery and personalized treatment are being made possible by the accurate identification, characterization, and improvement of bioactive substances made possible by developments in biotechnology, genomics, and metabolomics. Synthetic biology and Clustered Regularly Interspaced Short Palindromic Repeats technology-based gene editing provide methods to maximize metabolite synthesis and get around the drawbacks of conventional agriculture. In addition, further standardization and clinical validation of herbal products are being promoted by the development of evidence-based phytotherapy and regulatory frameworks. Medicinal plants are set to become increasingly important in the wellness and preventative healthcare sectors as consumer demand for plant-based nutraceuticals, cosmeceuticals, and functional foods keeps growing. Their significance in influencing the direction of global health innovation is further increased by their incorporation into AI-driven screening, digital health platforms, and sustainable agriculture.

About the Editors

Arpan Modi has been the Director of Axenic Agritech Private Limited, Ahmedabad, India since July 2023 and heading a molecular diagnostic unit at Paws Universe Superspeciality Veterinary Hospital, Ahmedabad, India since March 2025. Dr. Modi is also serving as visiting faculty at Pandit Deendayal Energy University, Gandhinagar, India, since January 2026. As a director in Plant Tissue Culture Unit, he is leading the production of 250,000 tissue culture-raised banana plants every year. He has 5 years of experience as a post-doctoral fellow from Agricultural Research Organization, Volcani Center, Israel and 7 years of experience as a researcher in the area of plant biotechnology and biochemistry. In his research tenure, he had worked on diversified plant species like stevia, date palm, papaya, sandalwood, *Ficus microcarpa*, fig, and apple. He has edited a book on plant tissue culture and got acceptance of three books, two of which with CRC Press and one with CABI publisher. He has 16 research papers, 5 review papers, and more than 30 book chapters in his publication list with the cumulative impact factor of 30.55. He has delivered online and offline lectures in various workshops, webinars, and conferences.

Megha Barot is Assistant Professor (Research Cadre) in the Department of Environmental Science, Parul Institute of Applied Sciences, Parul University, India. She has over a decade of experience in environmental science, life science research, and clinical diagnostics. She has served as the National Coordinator of the Biotech Innovation Ignition School (BIIS) at SRISTI, in collaboration with DBT–BIRAC, Government of India, and as a Junior Scientist and NABL Internal Auditor at the Neuberg Centre for Genomic Medicine, Ahmedabad. A recipient of the Research Contribution Award from GCPC, she also attended a fully funded training programme on Marine Environmental Health and Safety at Xiamen University, China. Dr. Barot has finalized seven academic books with CRC Press, Springer, and CABI, and has numerous research publications to her credit.

Sachidanand Singh is currently working as Associate Professor and Head Biotechnology, School of Energy and Technology, Pandit Deendayal Energy University, Gujarat, India with academic and administrative responsibilities. His education comes from an engineering background, that is, BTech Biotechnology, MTech. Bioinformatics, PhD Bioinformatics and postdoctoral experience from Ohio State University, College of Pharmacy, USA. He has more than 16 years of experience in teaching, administration, and research. Work experience includes working in Karunya University, Shri Ramswaroop Memorial University, and Vignan University. His research area includes plant biotechnology, metagenomics, systems biology, network biology, microRNA-based studies, expression data analytics, next-generation sequencing, and statistical genetics. His present focus also includes understanding the canonical and non-canonical pathways of miRNA biogenesis, dynamics and mechanisms of miRNA- mediated gene regulation. He teaches different subjects, especially genomics and proteomics, microarray and image analysis, drug design, and molecular modelling for graduate and postgraduate students. He has published more than 75 research and review articles in peer-reviewed journals and 12 patents published, among which 9 are granted. Out of the research and review publications, 40% of publications (book chapters, edited books, review articles, and research articles) are on plant-based studies in association to soil microbiome, herbal drug formulation, drug design, and microbe–microbe interactions and their expression-based studies. He has completed one SERB, Government of India-funded project and two university research grant projects. He has received several awards in his area of research and for his teaching strategies. He has organized different conferences and workshops and received government funds for the same. He has shared his area of expertise with different colleges and universities as invited speaker in different conferences and workshops. He has guided 36 BTech. and 10 MTech. projects. He has awarded 5 PhD students and currently 6 PhD students are working under him in the field of herbal products, drug design, and integrating with network biology. He has already edited three books under Springer and Taylor & Francis publishing groups. He has a collaboration with Mendel University Czech Republic, Europe, Lund University, Sweden, and Monash University Australia.

Azamal Husen has served as Full Professor and Head of the Department of Biology at the University of Gondar, Ethiopia. He is currently affiliated as a Foreign Delegate with Wolaita Sodo University, Ethiopia, and holds an Adjunct Professorship at Department of Biotechnology, Graphic Era (Deemed to be University) in Dehradun, Uttarakhand, India. His earlier academic roles include Full Professor at Sankalchand Patel University in Visnagar, India, and Visiting Faculty at the Forest Research Institute and Doon College of Agriculture and Forest in Dehradun, India. With more than 25 years of experience in academia and research, he is highly regarded for his expertise in nanobiotechnology, plant physiology, environmental botany, medicinal plants, and plant propagation. His research has been supported by prominent national and international organizations, including the World Bank, the National Agricultural Technology Project, the Indian Council of Agricultural Research, the Indian Council of Forestry Research and Education, and the Japan Bank for International Cooperation. He is a prolific scholar, having published more than 350 academic works, including research articles, books, and book chapters. His editorial contributions are equally noteworthy; he serves on the editorial boards and peer review panels of numerous prestigious journals published by Elsevier, Frontiers Media, Springer Nature, Taylor & Francis, RSC, Oxford University Press, the Royal Society, CSIRO, MDPI, PLOS, Sciendo, John Wiley & Sons, and UPM journals. He is also a member of the advisory board of Cambridge Scholars Publishing, United Kingdom. He is a fellow of the Plantae group under the American Society of Plant Biologists and holds membership in several scientific organizations, including the International Society of Root Research, the Asian Council of Science Editors, and the International Natural Product Sciences Taskforce. Throughout his career, he has delivered invited lectures, keynote addresses, and technical presentations at numerous international conferences, symposia, and workshops. In the field of academic development, he has made significant contributions through the design and delivery of orientation programmes for university faculty, the development of educational modules, and the formulation of curricula at the undergraduate, postgraduate, and doctoral levels. In recognition of his contributions to teaching, research, and community service, he was honoured with a special award from the University of Gondar. He currently serves as Editor-in-Chief of the *American Journal of Plant Physiology* and is Series Editor for several major academic collections, including *Exploring Medicinal Plants* (Taylor & Francis, USA), *Plant Biology, Sustainability, and Climate Change* (Elsevier, USA), and *Smart Nanomaterials Technology* (Springer Nature, Singapore). He also leads the *Handbook Series on Nanotechnology and Its Applications* (Springer

Nature, Singapore). His outstanding contributions to science have earned him international acclaim, including the prestigious distinction of being named among the "World's Top 2% Scientists" for three consecutive years (2020 onwards) by Stanford University, a recognition published in the Elsevier Data Repository. In addition to his regular academic responsibilities, he has served as Guest Editor for several high-impact special issues, including *Advances in Plant Metabolomics: Insights, Applications, and Integration* (*Plant Physiology and Biochemistry*), *Exploring Nanotechnology in Medicinal Plants* (*Journal of Herbal Medicine*), and *Integrating Nanotechnology with Plant Compounds: Enhancing Abiotic Stress Resilience and Crop Improvement via Signaling Molecules* (*Frontiers in Plant Science*), *Nanobiotechnology for Sustainable Agriculture and Crop Improvement (Discover Applied Sciences),* among others. His work continues to shape contemporary research at the intersection of plant science, biotechnology, and sustainable development. His academic leadership, research excellence, and international collaborations have positioned him as a prominent figure in global scientific circles.

Contributors

Mohammad Sirazuddin Ansari
Department of Botany
Banaras Hindu University
Varanasi, Uttar Pradesh, India

Shreya Arora
Department of Biotechnology
Dr. Ambedkar Institute of Technology for Divyangjan
Kanpur, Uttar Pradesh, India

Megha Barot
Research and Development Cell and Department of Environmental Science
Parul Institute of Applied Sciences
Parul University
Vadodara, Gujarat, India

Raviraj Bharatbhai Barot
Department of Biotechnology
School of Energy Technology
Pandit Deendayal Energy University
Gandhinagar, Gujarat, India

Susan Maria Binu
Division of Biotechnology
Karunya Institute of Technology and Sciences
Coimbatore, Tamil Nadu, India

Pritom Biswas
Department of Biotechnology
Central University of South Bihar
Gaya, Bihar, India

Sohilkhan Chauhan
Department of Chemistry
Gokul Science College
Gokul Global University
Siddhpur, Gujarat, India

Surojeet Das
Department of Biotechnology
School of Energy Technology
Pandit Deendayal Energy University
Gandhinagar, Gujarat, India

Azamal Husen
Wolaita Sodo University
Wolaita Sodo, South Ethiopia Regional State, Ethiopia;
Department of Biotechnology
Graphic Era (Deemed to be University)
Dehradun, Uttarakhand, India

Anu Jacob
Division of Biotechnology
Karunya Institute of Technology and Sciences
Coimbatore, Tamil Nadu, India

Tanvi Jain
Shri Ramswaroop Memorial University
Barabanki, Uttar Pradesh, India

Anshul Jaiswal
Department of Botany
Banaras Hindu University
Varanasi, Uttar Pradesh, India

Kalpana Katiyar
Department of Biotechnology
Dr. Ambedkar Institute of Technology for Divyangjan
Kanpur, Uttar Pradesh, India

Noopur Khare
Bhai Gurdas Institute of Engineering and Technology
Sangrur, Punjab, India;
Faculty of Biotechnology
Institute of Biosciences and Technology
Shri Ramswaroop Memorial University
Barabanki, Uttar Pradesh, India

Pragati Khare
BPS Educational Institution of Pharmacy
Etah, Uttar Pradesh, India

Nitish Kumar
Department of Biotechnology
Central University of South Bihar
Gaya, Bihar, India

Ankita Kumari
Department of Biotechnology
Central University of South Bihar
Gaya, Bihar, India

Antra Kumari
Department of Microbiology
JECRC University
Jaipur, Rajasthan, India

Kajal Kumari
Department of Biotechnology
Central University of South Bihar
Gaya, Bihar, India

Nalinee Kumari
Department of Microbiology
School of Life Sciences
Starex University
Gurugram, Haryana, India

Rebecca Jacey Kurian
Division of Biotechnology
Karunya Institute of Technology and Sciences
Coimbatore, Tamil Nadu, India

Arpan Modi
Axenic Agritech Private Limited
Ahmedabad, Gujarat, India

Tanvi Nayak
M. B. Patel Science College
Anand, Gujarat, India

Aakansha Pandey
Department of Botany
Banaras Hindu University
Varanasi, Uttar Pradesh, India

Gareema Pandey
Western Sydney University
Richmond, New South Wales, Australia

Dipika Pandya
Department of Microbiology
Shri Sarvajanik Science College (PG)
Mehsana, Gujarat, India

Anjali Pathak
Department of Zoology
University of Delhi
New Delhi, India

Amanullakhan Pathan
Department of Chemistry
Shri Sarvajanik Science College (PG)
Mehsana, Gujarat, India

Reut Peer
Institute of Plant Sciences
Agricultural Research Organization
Volcani Center
Rishon LeZiyyon, Israel

Ravi Kant Rahi
Department of Microbiology
JECRC University
Jaipur, Rajasthan, India

Richa Rawat
Department of Botany
Banaras Hindu University
Varanasi, Uttar Pradesh, India

Mahak Sachdev
BioNest, VentureStudio
Ahmedabad University
Ahmedabad, Gujarat, India

R. Sagar
Department of Botany
Banaras Hindu University
Varanasi, Uttar Pradesh, India

Hiral Sanghavi
Department of Biotechnology
School of Energy Technology
Pandit Deendayal Energy University
Gandhinagar, Gujarat, India

Divyang Ghanshyambhai Savaliya
Department of Biotechnology
School of Energy Technology
Pandit Deendayal Energy University
Gandhinagar, Gujarat, India

Ravi Shah
Research and Development
Living Carbon PBC
Hayward, California, USA

Ananya Sharma
Department of Botany and Microbiology
Hemwati Nandan Bahuguna Garhwal University
Srinagar, Uttarakhand, India

Sachidanand Singh
Department of Biotechnology
School of Energy Technology
Pandit Deendayal Energy University
Gandhinagar, Gujarat, India

Deepankar Yadav
Department of Botany
Banaras Hindu University
Varanasi, Uttar Pradesh, India

Manu Yadav
Division of Genetics
ICAR – Indian Agricultural Research Institute
New Delhi, India

Anna Zilov
ReaGenics Private Limited
Central District
Rehovot, Israel

1 Introduction to Molecular Markers

Ravi Kant Rahi, Nalinee Kumari, and Antra Kumari

1.1 INTRODUCTION

Molecular markers are defined as specific and identifiable sequences of DNA or RNA that serve as powerful tools for genetic analysis. These markers function as landmarks within the genome, enabling researchers to detect genetic variations, analyse evolutionary patterns, and study trait inheritance at a molecular level. The precision and reliability of molecular markers surpass traditional phenotypic markers, which are often influenced by environmental factors (Chamberlin and Lawit, 2017; Datta et al., 2024). Molecular markers have become indispensable in diverse fields, including genetics, evolutionary biology, conservation, and breeding programmes, particularly in the context of medicinal plants.

The foundation of molecular markers lies in the concept of genetic variation. Genetic variation encompasses the differences in nucleotide sequences among individuals within a population or between species, forming the foundation for evolutionary adaptation, population genetics. These variations occur naturally and form the basis of evolutionary adaptation and species diversity. Molecular markers exploit these variations, which manifest in forms such as single nucleotide polymorphisms (SNPs), microsatellites, Indels. For example, SNPs are highly abundant and useful for precise genetic mapping and diversity studies. Microsatellites, or simple sequence repeats (SSRs), consist of short tandem repeats which have a high mutation rate, ideal for population genetics and evolutionary relationships studies (Joshi and Nguyen, 1993).

Unlike traditional phenotypic markers, which rely on observable characteristics such as shoot height, flowering, and leaf shape, molecular markers operate at the genetic level. Traditional phenotypic markers are prone to inaccuracies because of environmental conditions, interactions with other organisms, or developmental stages. For instance, the expression of traits like flower size or secondary metabolite production may vary significantly under different environmental conditions, making it challenging to determine the genetic basis of these traits. In contrast, molecular markers provide a consistent and environment-independent method for identifying genetic variation. This prominent level of precision has made molecular markers an essential tool in marker-assisted selection (MAS), facilitating the improvement of both crop species and medicinal plants (Winter and Kahl, 1995; Sao et al., 2022).

Technological advancements have enhanced the applications of molecular markers. The introduction of polymerase chain reaction (PCR) technology has markedly improved the efficiency, sensitivity, and accessibility of various molecular marker techniques, including random amplified polymorphic DNA (RAPD), amplified fragment length polymorphism (AFLP), and SSRs. High-throughput sequencing technologies, such as next-generation sequencing (NGS), have further

DOI: 10.1201/9781003542681-1

expanded the scope of molecular marker research by enabling genome-wide marker discovery and genotyping at an unprecedented scale.

The significance of molecular markers in conservation genetics is especially remarkable. They have been extensively utilized to evaluate genetic diversity among rare and endangered medicinal plant species, providing valuable insights into their genetic structure. Maintaining genetic diversity is essential for ensuring population resilience, adaptability, and long-term survival. For instance, studies using SSR markers in endangered medicinal plant species have provided valuable insights into population structure and genetic bottlenecks, thereby aiding in the development of effective and sustainable conservation strategies. Additionally, molecular markers have been used to authenticate medicinal plant species and detect adulteration in herbal products, ensuring the safety and efficacy of traditional medicine (Semagn et al., 2006).

Molecular markers have thus become essential tools in modern genetics, offering unmatched precision in detecting genetic variations. They provide an accurate and reliable alternative to traditional phenotypic markers and have found extensive applications in plant breeding, genetic diversity studies, evolutionary biology, and conservation genetics. Their ability to analyse genetic differences independent of environmental influences makes them a cornerstone of molecular biology research (Nakagawa and Ebina, 2001).

1.2 IMPORTANCE OF MOLECULAR MARKERS

Molecular markers serve as indispensable tools in modern genetics, playing a crucial role in the analysis and understanding of genetic diversity. They enable the detection of polymorphisms at the DNA level, providing valuable insights into the patterns of genetic variation both within and between populations. This aspect is particularly critical for medicinal plants, where genetic diversity ensures adaptability, resilience, and the preservation of bioactive potential. For example, a study on *Rauvolfia serpentina*, a medicinal species renowned for its antihypertensive properties, utilized inter-simple sequence repeat (ISSR) markers to analyse genetic diversity across multiple populations. The findings revealed significant genetic variation, emphasizing the importance of conserving diverse genotypes to maintain the species' therapeutic potential (Collard and Mackill, 2008).

In the realm of medicinal plants, molecular markers are instrumental in identifying and characterizing species with pharmacological significance. They facilitate the authentication of plant materials, ensuring that the correct species and even specific genotypes are used in medicinal preparations. For instance, RAPD markers were employed to differentiate between two morphotypes of *Ocimum sanctum* (holy basil), a plant revered in traditional medicine. The study successfully distinguished the morphotypes at the genetic level, thereby aiding in the selection of superior chemotypes for cultivation and therapeutic use (Joshi et al., 2000).

Molecular markers are fundamental to contemporary breeding programmes aimed at improving medicinal plants. By identifying markers linked to key traits such as enhanced production of bioactive compounds or increased disease resistance, breeders can accurately select parental lines capable of transmitting these desirable attributes to their progeny. This marker-assisted approach significantly accelerates the development of superior cultivars with improved medicinal potential. In *Withania somnifera* (Ashwagandha), a medicinal plant valued for its withanolides content, SSR markers have been employed to identify genotypes exhibiting elevated withanolides levels, thereby facilitating the selection of elite lines for breeding programmes focused on developing high-yielding and therapeutically potent cultivars (Gupta et al., 1999).

The ability of molecular markers to precisely link phenotypic traits with specific genes represents a major advancement over conventional breeding methods. Quantitative trait locus (QTL) mapping, in particular, utilizes molecular markers to identify genomic regions associated with complex quantitative traits. In *Artemisia annua*, the natural source of the antimalarial compound artemisinin, QTL mapping has successfully identified genomic regions correlated with elevated artemisinin production. Such insights are invaluable for guiding genetic improvement strategies aimed at enhancing artemisinin yield and optimizing the plant's pharmaceutical potential (Pant et al., 2025).

1.3 HISTORICAL EVOLUTION

The history of molecular markers spans over six decades, beginning with the use of protein-based markers and evolving into highly advanced DNA-based technologies. Molecular markers were first introduced in the 1960s with the development of isozyme markers. These protein-based markers relied on the electrophoretic separation of enzymes to identify genetic differences among individuals. While isozyme markers provided early insights into population genetics, their low resolution and limited variability restricted their application (Tautz, 1989).

The 1970s witnessed a major breakthrough in molecular genetics with the development of DNA-based markers. The introduction of Restriction Fragment Length Polymorphism (RFLP) technology enabled researchers to analyse specific regions of DNA, providing a powerful tool for studying genetic variation, linkage analysis, and genome mapping. RFLP technology, based on restriction enzyme digestion and hybridization, enabled high-resolution genetic mapping for the first time. This innovation revolutionized genetic research and became the cornerstone of early molecular marker studies (Powell et al., 1996).

The next leap occurred in the 1980s and 1990s. Advent of PCR-based markers dramatically reduced the time and cost required for genetic analysis. RAPD utilized short random primers to amplify regions of DNA, while AFLP combined restriction digestion and selective PCR amplification, offering greater reproducibility and sensitivity. These technologies gained widespread adoption because of their methodological simplicity and effectiveness in detecting polymorphisms, even in non-model organisms (Becher and Heun, 1998).

Microsatellites, also known as SSRs, emerged in the late 1990s as highly polymorphic markers that advanced genetic diversity studies. Comprising tandemly repeated short nucleotide sequences, microsatellites became a preferred tool in population genetics due to their high mutation rates, abundance, and co-dominant inheritance patterns. In the early 2000s, the advent of SNPs further revolutionized molecular genetics by offering exceptional resolution and precision. SNPs, which represent single-base substitutions in DNA sequences, are highly abundant and amenable to large-scale analysis, making them ideal for genome-wide studies. Subsequent advancements in NGS technologies have dramatically expanded the scope and efficiency of molecular marker research. NGS platforms enable high-throughput genotyping and the simultaneous discovery of thousands of genetic markers within a single experiment. These innovations have facilitated genome-wide association studies (GWASs) and refined QTL mapping, thereby transforming genetic research and accelerating the understanding of complex traits across biological disciplines (Mace and Godwin, 2002).

The initial use of molecular markers was confined to understanding genetic diversity and evolutionary biology in model organisms. Early studies using isozyme and RFLP markers provided insights into population structure and phylogenetics. However, these techniques were primarily limited to basic genetic research due to the complexity and cost of the methods involved. As marker technologies advanced, their applications extended to economically important crops and medicinal plants. This is particularly important because adulteration and substitution are familiar challenges in the trade of medicinal plants. DNA fingerprinting, an application of molecular markers, has been used to identify species accurately, ensuring the safety and efficacy of herbal medicines. For instance, RAPD markers have been successfully employed to authenticate *Curcuma longa* (turmeric) and distinguish it from closely related species (Vos et al., 1995).

Molecular markers have also played a pivotal role in characterizing genetic diversity both within and between populations of medicinal plants. They provide precise and reproducible insights into genetic variation, enabling researchers to assess population structure, identify unique genotypes, and understand evolutionary relationships. Studies using SSR markers in *R. serpentina* have revealed significant genetic variation, highlighting the importance of conserving diverse genotypes to maintain therapeutic efficacy (Singh et al., 2017). Breeding programmes for medicinal plants have benefited immensely from the integration of molecular markers. MAS enables the identification of genotypes with desirable traits, such as high concentrations of secondary metabolites or resistance to pests and diseases. In *W. somnifera* (Ashwagandha), SSR markers have been used to select high-yielding genotypes for withanolides production, demonstrating the practical applications of molecular markers in improving medicinal plant quality and yield (Weising et al., 2005).

1.4 CHARACTERISTICS OF AN IDEAL MOLECULAR MARKER

Molecular markers are fundamental tools in genetic research, and their effectiveness depends on several key characteristics. These attributes determine their reliability, reproducibility, and suitability across various applications, including genetic diversity analysis, breeding programmes, and conservation genetics.

1.4.1 High Polymorphism for Effective Differentiation

Polymorphism refers to the presence of multiple alleles at a locus, allowing differentiation between individuals or populations. Markers with high polymorphism can detect extensive genetic variation, which is critical for studying genetic diversity and evolutionary relationships. For example, SSRs are highly polymorphic due to their tendency for variation in the number of repeat units. This feature makes SSRs invaluable for analysing population structure. A study on *Vigna radiata* (mung bean) used SSR markers to reveal substantial genetic diversity among accessions, enabling the identification of unique genotypes for breeding purposes (Xu, 2013). High polymorphism not only enhances resolution but also facilitates the identification of rare alleles, which are crucial for maintaining genetic resources.

1.4.2 Co-Dominant Inheritance for Clearer Genetic Analysis

Co-dominant inheritance enables molecular markers to differentiate between homozygous and heterozygous genotypes, offering a more detailed understanding of genetic variation. This feature provides a significant advantage over dominant markers, which can only reveal the presence or absence of a particular allele without distinguishing zygosity. SNPs are exemplary co-dominant markers. In a study on *Camellia sinensis* (tea), SNP markers effectively identified heterozygous loci associated with caffeine content, aiding in the selection of high-quality cultivars (Tanksley et al., 1997). The ability to detect both alleles at a locus is particularly beneficial in QTL mapping, where heterozygosity often correlates with complex trait inheritance.

1.4.3 Reproducibility and Consistency across Laboratories

Reproducibility is a fundamental requirement for any molecular marker system. Consistency ensures that results obtained from different laboratories and experimental setups are comparable. Reliable markers produce consistent amplification patterns across varying conditions, making them suitable for large-scale and collaborative studies. ISSR markers are known for their reproducibility when used under optimized conditions. A comparative study of ISSR markers across multiple laboratories in the genetic analysis of *Zingiber officinale* (ginger) demonstrated their robustness, highlighting their value in establishing standardized protocols for plant genetic studies (Kalia et al., 2011). Consistency in results bolsters confidence in marker-based conclusions and enhances their utility in applied research.

1.4.4 Cost-Effectiveness and Ease of Application

The cost and practicality of molecular markers significantly influence their adoption, particularly in resource-constrained settings. Markers that require minimal technical expertise, affordable reagents, and standard laboratory equipment are preferred for routine applications. Kompetitive Allele-Specific PCR (KASP) markers have gained prominence due to their cost-effectiveness and scalability. A study on *Triticum aestivum* (wheat) demonstrated the utility of KASP markers in genotyping a large number of samples with minimal cost and time, proving their suitability for high-throughput breeding programmes (Prasad et al., 2000). The simplicity of KASP markers makes them accessible to a broad range of researchers, from academic institutions to agricultural industries.

1.5 TYPES OF MOLECULAR MARKERS

1.5.1 DNA-Based Markers

Molecular markers play an essential role in genetic research and breeding programmes by providing valuable insights into genetic diversity, population structure, and trait inheritance. Among these, DNA-based markers are the most widely utilized, as they target specific regions within an organism's genome with high precision and reliability. Their versatility and accuracy make them indispensable tools across a wide range of genetic and genomic applications. Comparative studies evaluating molecular marker systems such as RFLP, RAPD, AFLP, and SSR have demonstrated differences in polymorphism detection and efficiency in genetic analysis (Powell et al., 1996).

1.5.1.1 Restriction Fragment Length Polymorphism

RFLP was one of the earliest DNA-based marker systems developed and relied on detecting polymorphisms in DNA fragment lengths resulting from restriction enzyme digestion. The method involves cleaving genomic DNA with restriction enzymes, separating the fragments by gel electrophoresis, and hybridizing them with labelled DNA probes. Variations in fragment lengths arise due to mutations in restriction sites or insertions/deletions in DNA, enabling the identification of genetic differences (Pootakham et al., 2020).

RFLP markers played a foundational role in the preliminary stages of genetic mapping and linkage analysis. For instance, a study on *Oryza sativa* (rice) employed RFLP markers to develop high-density genetic maps, yielding valuable insights into QTL associated with grain yield and drought tolerance (Bhat and Lakhanpaul, 2000). Although RFLP markers offer high accuracy and co-dominant inheritance, their application is limited today due to their labour-intensive nature and the requirement for enormous quantities of high-quality DNA.

1.5.1.2 Random Amplified Polymorphic DNA

RAPD is a PCR-based marker system that uses short, random primers to amplify anonymous regions of the genome. This method does not require prior knowledge of DNA sequences, which makes it a rapid, cost-effective, and widely accessible tool for researchers, particularly in preliminary genetic diversity and population studies. Polymorphisms are detected as differences in the presence or absence of amplified DNA bands, which result from variations in primer binding sites.

RAPD has been widely used in genetic diversity studies and the authentication of medicinal plants. A study on *O. sanctum* (holy basil) employed RAPD markers to differentiate between morphotypes, aiding in their proper classification and potential use in traditional medicine (Akkaya et al., 1992). Despite its simplicity, RAPD markers are often criticized for their lack of reproducibility, which can be mitigated by optimizing reaction conditions.

1.5.1.3 Amplified Fragment Length Polymorphism

AFLP combines the strengths of RFLP and PCR to provide a highly sensitive and reproducible marker system. The technique involves digesting genomic DNA with specific restriction enzymes, ligating adaptors to the resulting fragments, and selectively amplifying targeted subsets of these fragments through PCR. AFLP generates numerous polymorphic markers in a single reaction, making it ideal for genetic diversity studies, QTL mapping, and phylogenetic analysis. AFLP markers have been successfully applied in analysing genetic variation in crop plants. AFLP's high-throughput capability and reliability make it a preferred choice for large-scale studies (Adhikari et al., 2022).

1.5.1.4 Simple Sequence Repeats

SSRs, also known as microsatellites, are short tandemly repeated DNA sequences consisting of 2–6 base pairs distributed throughout the genome. Owing to their high mutation rates, SSRs display a high degree of polymorphism, making them highly informative molecular markers. Their co-dominant inheritance pattern enables the distinction between homozygous and heterozygous genotypes, providing detailed genetic information.

SSRs have been widely utilized in genetic mapping, diversity assessment, and population structure analysis. For example, in *Vitis vinifera* (grapevine), SSR markers were employed to examine genetic diversity among cultivars, facilitating the identification of distinct genotypes valuable for breeding programmes (Choudhary et al., 2013). Although the use of SSR markers requires prior knowledge of flanking sequences for primer design, their robustness, reproducibility, and high discriminatory power make them indispensable tools in modern genetic research.

1.5.1.5 Single Nucleotide Polymorphism

SNPs represent single base-pair substitutions within the genome and constitute the most abundant and stable form of genetic variation. Their widespread distribution and stability make them highly suitable for high-throughput genotyping platforms, enabling applications in GWASs and MAS. SNPs offer exceptional resolution for detecting genetic differences and, due to their co-dominant nature, allow precise identification of allelic variants in non-model plant species (Williams et al., 2010). SNP markers have become a cornerstone of modern molecular genetics and crop improvement research. For instance, in *Glycine max* (soybean), SNP-based analyses identified loci associated with salinity tolerance, thereby supporting the development of salt-tolerant cultivars through MAS, a reliable alternative to stress mitigation strategies (Joshi et al., 2013; Patani et al., 2023). The scalability, cost-efficiency, and accuracy of SNP genotyping systems, such as KASP, have further enhanced their utility across diverse breeding and genomics programmes (Kalendar et al., 2022).

1.5.2 RNA-Based and Protein-Based Markers

RNA-based and protein-based markers have significantly contributed to the advancement of functional genomics, particularly in exploring gene expression dynamics and protein functionality. Although their use in medicinal plant studies has been comparatively limited relative to DNA-based markers, recent investigations have demonstrated their considerable potential in elucidating complex biosynthetic pathways responsible for the production of bioactive compounds (Schubert and Müller-Starck, 2002).

RNA-based markers focus on the transcriptome, which encompasses the complete set of RNA transcripts synthesized by the genome under specific physiological or environmental conditions. Advanced techniques such as RNA sequencing (RNA-Seq) allow for comprehensive profiling of gene expression, enabling the identification of genes involved in the biosynthesis of pharmacologically active compounds. For example, RNA-Seq has been applied to characterize the transcriptomic landscape of *Tinospora cordifolia*, a medicinal plant valued for its therapeutic properties, thereby enhancing understanding of its metabolic and biosynthetic pathways (Rani and Singh, 2013). Furthermore, RNA interference (RNAi) technologies have been employed to silence target genes, providing valuable insights into gene function and regulatory mechanisms underlying secondary metabolite production (Chaudhary et al., 2024).

1.5.2.1 Protein-Based Markers

Protein-based markers focus on the proteome, which encompasses the complete set of proteins expressed by a genome, cell, tissue, or organism. Proteomic analyses, employing techniques such as two-dimensional gel electrophoresis and mass spectrometry, enable the identification, characterization, and quantification of proteins involved in various biosynthetic and metabolic pathways. In *Salvia miltiorrhiza*, a well-known medicinal herb, proteomic approaches have been utilized to identify proteins associated with the biosynthesis of tanshinones, a class of compounds with notable pharmacological properties (Ghafoor et al., 2002). Such studies provide valuable insights into the functional roles of proteins in medicinal plants, advancing the understanding of their complex metabolic networks and supporting efforts in metabolic engineering and drug discovery.

1.5.2.2 Integration in Functional Genomics

The integration of RNA-based and protein-based markers within functional genomics offers a comprehensive understanding of gene function and regulatory mechanisms in medicinal plants. By combining transcriptomic and proteomic analyses, researchers can develop detailed models of metabolic pathways, enabling the identification of novel bioactive compounds and the genetic enhancement of medicinal plant species through molecular breeding (Mondini et al., 2009). This multi-omics approach provides a holistic perspective on the intricate interactions among genes, proteins, and metabolites, thereby advancing the frontiers of plant-based medicine and supporting the development of more effective therapeutic resources.

1.6 ROLE OF MOLECULAR MARKERS IN MEDICINAL PLANTS RESEARCH

Genetic diversity in medicinal plants is fundamental to their adaptability, resilience, and long-term survival, as well as to the preservation of their therapeutic potential. This diversity ensures that medicinal plants can thrive under varying environmental conditions, resist diseases, and produce a wide array of bioactive compounds essential for their medicinal properties (Jain et al., 2012; Jain and Brar, 2009). Genetic variation within and among populations also underpins the evolutionary adaptability of medicinal plants, allowing them to respond effectively to climate change, habitat alterations, and other environmental stressors.

1.6.1 Assessing Genetic Diversity Using Molecular Markers

The assessment of genetic diversity in medicinal plants has been enhanced through the application of molecular marker technologies. Techniques such as RAPD, AFLP, SSR, and NGS are widely utilized to investigate genetic variation at the molecular level. These approaches provide precise, reproducible, and comprehensive insights into the genetic structure and evolutionary relationships of medicinal plant populations. For example, an investigation of *W. somnifera* populations using AFLP markers revealed substantial genetic variation among populations from diverse ecological zones in India, underlining the importance of geographic and environmental influences on genetic diversity (Nadeem et al., 2018). Similarly, SSR markers were used to assess genetic variation in *R. serpentina*, a plant widely known for its antihypertensive properties, which highlighted significant intra-species genetic diversity critical for breeding and conservation programs.

1.6.2 Identification of Bioactive Compound-Related Traits

Genetic studies have also facilitated the identification of traits linked to bioactive compound production. Advances in transcriptomics and metabolomics have provided insights into the biosynthetic pathways of key phytochemicals. For instance, research on *Curcuma longa* (turmeric) identified genetic markers associated with the production of curcumin, a compound with strong anti-inflammatory and antioxidant properties (Varshney et al., 2012). These discoveries have direct applications in breeding programmes designed to enhance the concentration of bioactive compounds in medicinal plants. Similarly, in *Nigella sativa* (black cumin), transcriptomic analyses have identified genes involved in the biosynthesis of thymoquinone, a bioactive compound recognized for its potent anticancer and antimicrobial activities.

MAS and GWASs are increasingly used to link specific genetic loci to bioactive compound production. This has enabled researchers to develop high-yielding varieties of medicinal plants with enhanced therapeutic properties while maintaining genetic diversity.

1.6.3 Conservation Strategies for Medicinal Plants

Conserving genetic diversity is crucial for the sustainability of medicinal plant species. In situ conservation, which involves protecting plants within their natural habitats, allows for the maintenance of ecological interactions and the continuous evolution of genetic traits. This method is especially vital for species that rely on specific environmental factors for their survival. However, habitat destruction, overharvesting, and climate change pose significant threats to in situ conservation efforts. Ex situ conservation methods, such as seed banking, cryopreservation, and tissue culture, provide alternative strategies to safeguard genetic material. For example, seed banks have been established for *Tinospora cordifolia*, ensuring the preservation of its genetic diversity for future research and medicinal use (Ipek et al., 2020).

1.6.4 Role of Biotechnological Advancements

Biotechnological tools have expanded the scope of conserving and utilizing genetic diversity in medicinal plants. Techniques like CRISPR/Cas9 genome editing have enabled researchers to enhance the production of bioactive compounds, reducing dependency on wild harvesting and preserving genetic resources. In *Catharanthus roseus*, the CRISPR/Cas9 genome-editing system has been employed to precisely target genes associated with alkaloid biosynthesis, resulting in enhanced production of key anticancer compounds, including vincristine and vinblastine (Soriano, 2020). This advancement demonstrates the potential of gene-editing technologies in optimizing the yield and therapeutic efficacy of medicinally important phytochemicals.

1.7 ADVANTAGES OF MOLECULAR MARKERS

1.7.1 High Specificity and Precision in Genetic Studies

Molecular markers provide unparalleled specificity and precision in genetic research by allowing the direct detection and analysis of DNA sequence variations. Unlike traditional markers that rely on observable phenotypic traits, molecular markers can identify polymorphisms at the molecular level, such as SNPs and microsatellites. This high resolution enables researchers to construct detailed genome maps, identify genes associated with desirable traits, and analyze genetic structure and diversity within populations (Gitler et al., 2003). For instance, the application of SNP markers has advanced the fine mapping of disease-resistance genes in *Oryza sativa* (rice), facilitating the development of cultivars with improved resistance to bacterial blight.

The specificity of molecular markers also plays a vital role in the identification of QTLs in plants. QTL mapping using marker systems such as AFLP and SSR has identified genomic regions linked to important agronomic traits, including drought tolerance in wheat and pest resistance in cotton (Hormi-Carver & Souza, 2009). By enabling precise localization of these genetic regions, molecular markers have become indispensable tools in MAS, significantly reducing the time and resources needed to develop superior plant varieties with enhanced performance and resilience.

1.7.2 Independence from Environmental Influences

One of the most significant advantages of molecular markers is their independence from environmental factors, which often complicate phenotypic studies. Unlike morphological or biochemical markers, which are often influenced by environmental factors and developmental stages, molecular markers are derived from stable DNA sequences that remain unaffected by external conditions (Amiteye, 2021). This stability ensures consistent and reliable results, making molecular markers superior tools for accurate genetic identification and analysis. This reliability ensures consistent results across diverse environmental settings and makes molecular markers invaluable for studying traits that are difficult to evaluate phenotypically, such as disease resistance or secondary metabolite production.

For example, the assessment of genetic diversity in *W. somnifera* using ISSR markers revealed elevated levels of genetic variation, unaffected by the plant's growing conditions. This independence is particularly beneficial in identifying genetic material for conservation and breeding purposes, as it enables the selection of superior genotypes regardless of external environmental pressures (Valdman, 2003).

1.7.3 Applicability across Plant Species

Molecular markers are highly versatile tools that can be applied to a wide range of plant species, including those with complex, polyploid, or poorly characterized genomes. They have been successfully used in both model plants like *Arabidopsis thaliana* and non-model crops such as banana, cassava, and sorghum. This wide applicability stems from the availability of universal marker systems, such as SSR and SNP, which can be customized for distinct species without the need for extensive genome sequencing (Storer and Keyes, 2016).

For instance, Diversity Arrays Technology (DArT) has been employed in wheat and barley to identify genetic loci associated with yield and stress tolerance, even in the absence of reference genomes. Similarly, SNP arrays have facilitated genomic selection in perennial crops like grapes, accelerating breeding cycles and improving fruit quality traits (Hoerder, 2007). The ability to apply molecular markers across diverse plant taxa has made them indispensable in addressing global challenges such as food security and climate resilience.

1.8 LIMITATIONS OF MOLECULAR MARKERS

1.8.1 Cost and Technological Barriers in Some Regions

The application of molecular markers often requires substantial financial investment, posing a significant challenge in resource-limited regions. The inflated cost of reagents, such as DNA extraction kits, primers, and enzymes, coupled with the need for advanced laboratory infrastructure, restricts widespread use. For example, NGS technologies, which are increasingly used for genomic studies, demand high-performance computing and substantial storage capacities, both of which are expensive (Panahi et al., 2024). The cost of consumables and instrument maintenance further exacerbates the financial burden, making molecular marker technologies inaccessible in many developing countries. A study analysing the genetic diversity of medicinal plants in sub-Saharan Africa found that inadequate funding for molecular biology research limited the use of cutting-edge techniques like SNP arrays and NGS (Tar'an, 2000).

Collaborative efforts between developed and developing nations have attempted to bridge this gap, but the cost remains a persistent barrier. Subsidies or technology-sharing initiatives are often insufficient to cover the full range of expenses, leaving many researchers reliant on outdated or less accurate methods.

1.8.2 Requirement for Specialized Equipment and Expertise

The successful application of molecular marker technologies necessitates specialized equipment, such as thermal cyclers for PCR, electrophoresis apparatus for DNA fragment analysis, and sequencing machines for genome-wide studies. These instruments not only are costly but also require regular calibration and maintenance to ensure precision. In many cases, these facilities are concentrated in urban research institutions, leaving rural or less-developed areas underserved (Singh and Rameshwar, 2016). Equally important is the need for technical expertise. Researchers must be skilled in various molecular techniques, including DNA extraction, amplification, and sequencing. Analytical tools such as bioinformatics software for interpreting large datasets generated by molecular markers add an additional layer of complexity. Training programmes and capacity-building initiatives can help overcome these limitations; however, the substantial time, infrastructure, and financial resources needed to develop skilled professionals often result in delayed progress and slower implementation of advanced molecular techniques.

1.8.3 Limited Availability of Reference Genomes for Some Medicinal Plants

The availability of high-quality reference genomes is critical for maximizing the utility of molecular markers, as these genomes provide a framework for identifying genetic variations, mapping genes, and understanding complex traits. Unfortunately, many medicinal plants lack complete reference genomes, particularly those that are underrepresented in genomic research. This limitation hinders the development of species-specific markers and makes it challenging to conduct detailed genetic analyses (Livneh and Vardi, 1998).

For instance, while extensive genomic data exist for model plants such as rice and *A. thaliana*, medicinal plants like *W. somnifera* and *O. sanctum* have incomplete or fragmented genomic information. This gap limits studies on their genetic diversity, conservation strategies, and the biosynthetic pathways of bioactive compounds. Researchers working on *R. serpentina* faced similar challenges, as the lack of a comprehensive genome constrained their ability to identify and map loci associated with its valuable alkaloids (Dumitrescu, 2012). The absence of reference genomes also impacts comparative genomics, which is essential for identifying conserved genes and evolutionary patterns among medicinal plant species. This limitation reduces the effectiveness of molecular markers in exploring traits like drought tolerance, pest resistance, and secondary metabolite production.

1.9 CONCLUSION

Molecular markers have transformed the landscape of genetic research, offering precise and reliable methods for studying genetic variation, mapping genomes, and identifying traits linked to agronomic and medicinal importance. Their ability to directly analyse DNA sequences provides unmatched specificity, allowing for the identification of genes and QTLs associated with complex traits such as drought tolerance, disease resistance, and bioactive compound production. Unlike traditional methods that depend on phenotypic observations, molecular markers operate independently of other environmental factors, ensuring consistent results across diverse conditions. Their wide applicability across plant species, including medicinal plants, has significantly enhanced our understanding of plant genetics and improved breeding programmes aimed at addressing global challenges such as food security, climate change, and health.

Despite these advantages, the use of molecular markers is not without challenges. The excessive cost of reagents, equipment, and laboratory infrastructure continues to be a significant barrier in developing countries, limiting their accessibility to many researchers. Additionally, the requirement for specialized expertise in molecular biology and bioinformatics often restricts their application to well-equipped laboratories, leaving rural and underfunded institutions unable to adopt these technologies. The limited availability of reference genomes for many medicinal plants further constrains the development of species-specific markers and hinders detailed genetic studies. This gap in genomic resources makes it difficult to fully harness the potential of molecular markers for studying underrepresented species critical to traditional medicine and pharmaceutical industries.

The significance of molecular markers in medicinal plant research cannot be overstated. By revealing genetic diversity, these tools provide insights into the evolutionary processes that shape plant populations and inform conservation strategies to protect genetic resources. For example, studies using molecular markers have uncovered genetic variations in *W. somnifera* and *R. serpentina*, leading to more targeted breeding programmes for these medicinally important species. Additionally, molecular markers have facilitated the identification of genes involved in the biosynthesis of bioactive compounds, accelerating the development of high-yielding varieties with enhanced medicinal properties.

As the field continues to evolve, sustained research and technological innovation remain essential to overcoming existing challenges and maximizing the potential of molecular markers. Advancements in NGS technologies, such as whole-genome sequencing and transcriptomics, are significantly lowering costs and enhancing the availability of comprehensive genomic data. Concurrently, progress in bioinformatics is streamlining data processing and interpretation, making complex analyses more accessible to researchers with varying levels of expertise. Moreover, precise genome-editing tools like CRISPR/Cas9 are revolutionizing marker-assisted breeding by improving accuracy and efficiency, thereby facilitating the development of crops with superior traits while preserving critical genetic diversity.

The integration of molecular markers into medicinal plant genetics and improvement programmes remains critical for the sustainable development of these resources. By bridging the gaps in genomic knowledge and addressing technological and financial barriers, molecular markers can unlock the full potential of medicinal plants in addressing global health and biodiversity challenges. Their application in breeding, conservation, and pharmaceutical research ensures that the benefits of genetic advancements are accessible and impactful, thereby contributing to the long-term sustainability of medicinal plant resources.

REFERENCES

Adhikari, S., Biswas, A., Saha, S., Bandyopadhyay, T.K., & Ghosh, P. (2022). AFLP-based assessment of genetic variation in certain Indian elite cultivars of Cymbopogon species. *Journal of Applied Research on Medicinal and Aromatic Plants*, 29, 100372.

Akkaya, M. S., Bhagwat, A. A., & Cregan, P. B. (1992). Length polymorphisms of simple sequence repeat DNA in soybean. *Genetics*, 132(4), 1131–1139. https://doi.org/10.1093/genetics/132.4.1131

Amiteye, S. (2021). Basic concepts and methodologies of DNA marker systems in plant molecular breeding. *Heliyon*, *7*(10), e08093.

Becher, S. A., & Heun, M. (1998). SSR markers for barley: use in genetic diversity analyses. *Plant Breeding*, 117(4), 340–342. https://doi.org/10.1111/j.1439-0523.1998.tb01965.x

Bhat, K. V., & Lakhanpaul, S. (2000). RAPD markers for genetic diversity study in Indian mungbean [*Vigna radiata (L.)* Wilczek] germplasm. *Indian Journal of Genetics and Plant Breeding*, 60(3), 287–293.

Chamberlin, M. A., & Lawit, S. J. (2017). Development and observation of mature megagametophyte cell-specific fluorescent markers. In M. A. Germana & M. Lambardi (Eds.), *In Vitro Embryogenesis in Higher Plants* (pp. 237–248). Humana Press. https://doi.org/10.1007/978-1-4939-7286-9_5

Chaudhary, D., Jeena, A. S., Rohit, Gaur, S., Raj, R., Mishra, S., Kajal, Gupta, O. P., & Meena, M. R. (2024). Advances in RNA interference for plant functional genomics: unveiling traits, mechanisms, and future directions. *Applied Biochemistry and Biotechnology*, 196(9), 5681–5710.

Choudhary, S., Gaur, R. K., & Sharma, P. (2013). Molecular markers and their applications in genetic diversity assessment. *International Journal of Biotechnology and Bioengineering Research*, 4(4), 341–354.

Collard, B. C. Y., & Mackill, D. J. (2008). Marker-assisted selection: an approach for precision plant breeding in the twenty-first century. *Philosophical Transactions of the Royal Society B: Biological Sciences*, 363(1491), 557–572. https://doi.org/10.1098/rstb.2007.2170

Datta, R., Kapadia, C., Danish, S. and Singh, S. (Eds.). (2024). *Non-Coding RNA in Plants: Modulation and Stress Responses*. Elsevier.

Dumitrescu, R. G. (2012). Epigenetic markers of early tumor development. In R. Dumitrescu & M. Verma (Eds.), *Methods in Molecular Biology* (pp. 3–14). Humana Press. https://doi.org/10.1007/978-1-61779-612-8_1

Ghafoor, A., Ahmad, Z., Qureshi, A. S., & Bashir, M. (2002). Genetic relationship in *Vigna mungo* (L.) Hepper and *V. radiata* (L.) Wilczek based on morphological traits and RAPD markers. *Plant Genetic Resources Newsletter*, 130, 1–6.

Gitler, A. D., Lu, M. M., Jiang, Y. Q., Epstein, J. A., & Gruber, P. J. (2003). Molecular markers of cardiac endocardial cushion development. *Developmental Dynamics*, 228(4), 643–650. https://doi.org/10.1002/dvdy.10418

Gupta, P. K., Varshney, R. K., Sharma, P. C., & Ramesh, B. (1999). Molecular markers and their applications in wheat breeding. *Plant Breeding*, 118(5), 369–390. https://doi.org/10.1046/j.1439-0523.1999.00401.x

Hoerder, A. (2007). *Mouse cortical subplate neurones: molecular markers, connectivity and development* (Doctoral dissertation, University of Oxford). http://ethos.bl.uk/OrderDetails.do?uin=uk.bl.ethos.442449

Hormi-Carver, K., & Souza, R. F. (2009). Molecular markers and genetics in cancer development. *Surgical Oncology Clinics of North America*, 18(3), 453–467. https://doi.org/10.1016/j.soc.2009.03.002

Ipek, M., Ipek, A., & Simon, P. W. (2020). Molecular markers development and application in garlic. *Acta Horticulturae*, 1297, 653–658. https://doi.org/10.17660/ActaHortic.2020.1297.86

Jain, M., Moharana, K. C., Shankar, R., & Kumari, S. (2012). Transcriptome sequencing to develop gene-based markers in chickpea. *Plant Biotechnology Journal*, 10(6), 720–732. https://doi.org/10.1111/j.1467-7652.2012.00710.x

Jain, S. M., & Brar, D. S. (Eds.). (2009). *Molecular Techniques in Crop Improvement* (2nd ed.). Springer. https://doi.org/10.1007/978-90-481-2967-6

Joshi, C. P., & Nguyen, H. T. (1993). Application of molecular markers in plant biology. *Current Science*, 64(8), 616–622.

Joshi, M. A., & Jain, R. K. (2013). Development of SSR markers from the expressed sequence tags of sunflower (*Helianthus annuus* L.) and their application in diversity analysis. *Indian Journal of Biotechnology*, 12, 186–193.

Joshi, S. P., Gupta, V. S., Aggarwal, R. K., Ranjekar, P. K., & Brar, D. S. (2000). Genetic diversity and phylogenetic relationship as revealed by inter simple sequence repeat (ISSR) polymorphism in the genus Oryza. *Theoretical and Applied Genetics*, 100(8), 1311–1320. https://doi.org/10.1007/s001220051440

Kalendar, R., Shustov, A. V., Akhmetollayev, I., & Kairov, U. (2022). Designing allele-specific competitive-extension PCR-based assays for high-throughput genotyping and gene characterization. *Frontiers in molecular biosciences*, 9, 773956.

Kalia, R. K., Rai, M. K., Kalia, S., Singh, R., & Dhawan, A. K. (2011). Microsatellite markers: an overview of the recent progress in plants. *Euphytica*, 177(3), 309–334. https://doi.org/10.1007/s10681-010-0286-9

Livneh, O., & Vardi, E. (1998). Molecular genetic markers. In C. Kole (Ed.), *Hybrid Cultivar Development* (pp. 201–220). Springer. https://doi.org/10.1007/978-3-662-07822-8_8

Mace, E. S., & Godwin, I. D. (2002). AFLP analysis of genetic diversity in taro (*Colocasia esculenta*) in Vanuatu. *Plant Genetic Resources*, 1(2–3), 149–158. https://doi.org/10.1079/PGR200236

Mondini, L., Noorani, A., & Pagnotta, M. A. (2009). Assessing plant genetic diversity by molecular tools. *Diversity*, 1(1), 19–35. https://doi.org/10.3390/d1010019

Nadeem, M. A., Nawaz, M. A., Shahid, M. Q., Doğan, Y., Comertpay, G., Yıldız, M., ... & Baloch, F. S. (2018). DNA molecular markers in plant breeding: current status and recent advancements in genomic selection

and genome editing. *Biotechnology & Biotechnological Equipment*, 32(2), 261–285. https://doi.org/10.1080/13102818.2017.1400401
Nakagawa, H., & Ebina, M. (2001). Development of molecular markers for the analysis of apomixis. In G. Spangenberg (Ed.), *Developments in Plant Breeding* (pp. 161–173). Springer. https://doi.org/10.1007/978-94-015-9700-5_9
Panahi, B., Jalaly, H. M., & Hamid, R. (2024). Using next-generation sequencing approach for discovery and characterization of plant molecular markers. *Current Plant Biology*, 40, 100412.
Pant, M., Sohrab, S. S., & Husen, A. (2025) *Genetic Improvement and Conservation Practice of Medicinal Plants*. Taylor & Francis Group. https://doi.org/10.1201/9781003482659
Patani, A., Prajapati, D., Singh, S., Enakiev, Y., Bozhkov, S., Jabborova, D., & Joshi, C. (2023). Beneficial effects of biochar application on mitigating the drought and salinity stress implications on plants. *Plant Science Today*, 10, 188–193.
Pootakham, W., Shearman, J. R., & Tangphatsornruang, S. (2020). Development of molecular markers in *Hevea brasiliensis* for marker-assisted breeding. In M. Matsui & K. S. Chow (Eds.), *The Rubber Tree Genome* (pp. 67–79). Springer. https://doi.org/10.1007/978-3-030-42258-5_5
Powell, W., Morgante, M., Andre, C., Hanafey, M., Vogel, J., Tingey, S., & Rafalski, A. (1996). The comparison of RFLP, RAPD, AFLP and SSR (microsatellite) markers for germplasm analysis. *Molecular Breeding*, 2(3), 225–238. https://doi.org/10.1007/BF00564200
Prasad, M., Varshney, R. K., Roy, J. K., Balyan, H. S., & Gupta, P. K. (2000). The use of microsatellite markers for detecting DNA polymorphism, genotype identification and genetic diversity in wheat. *Theoretical and Applied Genetics*, 100(3–4), 494–502. https://doi.org/10.1007/s001220050067
Rani, K., & Singh, R. (2013). Genetic diversity analysis of cowpea (*Vigna unguiculata*) genotypes using RAPD markers. *Indian Journal of Agricultural Sciences*, 83(2), 191–195.
Sao, P., Parveen, R., Khattri, A., Sharma, S., Tiwari, N., & Singh, S. (2022). Understanding the metabolomics of medicinal plants under environmental pollution: challenges and opportunities. In A. Husen (Ed.), *Environmental Pollution and Medicinal Plants* (pp. 193–212). CRC Press/Taylor & Francis.
Schubert, R., & Müller-Starck, G. (2002). Development and application of molecular markers in conifers. In S. M. Jain, D. S. Brar, & B. S. Ahloowalia (Eds.), *Molecular Techniques in Crop Improvement* (pp. 139–160). Springer. https://doi.org/10.1007/978-94-017-2356-5_5
Semagn, K., Bjørnstad, Å., & Ndjiondjop, M. N. (2006). An overview of molecular marker methods for plants. *African Journal of Biotechnology*, 5(25), 2540–2568.
Singh, S., Kumari, K., Chaturvedi, S., Pandey, N., & Varghese, A. (2017). Molecular characterization of *Anogeissus acuminata* genotypes employing RAPD markers. In K. Mukhopadhyay, A. Sachan, & M. Kumar (Eds.), *Applications of Biotechnology for Sustainable Development* (pp. 7–14). Springer. https://doi.org/10.1007/978-981-10-5538-6_2
Singh, S. R., & Rameshwar, P. (2016). *MicroRNA in Development and in the Progression of Cancer*. Springer.
Soriano, J. M. (2020). Molecular marker technology for crop improvement. *Agronomy*, 10(10), 1462. https://doi.org/10.3390/agronomy10101462
Storer, M., & Keyes, W. M. (2016). Detection of senescence markers during mammalian embryonic development. In M. Nikiforov (Ed.), *Methods in Molecular Biology* (pp. 199–210). Springer. https://doi.org/10.1007/978-1-4939-6670-7_19
Tanksley, S. D., & McCouch, S. R. (1997). Seed banks and molecular maps: unlocking genetic potential from the wild. *Science*, 277(5329), 1063–1066. https://doi.org/10.1126/science.277.5329.1063
Tar'an, B. (2000). *Development and application of molecular markers in common bean breeding* (Doctoral dissertation, National Library of Canada). www.collectionscanada.ca/obj/s4/f2/dsk1/tape2/PQDD_0015/NQ47413.pdf
Tautz, D. (1989). Hypervariability of simple sequences as a general source for polymorphic DNA markers. *Nucleic Acids Research*, 17(16), 6463–6471. https://doi.org/10.1093/nar/17.16.6463
Valdman, A. (2003). *Molecular genetic markers of prostate cancer development* (Doctoral dissertation, Karolinska Institutet). http://diss.kib.ki.se/2003/91-7349-618-9/
Varshney, R. K., Prasad, M., Kota, R., Sigmund, R., Börner, A., Scholz, U., ... & Graner, A. (2012). Functional molecular markers in barley: development and applications. *Czech Journal of Genetics and Plant Breeding*, 41, 128–133. https://doi.org/10.17221/6152-cjgpb

Vos, P., Hogers, R., Bleeker, M., Reijans, M., van de Lee, T., Hornes, M., ... & Zabeau, M. (1995). AFLP: a new technique for DNA fingerprinting. *Nucleic Acids Research*, 23(21), 4407–4414. https://doi.org/10.1093/nar/23.21.4407

Weising, K., Nybom, H., Wolff, K., & Meyer, W. (2005). *DNA Fingerprinting in Plants: Principles, Methods, and Applications*. CRC Press.

Williams, L. M., Ma, X., Boyko, A. R., Bustamante, C. D., & Oleksiak, M. F. (2010). SNP identification, verification, and utility for population genetics in a non-model genus. *BMC Genetics*, 11(1), 32.

Winter, P., & Kahl, G. (1995). Molecular marker technologies for plant improvement. *World Journal of Microbiology and Biotechnology*, 11(4), 438–448. https://doi.org/10.1007/BF00364604

Xu, Y. (2013). *Molecular Plant Breeding*. CAB International.

2 Advances in Biotechnological Interventions of Medicinal Plants and Herbal Formulations

Current Status and Regulatory Affairs

Pragati Khare, Noopur Khare, Tanvi Jain, Sachidanand Singh, Megha Barot, and Azamal Husen

2.1 INTRODUCTION

Medicinal plants have been a cornerstone of traditional medicine for centuries, providing essential bioactive compounds used in drug development. However, the sustainable production and quality control (QC) of herbal formulations remain a challenge. Recent advances in biotechnology offer promising solutions to enhance the yield, consistency, and therapeutic efficacy of medicinal plants. The integration of biotechnological tools such as plant tissue culture, genetic engineering, metabolic engineering, and nanotechnology has significantly improved the commercial viability of medicinal plants (Husen and Pant 2024; Pant et al., 2025). These techniques aid in the large-scale production of bioactive compounds, ensuring a steady supply of medicinal ingredients while conserving endangered plant species. Additionally, synthetic biology and bioinformatics have opened new avenues for optimizing metabolic pathways and improving the quality of herbal formulations (Mustafa et al., 2017).

Regulatory affairs are essential for maintaining the safety, efficacy, and quality of herbal medicines (Mosihuzzaman and Choudhary, 2008). International frameworks—such as World Health Organization (WHO) guidelines, Food and Drug Administration (FDA) regulations, and national policies governing traditional medicine—offer systematic approaches for the standardization and quality assurance of herbal formulations (Jain et al., 2019). However, challenges including variations in phytochemical composition, absence of uniform testing standards, and issues related to intellectual property rights must be effectively addressed to ensure the sustainable development of the herbal medicine sector (Raju and Das, 2024). This chapter discusses recent biotechnological advancements in medicinal plant research and herbal formulation, emphasizing their influence on modern healthcare and the regulatory challenges they present. The objective is to provide a comprehensive understanding of current innovations and outline the strategic measures required for the successful integration of biotechnology into traditional medicine systems.

DOI: 10.1201/9781003542681-2

2.2 BIOTECHNOLOGICAL INTERVENTIONS IN MEDICINAL PLANTS

2.2.1 Plant Tissue Culture and Micropropagation

Plant tissue culture has become one of the most efficient methods for propagating medicinal plants on a large scale. In vitro propagation techniques allow the production of high-quality plant material, ensuring the conservation of rare and endangered species (Hussain et al., 2012).

- *Micropropagation Techniques:* Callus culture, shoot tip culture, and somatic embryogenesis play a key role in rapid plant multiplication. Clonal propagation ensures the production of genetically uniform plants with desirable traits. Rooting and acclimatization methods enhance the survival rates of tissue-cultured plants (Chokheli et al., 2020).
- *Cryopreservation for Germplasm Conservation:* Long-term storage of medicinal plant cells and tissues in liquid nitrogen preserves genetic resources. Techniques like encapsulation—dehydration and vitrification enhance cryopreservation success rates (Stanwood, 2024).
- *Micropropagation of Medicinal Plants:* Micropropagation of *Rauwolfia serpentina* is done for reserpine production. In vitro propagation of *Withania somnifera* is carried out to enhance alkaloid yield (Lemma et al., 2020).

2.2.2 Genetic Engineering and Metabolic Engineering

Genetic engineering techniques allow the modification of medicinal plants to enhance the production of secondary metabolites with therapeutic properties (Halka et al., 2025).

- *Transgenic Approaches for Enhanced Secondary Metabolite Production:* Introduction of genes encoding key enzymes in metabolic pathways (e.g., artemisinin production in *Artemisia annua*). Overexpression of regulatory genes to boost alkaloid, flavonoid, and terpenoid synthesis (Satish et al., 2019).
- *CRISPR/Cas9 Genome Editing Applications:* Precision genome editing to knock out unwanted genes and improve phytochemical content. Development of stress-resistant medicinal plants through targeted mutations (Liu et al., 2017).
- *Agrobacterium-Mediated Transformation:* An effective method for gene transfer in medicinal plants to enhance secondary metabolite accumulation. Case studies on genetic transformation of *Catharanthus roseus* for improved vinblastine and vincristine production (Li et al., 2017).
- *Enhancing Biosynthetic Pathways through Metabolic Engineering:* Metabolic flux analysis to optimize phytochemical synthesis. Pathway engineering using synthetic promoters and transcription factors (Das et al., 2020).

2.2.3 Synthetic Biology and Bioinformatics in Medicinal Plants

Synthetic biology and bioinformatics tools are transforming the study and utilization of medicinal plants by enabling rational design and optimization of metabolic pathways.

- *Designing Synthetic Pathways for Bioactive Compound Production:*
  - Reconstruction of entire biosynthetic pathways in microbial hosts for scalable production.
  - Application of synthetic promoters to enhance gene expression in medicinal plants (Liu et al., 2024).
- *Computational Modelling of Metabolic Networks:*
  - Predictive modelling of secondary metabolite biosynthesis pathways.
  - Identification of key regulatory nodes for metabolic engineering (Kalyankar et al., 2012).

- *Genome Sequencing and Annotation of Medicinal Plants:*
  - Whole-genome sequencing of medicinal plants to uncover novel genes involved in bioactive compound synthesis.
  - Comparative genomics to identify conservation and variation in metabolic pathways (Hasnain et al., 2022).

2.2.4 Nanobiotechnology in Herbal Medicine

Nanobiotechnology, the interdisciplinary integration of nanotechnology and biotechnology, has become a ground-breaking advancement in the realm of herbal medicine. This emerging field focuses on improving the efficacy, bioavailability, and safety of herbal formulations through the design of nanoscale drug delivery systems. With the growing global preference for herbal medicines due to their therapeutic benefits and reduced side effects compared to synthetic drugs, the incorporation of nanobiotechnology offers significant opportunities as well as notable challenges (Dewi et al., 2022; Choudhary et al., 2011).

2.2.5 Current Status of Nanobiotechnology in Herbal Medicine

Nanobiotechnology focuses on manipulating materials at the nanoscale (1–100 nm) to develop novel drug delivery systems (NDDS). These advanced systems are designed to improve the solubility, stability, and bioavailability of herbal compounds, which frequently exhibit poor absorption resulting from their large molecular size or limited water solubility (Patra et al., 2018).

Nanotechnology has gained considerable importance in herbal medicine by improving its efficacy and therapeutic performance (Khare et al., 2025a). A key application lies in advanced drug delivery, where nanocarriers such as liposomes, solid lipid nanoparticles (SLNs), polymeric nanoparticles, and nano-emulsions are utilized to enhance the delivery of herbal extracts. These nano-systems encapsulate bioactive compounds, safeguarding them from degradation and facilitating targeted delivery to specific sites within the body. Another major advantage of nanotechnology is the improvement of bioavailability, as many herbal constituents possess poor solubility, resulting in limited absorption (Khare et al., 2025b). Nanoparticles can significantly enhance the solubility and stability of these compounds, thereby boosting their therapeutic effectiveness. For example, research has shown that nanoparticles derived from *Cuscuta chinensis* notably increased the absorption of its flavonoids and lignans. Additionally, nanotechnology supports controlled release mechanisms, enabling the sustained and gradual release of active ingredients. This not only ensures consistent therapeutic levels but also reduces dosing frequency and enhances patient compliance (Kumar et al., 2022).

Recent research in nanobiotechnology has been increasingly directed towards developing advanced nanostructured carrier systems for various herbal medicines (Khare et al., 2025a). Among these, polymeric nanoparticles have been effectively used to deliver anticancer agents derived from herbs such as *Curcuma longa* (turmeric) and *Momordica charantia* (bitter melon), demonstrating enhanced therapeutic effects compared to traditional formulations. Lipid-based systems, including liposomes and SLNs, have also shown promise by encapsulating herbal extracts from plants like *Panax ginseng*, resulting in improved absorption and increased bioactivity. Additionally, the plant-mediated synthesis of metallic nanoparticles is gaining significant attention for its antibacterial properties, especially when combined with traditional herbal extracts, highlighting the potential of these hybrid systems in combating microbial infections (Mittal et al., 2022). Looking ahead, the future of nanobiotechnology in herbal medicine hinges on expanding research and development to explore novel applications. One key area is targeted therapy, where efforts are underway to design delivery systems that can transport herbal compounds directly to diseased tissues or cells, thereby maximizing treatment efficacy and minimizing side effects. Another promising avenue is personalized medicine, where the integration of nanotechnology with individualized treatment

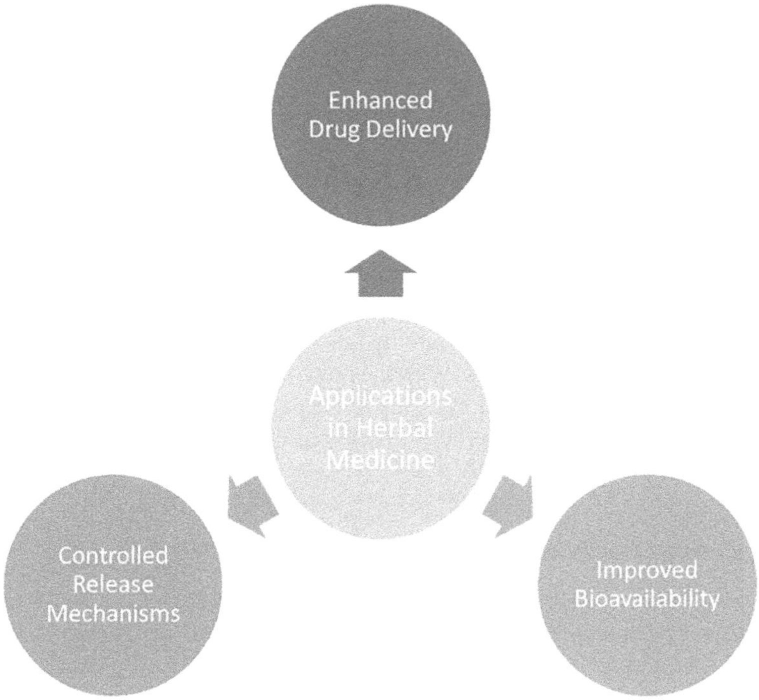

FIGURE 2.1 Applications for herbal medicine.

strategies could enable the development of customized herbal therapies based on a patient's genetic make-up (Bhardwaj et al., 2023).

As this innovative field progresses, it becomes increasingly important to establish robust regulatory frameworks. This includes creating standardised protocols for the production and characterization of nano-pharmaceuticals derived from herbal sources, ensuring consistency, safety, and efficacy. Safety assessments will be critical, with regulatory agencies tasked with formulating clear guidelines for evaluating the potential risks and benefits of these nanotechnological interventions. Moreover, the advancement of nanobiotechnology in herbal medicine will require collaboration across disciplines. Researchers, clinicians, regulatory authorities, and industry stakeholders must work together to facilitate knowledge sharing, align on best practices, and ensure regulatory compliance. Furthermore, public engagement plays a vital role in building trust; by educating the public on the benefits and safety of nanobiotechnological applications in herbal medicine (Figure 2.1), informed decisions can be encouraged, and widespread acceptance can be achieved (Ansari et al., 2012).

2.2.6 Challenges Ahead

Despite the promising prospects of nanobiotechnology in herbal medicine, several critical challenges must be addressed to fully realize its potential. One of the foremost concerns is safety, particularly the long-term effects of nanoparticles. Comprehensive toxicity studies are essential to evaluate any adverse health outcomes associated with the use of nanoparticles, especially when combined with traditional herbal formulations. Additionally, the environmental impact of manufacturing and disposing of nanomaterials must be carefully assessed to ensure sustainable and eco-friendly practices. Another significant hurdle is market acceptance. For nanobiotechnological approaches to gain widespread adoption in herbal medicine, consumer awareness must be improved. Educating the public about the benefits, safety, and scientific basis of nanotechnology-enhanced herbal products is vital to overcoming scepticism. Many consumers may hold misconceptions or harbour doubts about products that incorporate advanced technologies like nanotechnology, making clear and accessible communication a priority (Gupta et al., 2024).

Economic barriers also pose a challenge to the development and commercialization of these advanced formulations. The excessive costs associated with research, development, and regulatory compliance may discourage small-scale or emerging producers from investing in nanotechnology applications (Parveen et al., 2023). Furthermore, the competitive market landscape often favours larger, well-established pharmaceutical companies that have the resources to support the development and marketing of such innovations. Nevertheless, nanobiotechnology holds tremendous potential to revolutionize herbal medicine by enhancing drug delivery systems, improving the bioavailability of active compounds, and enabling targeted therapeutic effects. While scientific advancements in this area continue to evolve, addressing the intertwined challenges of safety, regulation, market perception, and economic viability is crucial. Through interdisciplinary collaboration and increased public engagement, stakeholders can work collectively to unlock the full benefits of nanobiotechnology, enhancing traditional herbal therapies and contributing to better health outcomes (Xuan et al., 2023).

2.3 STANDARDIZATION AND QC OF HERBAL FORMULATIONS

2.3.1 Importance of Standardization in Herbal Medicine

Herbal formulations are complex mixtures of bioactive compounds, and their efficacy depends on consistency and quality. Standardization is essential for ensuring batch-to-batch uniformity, reproducibility of therapeutic effects, and safety of herbal products (Sachan et al., 2016).

- *Factors Affecting Herbal Formulation Quality:*
  - Variability in phytochemical composition due to genetic and environmental factors.
  - Differences in harvesting, processing, and storage conditions.
  - Challenges in maintaining consistent active ingredient concentrations.
- *Global Standardization Efforts:*
  - WHO guidelines for QC of herbal medicines.
  - Role of pharmacopoeias (e.g., USP, BP, ayurvedic pharmacopoeia) in setting quality standards.
  - Implementation of Good Agricultural and Collection Practices (GACP) for medicinal plants.

2.3.2 Analytical Techniques for Herbal Drug QC

Herbal medicines have been used for centuries, and their popularity has grown significantly in recent years as consumers increasingly turn to natural alternatives to conventional pharmaceuticals. Despite this growing acceptance, ensuring the safety, efficacy, and quality of herbal products remains a critical concern (Patani et al., 2023). QC plays a vital role in verifying that herbal medicines adhere to established standards of identity, purity, potency, and safety. This chapter examines the various analytical techniques applied in the QC of herbal drugs, highlighting their significance in preserving product integrity and ensuring consumer safety (Wang et al., 2023).

2.3.3 Importance of QC in Herbal Medicines

QC in herbal medicines is essential for ensuring their safety, efficacy, and overall credibility (Figure 2.2). A primary aspect of QC is consumer safety—comprehensive testing ensures that herbal formulations are free from harmful contaminants such as heavy metals, pesticides, and

FIGURE 2.2 Importance of QC in herbal medicines.

microbial toxins, while also verifying that they contain the correct active ingredients in appropriate concentrations (Wang et al., 2023). This is vital for minimizing health risks posed by adulterated or substandard products. Moreover, QC enhances efficacy by promoting the standardization of bioactive compounds, ensuring consistent and reliable therapeutic outcomes across production batches (Khare et al., 2025a). Regulatory compliance is equally important, as many nations enforce stringent standards and guidelines governing the manufacture and distribution of herbal products, necessitating the adoption of rigorous QC procedures. In addition, effective quality assurance safeguards market integrity by preventing adulteration and mislabelling, thereby strengthening consumer trust and confidence in the safety, authenticity, and effectiveness of herbal medicines (Ekor, 2014).

2.3.4 Key Analytical Techniques for QC

2.3.4.1 Chromatographic Techniques

Chromatography is one of the most widely employed analytical techniques in the evaluation of herbal medicines, as it enables the separation, identification, and quantification of various components within complex mixtures. Among the different chromatographic methods, high-performance liquid chromatography (HPLC) stands out for its high resolution and sensitivity. HPLC is extensively used to separate and quantify active compounds in herbal extracts, making it particularly suitable for analysing complex matrices. It is commonly applied to determine the concentration of specific bioactive ingredients such as flavonoids, alkaloids, and other phytochemicals. The technique offers rapid analysis with minimal sample preparation and can be enhanced using various detectors, including ultraviolet (UV) and fluorescence, for improved sensitivity. Another important method is gas chromatography (GC), which is especially effective for analysing volatile compounds and essential oils found in many medicinal herbs. GC is typically used to detect and quantify terpenes, fatty acids, and other volatile constituents. It offers high sensitivity and resolution; however, it is

FIGURE 2.3 Key analytical techniques for QC.

limited to compounds that are naturally volatile or can be chemically modified (derivatized) to achieve volatility. Additionally, high-performance thin-layer chromatography (HPTLC) is gaining popularity due to its ability to combine the simplicity of thin-layer chromatography with enhanced performance capabilities. HPTLC is widely used for both qualitative and quantitative analysis of multiple components in a single run. Its main advantages include the ability to process a large number of samples efficiently and its cost-effectiveness, making it an ideal choice for routine QC in herbal drug analysis (Figure 2.3) (Kumari et al., 2022).

2.3.4.2 Spectroscopic Techniques

Spectroscopy, which studies the interaction of light with matter, provides critical insights into the molecular structure and composition of various substances. Among the most commonly used spectroscopic methods, ultraviolet–visible spectrophotometry (UV–Vis) is frequently applied to quantify active constituents in herbal extracts by analysing their absorbance properties. This technique is particularly effective for determining phenolic and flavonoid content and is valued for its simplicity, rapid results, and minimal sample preparation requirements. Fourier transform infrared spectroscopy (FTIR) is another prominent method, used primarily to identify functional groups in organic compounds through infrared absorption spectra. FTIR is highly effective in assessing the chemical composition of herbal materials and detecting adulteration. Its advantages include quick analysis, ease of operation, and minimal sample handling, making it ideal for routine QC. Nuclear magnetic resonance (NMR) spectroscopy serves as a more advanced analytical tool, offering detailed structural and dynamic information about molecules. It is especially beneficial for elucidating the structures of complex mixtures found in herbal medicines. A major advantage of NMR is its non-destructive nature, combined with its ability to provide comprehensive molecular insights, making it indispensable for advanced research and the development of high-quality herbal formulations (Sinha et al., 2023).

2.3.4.3 Mass Spectrometry

Mass spectrometry (MS) is often employed in conjunction with chromatographic techniques such as gas chromatography–mass spectrometry (GC-MS) and liquid chromatography–mass spectrometry (LC-MS) to analyse the complex mixtures typically present in herbal formulations. This combined approach enhances the ability to identify, separate, and quantify bioactive compounds with high sensitivity and precision, making it indispensable for the comprehensive chemical profiling and quality assessment of herbal products. This coupling enhances the ability to separate, identify, and quantify individual components within intricate herbal formulations. The key advantages of MS include its high sensitivity and specificity, which make it particularly effective for detecting trace amounts of active compounds or potential contaminants, thereby ensuring the quality and safety of herbal medicines (Lai and Wang, 2023).

2.3.4.4 Biological Assays

Biological assays are essential tools for evaluating the pharmacological activity of herbal medicines through either in vitro or in vivo testing methods. These assays are commonly used to assess various therapeutic properties of herbal extracts, including antioxidant activity, antimicrobial effects, and cytotoxicity. By directly measuring the biological response to herbal compounds, these tests provide concrete evidence of their efficacy. However, while biological assays offer valuable insights, they often require extensive validation to ensure the accuracy, reliability, and reproducibility of the results, which is crucial for the development and standardisation of effective herbal therapies (Barba-Ostria et al., 2022).

2.3.4.5 DNA Barcoding

DNA barcoding is a technique that involves sequencing specific regions of DNA to accurately identify plant species. This method is particularly useful in the authentication of herbal materials, as it allows for the confirmation of plant identity at the genetic level. DNA barcoding is especially valuable for distinguishing species that may appear morphologically similar but possess different therapeutic properties. The key advantage of this approach is its reliability in providing a precise and unambiguous identification of plants, which is crucial for ensuring the safety and effectiveness of herbal medicines (Antil et al., 2023).

2.3.4.6 Chemometric Techniques

Chemometrics are particularly useful for enhancing the interpretation of data from chromatographic fingerprints or spectroscopic profiles of herbal medicines. By applying chemometric methods, researchers can uncover hidden patterns or correlations within the data that may not be easily detected through traditional analysis methods. The main advantage of chemometrics is its ability to provide deeper insights into the composition and quality of herbal products, thereby improving data analysis efficiency and accuracy (Hurkul et al., 2024).

2.3.5 Challenges in QC

Despite notable progress in analytical technologies, several challenges persist in maintaining the quality and consistency of herbal medicines. A major obstacle is the inherent complexity of herbal formulations, which often contain numerous active constituents, making it difficult to achieve standardization using a single marker compound. Variability in raw materials further complicates quality assurance, as factors like geographical origin, cultivation conditions, harvesting time, and processing techniques can significantly influence the chemical composition and therapeutic potency of the final product. The absence of standardized testing protocols and globally accepted quality guidelines also hinders regulatory compliance and leads to inconsistencies among manufacturers. Additionally, the issue of adulteration—such as contamination with synthetic chemicals or substitution with inferior plant materials—continues to pose serious risks to consumer safety and undermines the credibility of herbal medicines (Klein et al., 2021).

2.4 REGULATORY FRAMEWORKS

Regulatory bodies play a vital role in ensuring the QC of herbal medicines, especially as the global market for these products has expanded significantly due to growing consumer demand for natural and alternative therapies. With this increasing popularity, concerns about safety, efficacy, and quality of herbal medicines have also risen (Zhang et al., 2012). In response, regulatory agencies are essential in establishing and enforcing standards that guarantee the quality of these products. These agencies oversee various aspects of herbal medicine production, from the sourcing of raw materials to the final formulation, ensuring that the products are safe for consumption and effective in their therapeutic use (Edo et al., 2024). Herbal medicines, derived from plant sources, are commonly regarded as safer and more natural alternatives to synthetic pharmaceuticals. They are extensively used to manage a wide range of health conditions, including digestive disorders, respiratory problems, and chronic diseases. According to the WHO, nearly 80% of the global population depends on herbal medicines for some aspects of primary healthcare. However, despite their widespread use and perceived safety, inadequate QC poses serious health risks, including adverse effects, toxicity, and reduced therapeutic efficacy (Parvin et al., 2023). Historically, herbal medicines have been integral to traditional medicine systems such as Traditional Chinese Medicine (TCM), Ayurveda, and Native American healing practices. These systems have developed extensive knowledge regarding the therapeutic uses of various plants. However, as herbal products enter modern healthcare systems, it becomes essential to validate their efficacy and safety through rigorous QC measures (Pankaj et al., 2024).

2.4.1 Importance of QC

Herbal medicines have been utilized for centuries in diverse cultural traditions for their therapeutic benefits. As global interest in natural and plant-based remedies continues to rise, the demand for herbal products has grown substantially. However, this increasing demand underscores the urgent need for robust QC measures to guarantee the safety, efficacy, and consistency of these products. QC involves a comprehensive set of procedures designed to monitor and maintain the quality of herbal medicines throughout their entire lifecycle—from cultivation and processing to packaging and consumption. This chapter highlights the significance of QC in herbal medicines, emphasizing its role in ensuring consumer safety, achieving regulatory compliance, promoting standardization, and preventing adulteration and contamination (Edo et al., 2024). Ensuring consumer safety is of paramount importance when it comes to herbal medicines. QC measures play a crucial role in mitigating risks associated with contaminants and adulterants that may be present in these products. Herbal products can be contaminated with heavy metals, such as lead and arsenic, pesticides, or microbial pathogens during cultivation or processing. QC protocols involve testing for these contaminants to ensure that products are safe for consumption. Adulteration is another significant concern, occurring when inferior or unrelated plant materials are mixed with genuine herbs to increase profit margins. This practice not only compromises product quality but also poses serious health risks to consumers. Maintaining efficacy is another vital aspect of QC in herbal medicines. It ensures that herbal products contain consistent levels of active compounds responsible for their therapeutic effects. Variability in active ingredient concentrations can lead to inconsistent therapeutic outcomes, which is why standardization is essential. Standardizing the extraction and formulation processes helps maintain consistent potency across different batches by identifying key active compounds and setting benchmarks for their concentrations. QC measures also ensure batch-to-batch consistency, minimizing variability between batches by implementing rigorous testing protocols to evaluate the composition and potency of each batch before it reaches consumers (Shaban et al., 2016).

Regulatory compliance is an essential component of QC for herbal medicines. Regulatory bodies oversee the manufacturing and marketing of herbal products, ensuring that they meet established guidelines and standards. Organizations such as the WHO, FDA, and European Medicines Agency (EMA) provide guidelines on good manufacturing practices (GMP) and quality assurance protocols, ensuring that herbal products are produced and marketed safely. Manufacturers must obtain licences and register their products with regulatory agencies, demonstrating compliance with safety and efficacy standards through rigorous testing before products can reach the market. Preventing adulteration and contamination is another key focus of QC measures in herbal medicines. Analytical techniques, such as chromatography (HPLC, GC) and spectroscopy (UV–Vis), are employed to identify and quantify active compounds while detecting potential adulterants. Authenticity testing methods, such as DNA barcoding and chemical profiling, are used to authenticate herbal ingredients, ensuring that consumers receive the correct species with known therapeutic properties (Mollah et al., 2021).

2.4.2 Components of QC

QC in herbal medicines encompasses several key components to ensure the safety, efficacy, and consistency of the products. Accurately identifying and authenticating the raw materials used in herbal products is critical for ensuring the authenticity and therapeutic potential of the herbs. Standardization processes are essential for maintaining consistent quality across different batches of herbal products. Identifying key active compounds helps understand the therapeutic properties of the herbs, while establishing reference standards provides benchmarks for desired levels of active compounds during QC testing (Mukherjee, 2019).

MS, when coupled with chromatography, provides detailed information about molecular weight and structure, further aiding in the comprehensive analysis of herbal products. Microbial testing is another crucial aspect of QC, as microbial contamination can pose significant health risks. Testing for microbial load ensures that herbal products meet safety standards for pathogens, and endotoxin testing is essential for injectable herbal formulations to prevent adverse reactions in consumers (Kaushik et al., 2014).

2.4.3 Challenges in QC

Despite its importance, several challenges hinder effective QC in herbal medicine. One significant issue is the variability in plant materials, which can lead to inconsistencies in product composition. Geographical differences, such as variations in soil composition, climate conditions, and cultivation practices, can all affect the concentration of active compounds in the plants. The lack of standardized protocols is another challenge that complicates regulatory compliance. Inconsistent manufacturing practices, such as variability in extraction methods or formulation processes among different manufacturers, can lead to differences in product quality. Furthermore, harmonization issues arise due to disparities between national regulations, which create obstacles for international trade in herbal medicines and hinder the establishment of global standards. Limited research funding also presents a barrier to advancing QC in herbal medicines. Insufficient scientific data on the safety and efficacy of many herbal products limits their acceptance within mainstream medicine (Van Wyk et al., 2020).

2.4.4 Future Directions in QC

To enhance QC measures for herbal medicines, several strategies can be pursued. One key approach is strengthening regulatory frameworks (Khare et al., 2025a). Regulatory agencies should work towards establishing more comprehensive regulations governing the production and sale of herbal medicines. Collaborative efforts between countries can help harmonize regulations related to QC practices, ensuring consistency across borders. Additionally, building the capacity of regulatory personnel through training programmes will enhance their ability to effectively oversee compliance and maintain lofty standards. Investing in research is another important strategy to address existing gaps in herbal medicine QC. Increased funding for clinical trials will provide valuable data on the safety and efficacy of herbal products, helping to validate their therapeutic uses. Furthermore, developing standardized testing methods will improve consistency across laboratories conducting quality assessments, ensuring that herbal medicines meet established quality benchmarks (Dubale et al., 2025; Chisholm and Critchley, 2023).

Enhancing public awareness about the importance of QC is vital for strengthening consumer confidence in herbal medicines. Educational campaigns can help inform the public about safe usage practices, while transparency initiatives that clearly communicate regulatory procedures can further build trust in quality assurance systems. Implementing rigorous QC measures is essential to ensure the safety, efficacy, and consistency of herbal medicines. With global consumer demand steadily increasing, such measures must be enforced throughout every stage of production—from raw material sourcing to the final product's distribution. Addressing key challenges such as variability in plant materials, lack of standardization, limited research funding, and insufficient public awareness can significantly improve the credibility and integrity of herbal medicines while protecting public health (Khare et al., 2024). Furthermore, integrating traditional knowledge with modern scientific innovations can strengthen QC frameworks. Ultimately, comprehensive and transparent quality assurance practices will foster greater consumer trust in herbal products and promote their responsible integration into healthcare systems worldwide (Gatt et al., 2024).

2.4.5 Regulatory Frameworks for Herbal Medicines

In the United States, the Food and Drug Administration (FDA) regulates dietary supplements, including herbal products, under the Dietary Supplement Health and Education Act (DSHEA) of 1994. The FDA mandates that herbal product labels must accurately reflect their contents and not make misleading claims about health benefits. Manufacturers are required to adhere to GMP guidelines to ensure product safety and quality. The FDA also monitors adverse event reports related to dietary supplements to identify potential safety issues. The EMA oversees the regulation of herbal medicinal products within the European Union. The Traditional Herbal Medicinal Products Directive (THMPD) provides a framework for the registration and approval of traditional herbal medicines (THMs) based on long-standing use. The EMA sets stringent quality standards for herbal products, requiring evidence of both safety and efficacy before approval. Furthermore, the EMA emphasizes ongoing post-market surveillance to ensure the continued safety of herbal products after they enter the market (Bailey et al., 2020).

In addition to global and regional regulatory bodies, many countries have established their own regulatory authorities to oversee herbal medicines. For example, India's Department of AYUSH

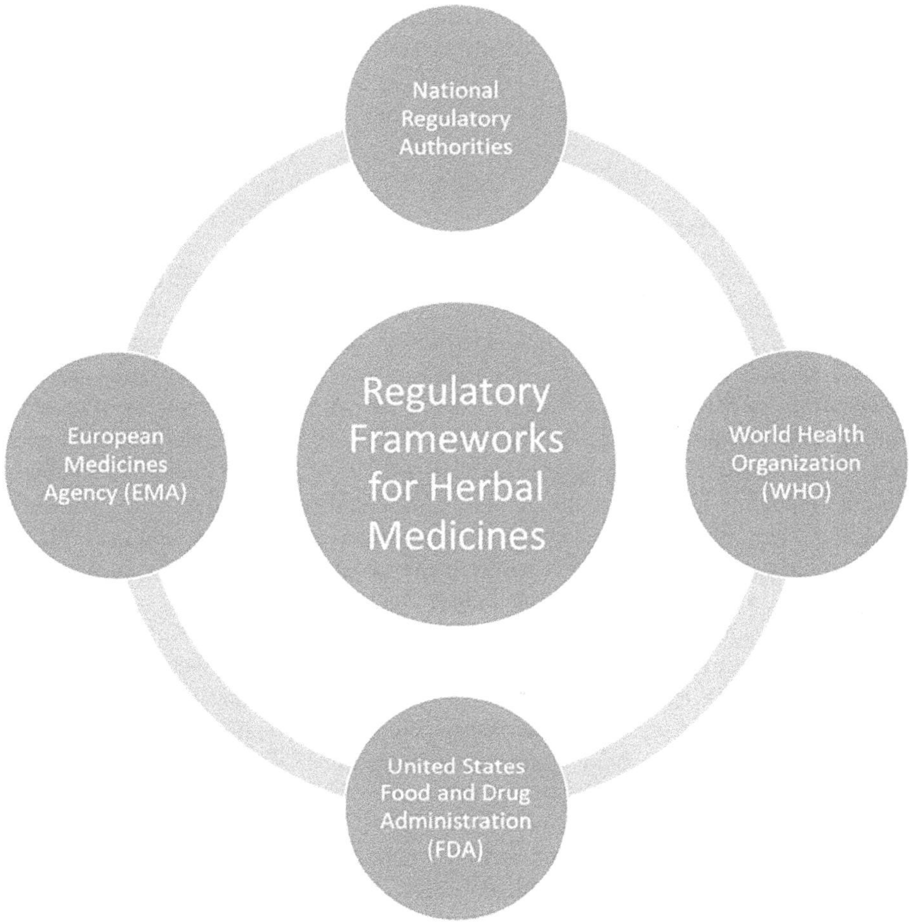

FIGURE 2.4 Regulatory frameworks for herbal medicines.

regulates Ayurvedic, Unani, Siddha, and Homoeopathic systems, enforcing compliance with quality standards through licensing and monitoring. Australia's Therapeutic Goods Administration (TGA) regulates complementary medicines, including herbal products, ensuring they meet safety and efficacy standards. In Canada, the Natural Health Products Directorate (NHPD) oversees natural health products, including herbs, requiring evidence of safety and effectiveness before they are approved for market entry (Figure 2.4) (Thakkar et al., 2020).

2.4.6 Key Functions of Regulatory Bodies

Regulatory bodies perform several critical functions to ensure the QC of herbal medicines, which are essential in safeguarding consumer health and maintaining the integrity of the herbal medicine industry. One of their key functions is establishing standards. Regulatory agencies develop guidelines that outline acceptable standards for various aspects of herbal medicine production. This includes QC protocols that establish methods for testing identity, purity, potency, and contaminants in herbal products. Licensing and registration are also essential components of regulatory oversight. Before an herbal product can be marketed, manufacturers are required to obtain licences or register their products with regulatory bodies. Manufacturers must demonstrate compliance with GMP to obtain a manufacturing license. In many authorities, pre-market approval or product registration is necessary, requiring manufacturers to submit evidence of safety and efficacy before their products can be sold to the public (Heinrich, 2015; Sahu et al., 2024).

Monitoring compliance is another vital role of regulatory agencies. These agencies are responsible for ensuring that herbal products adhere to established standards through regular inspections of manufacturing facilities. Inspections are conducted to ensure that manufacturers are complying with GMP and other regulatory requirements. Random sampling and testing of marketed products are also carried out to identify any non-compliance or adulteration issues, ensuring that products on the market meet the required quality standards. Finally, regulatory bodies establish systems for adverse event reporting. Post-market surveillance is crucial to monitor adverse reactions to herbal medicines and to identify potential safety concerns related to specific products or ingredients. When adverse events are reported, regulatory agencies evaluate the information to determine whether further action is needed, such as product recalls or issuing warnings to the public (Eruaga et al., 2024).

2.5 CHALLENGES IN SAFETY AND TOXICITY ASSESSMENTS

The safety evaluation of herbal medicines is fraught with several challenges that need to be addressed to ensure consumer health protection. One major issue is the complexity of herbal formulations. These medicines often contain multiple active constituents that can interact in unpredictable ways. The effects of these interactions may differ significantly from those observed when individual components are studied in isolation. This complexity makes it difficult to assess the safety profiles of herbal products comprehensively, as the combined effects of various compounds may be unknown or difficult to predict. The absence of universally accepted methods for evaluating the safety and toxicity of herbal products creates variability in quality and efficacy. Differences in preparation methods, dosages, and plant sources can lead to inconsistent results, making it challenging to establish reliable safety standards. Inadequate toxicological data is also a critical obstacle. Many herbal products lack comprehensive and systematic toxicological studies, which makes it difficult to accurately evaluate their safety. Existing studies often focus on traditional uses of herbal remedies, but they typically lack modern, rigorous toxicological assessments that are necessary to determine the potential risks associated with these products (Muyumba et al., 2021).

2.5.1 Methodologies for Safety and Toxicity Assessment

Preclinical toxicity studies are an essential first step in evaluating the safety profile of herbal medicines. These studies involve both in vitro (cell-based) and in vivo (animal-based) assessments to determine potential harmful effects before human exposure. Acute toxicity testing is used to evaluate the effects of a single high dose of a substance over a fleeting period, helping identify immediate health risks. Sub-acute and chronic toxicity testing assess the effects of repeated doses over longer periods, which helps identify potential long-term health impacts such as organ damage or systemic toxicity. Clinical trials involving human participants are crucial for assessing the safety and efficacy of herbal medicines in real-world settings. These trials follow strict ethical guidelines and regulatory requirements to ensure participant safety and to gather data on the product's effectiveness in treating specific conditions. Clinical trials provide the most direct evidence of how herbal medicines interact with the human body and their potential therapeutic benefits or risks. Regulatory frameworks from various global bodies also play a vital role in overseeing the safety of herbal medicines. The regulatory guidelines are essential in maintaining consumer safety and ensuring that herbal medicines meet necessary standards of quality and efficacy (Mugale et al., 2024; Sahu et al., 2024).

2.5.2 Commonly Recognized Toxic Constituents in Herbal Medicines

Certain herbs contain well-documented toxic compounds that can pose significant risks when consumed. Pyrrolizidine alkaloids, found in plants like *Senecio* species, are known for their hepatotoxic effects, potentially leading to liver damage or failure when consumed over extended periods. Aristolochic acids, present in *Aristolochia* species, are associated with nephrotoxicity, causing kidney damage, and have been linked to cancer in some cases. Aconitine, which is found in *Aconitum* species, can cause severe cardiovascular toxicity, potentially leading to fatal arrhythmias or other heart-related issues even in small doses. Furan derivatives, which are present in certain plants used in traditional medicine, have been linked to hepatotoxicity, potentially causing liver damage and dysfunction. These toxic compounds emphasize the importance of rigorous safety assessments and QC in the use of herbal medicines (Moreira et al., 2018).

2.6 REGULATORY AFFAIRS IN HERBAL MEDICINE

Regulatory affairs in herbal medicine encompass a complex framework of legislation, guidelines, and practices intended to ensure the safety, efficacy, and quality of herbal products. As the demand for herbal remedies continues to grow globally, the need for robust regulatory systems becomes increasingly critical. This document explores the current state of regulatory affairs in herbal medicine, focusing on various countries' approaches, challenges faced, and potential improvements.

2.6.1 Regulatory Frameworks by Region

India has a rich tradition of herbal medicine, particularly through systems like Ayurveda and Siddha. The Indian government has established several initiatives to enhance the regulatory framework for herbal medicines. The Department of AYUSH, established in 2003, oversees the regulation and promotion of traditional medicine systems, including Ayurveda and Unani. The Traditional Herbal Medicines Act of 2006 aims to regulate the sale of THMs that lack proper licensing, mandating QC measures, and ingredient verification. However, despite these efforts, challenges remain regarding QC and the prevalence of unregulated herbal products in the market (Kala, 2017).

In the European Union, herbal medicines are regulated under specific directives that require them to be authorized for marketing if they meet certain criteria related to safety and efficacy. Directive 2004/24/EC outlines the requirements for registering traditional herbal medicinal products (THMPs), which must demonstrate safety and traditional use for at least 30 years. The Commission E evaluates herbal products and publishes monographs that provide guidance on their use. While the EU has a structured approach to regulation, issues such as high licensing fees and insufficient data on efficacy still pose challenges (Peschel and Alvarez, 2018).

The Office of Alternative Medicine, established within the NIH, aims to promote research on dietary supplements but does not evaluate products for safety or efficacy before they reach consumers. This regulatory approach has led to a market filled with unregulated products, raising public health concerns. Ethiopia faces significant challenges in regulating THM. The Ethiopian Food and Drug Authority is working to strengthen its legislative framework but lacks comprehensive regulations. One major issue is the absence of post-market surveillance, as there is no system in place to monitor the safety and efficacy of THM after they are sold. Additionally, effective regulation requires collaboration among traditional healers, regulatory bodies, and scientific communities (Hoffman, 2001).

Despite various frameworks in place globally, several common challenges hinder effective regulation of herbal medicines. These include QC issues, as many countries struggle with ensuring consistent quality across herbal products. In India, for instance, random testing has revealed contamination in many traditional preparations. There is also insufficient research on the safety and efficacy of many herbal products, which leads to regulatory gaps. In Ethiopia, 90% of experts identified insufficient research as a significant barrier. Furthermore, traditional healers often resist engaging with scientific communities or regulatory bodies, complicating efforts to standardize practices. To enhance the regulation of herbal medicines globally, several strategies can be implemented. Strengthening research initiatives is crucial, as investing in research can provide valuable data on the safety and efficacy of herbal products. Governments and the private sector should collaborate to fund studies that validate traditional uses with scientific evidence. Enhancing training programmes for both regulators and practitioners will improve understanding of QC measures and compliance requirements. Promoting international collaboration will allow countries to share best practices and harmonize regulations, potentially creating global standards for herbal product registration and monitoring. Finally, increasing public awareness through educational campaigns can inform consumers about the risks associated with unregulated herbal products and promote awareness about certified alternatives (Fontanarosa et al., 2003).

2.7 CONCLUSION

Biotechnological advancements are revolutionizing medicinal plant research and the development of herbal formulations. These technologies enhance the productivity, consistency, and therapeutic potential of medicinal plants, contributing to the creation of safer and more effective herbal products. However, the rapid pace of innovation necessitates the establishment of robust regulatory frameworks to ensure that these advancements meet stringent standards for safety, efficacy, and quality. Future research should emphasize the integration of cutting-edge biotechnological tools—such as tissue culture, genetic engineering, and elicitation techniques—with traditional medicinal knowledge to develop standardised, high-quality herbal products suitable for global healthcare applications. Although the prospects for biotechnological interventions in medicinal plants are highly promising, they are accompanied by challenges related to regulatory compliance, intellectual property rights, and market dynamics. Addressing these challenges through interdisciplinary collaboration among scientists, policymakers, industry stakeholders, and consumers will be essential to fostering sustainable innovation. A well-balanced framework that harmonizes technological progress with traditional wisdom can ensure the safe, effective, and ethical advancement of herbal medicine in the modern era.

REFERENCES

Ansari, S. H., Islam, F., & Sameem, M. (2012). Influence of nanotechnology on herbal drugs: A review. *Journal of Advanced Pharmaceutical Technology & Research, 3*(3), 142–146.

Antil, S., Abraham, J. S., Sripoorna, S., Maurya, S., Dagar, J., Makhija, S., ... & Toteja, R. (2023). DNA barcoding, an effective tool for species identification: A review. *Molecular Biology Reports, 50*(1), 761–775.

Bailey, R. L. (2020). Current regulatory guidelines and resources to support research of dietary supplements in the United States. *Critical Reviews in Food Science and Nutrition, 60*(2), 298–309.

Barba-Ostria, C., Carrera-Pacheco, S. E., Gonzalez-Pastor, R., Heredia-Moya, J., Mayorga-Ramos, A., Rodríguez-Pólit, C., & Guamán, L. P. (2022). Evaluation of biological activity of natural compounds: Current trends and methods. *Molecules, 27*(14), 4490.

Bhardwaj, A. K., Kant, A., Rehalia, A., Singh, V., & Sharma, R. (2023). A review on nanomaterials for drug delivery systems and application of carbon-based nanomaterials. *ES Materials & Manufacturing, 21*(2), 824.

Chisholm, O., & Critchley, H. (2023). Future directions in regulatory affairs. *Frontiers in Medicine, 9*, 1082384.

Chokheli, V. A., Dmitriev, P. A., Rajput, V. D., Bakulin, S. D., Azarov, A. S., Varduni, T. V., ... & Minkina, T. M. (2020). Recent development in micropropagation techniques for rare plant species. *Plants, 9*(12), 1733.

Choudhary, N., & Sekhon, B. S. (2011). An overview of advances in the standardization of herbal drugs. *Journal of Pharmaceutical Education and Research, 2*(2), 55.

Das, M., Patra, P., & Ghosh, A. (2020). Metabolic engineering for enhancing microbial biosynthesis of advanced biofuels. *Renewable and Sustainable Energy Reviews, 119*, 109562.

Dewi, M. K., Chaerunisaa, A. Y., Muhaimin, M., & Joni, I. M. (2022). Improved activity of herbal medicines through nanotechnology. *Nanomaterials, 12*(22), 4073.
Dubale, S., Usure, R. E., Mekasha, Y. T., Hasen, G., Hafiz, F., Kebebe, D., & Suleman, S. (2025). Traditional herbal medicine legislative and regulatory framework: A cross-sectional quantitative study and archival review perspectives. *Frontiers in Pharmacology, 16*, 1475297.
Edo, G. I., Obasohan, P., Makia, R. S., Abiola O. T., Umelo, E. C., Jikah, A. N., Yousif, E., Isoje, E. F., Igbuku, U. A., Opiti, R. A., & Essaghah, A. E. A. (2024). The use of quality control parameters in the evaluation of herbal drugs. A review. *Discover Medicine, 1*(1), 168.
Ekor, M. (2014). The growing use of herbal medicines: Issues relating to adverse reactions and challenges in monitoring safety. *Frontiers in Pharmacology, 4*, 177.
Eruaga, M. A., Itua, E. O., & Bature, J. T. (2024). The role of regulatory authorities in the regulation and control of herbal medicines: A case study of NAFDAC. *International Journal of Science and Research Archive, 11*(2), 207–211.
Fontanarosa, P. B., Rennie, D., & DeAngelis, C. D. (2003). The need for regulation of dietary supplements—Lessons from ephedra. *JAMA, 289*(12), 1568–1570.
Gatt, A. R., Vella Bonanno, P., & Zammit, R. (2024). Ethical considerations in the regulation and use of herbal medicines in the European Union. *Frontiers in Medical Technology, 6*, 1358956.
Gupta, A. P., Pathak, A., & Pandey, P. (2024). Challenges and future of nanotechnology in global herbal medicine practices. In S. C. Izah, M. C. Ogwu, M. Akram (Eds.), *Herbal medicine phytochemistry: Applications and trends* (pp. 1–27). Springer International Publishing.
Halka, J., Kowsalya, K., Anand, A. V., Anand, M., Ramalashmi, K., Begam, A. K. U., & Arun, M. (2025). Genetic engineering of medicinal plants for enhanced secondary metabolite production. In J.-T. Chen (Ed.), *Biotechnology, multiple omics, and precision breeding in medicinal plants* (pp. 146–159). CRC Press.
Hasnain, A., Naqvi, S. A., Ayesha, S. I., Khalid, F., Ellahi, M., Iqbal, S., Hassan, M. Z., Abbas, A., Adamski, R., Markowska, D., & Baazeem, A. (2022). Plants in vitro propagation with its applications in food, pharmaceuticals, and cosmetic industries; Current scenario and future approaches. *Frontiers in Plant Science, 13*, 1009395.
Heinrich, M. (2015). Quality and safety of herbal medical products: Regulation and the need for quality assurance along the value chains. *British Journal of Clinical Pharmacology, 80*(1), 62–66.
Hoffman, F. A. (2001). Regulation of dietary supplements in the United States: Understanding the dietary supplement Health and Education Act. *Clinical Obstetrics and Gynecology, 44*(4), 780–788.
Hurkul, M. M., Cetinkaya, A., Yayla, S., & Ozkan, S. A. (2024). Advanced sample preparation and chromatographic techniques for analyzing plant-based bioactive chemicals in nutraceuticals. *Journal of Chromatography Open, 5*, 100131.
Hussain, A., Qarshi, I. A., Nazir, H., & Ullah, I. (2012). Plant tissue culture: Current status and opportunities. In A. Leva & L. M. R. Rinaldi (Eds.), *Recent advances in plant in vitro culture* (pp. 1–16). IntechOpen.
Husen, H., & Pant, M. (2024) *Tissue culture techniques and medicinal plants: Enhancing propagation and production.* Taylor & Francis Group. https://doi.org/10.1201/b23374
Jain, S. D., Shrivastava, S. K., Agrawal, A., & Gupta, A. K. (2024). WHO guidelines for quality control of herbal medicines: From cultivation to consumption. *International Journal of Pharmaceutical Chemistry and Analysis, 11*, 212–225.
Kala, C. P. (2017). Traditional health care systems and herbal medicines. *European Journal of Environment and Public Health, 1*(1), 03.
Kalyankar, S. N., Kalyankar, N. V., & Thaseen, M. (2012). Computational modeling of metabolic networks. *International Journal of Machine Learning and Computing, 2*(5), 716.
Kaushik, D., Pandey, M. K., & Sharma, A. (2014). Current issues in authentication and quality control of natural products. *Research in Plant Biology, 4*(5).
Khare, N., Khare, P., & Singh, S. (2025a). Cancer drug development from the wild edible plants: Needs and current status. In V. Agarwal, S. Singh, & R. Datta (Eds.), *Exploring traditional wild edible plants* (pp. 144–159). CRC Press.
Khare, N., Khare, P., & Singh, S. (2025b). Molecular and physiological concepts: Macronutrients in crop plant growth and development. In A. Husen (Ed.), *Agricultural crop improvement* (pp. 148–164). CRC Press.
Khare, N., Khare, P., Desai, H. S., Trivedi, M., Johri, P., & Singh, S. (2024). Overview of nutraceuticals for management of lifestyle diseases. In M. Trivedi, S. Singh, P. Johri, & P. Lopez-Sanchez (Eds.), *Nutraceuticals inspiring the contemporary therapy for lifestyle diseases* (pp. 1–8). CRC Press.

Klein-Junior, L. C., de Souza, M. R., Viaene, J., Bresolin, T. M., de Gasper, A. L., Henriques, A. T., & Vander Heyden, Y. (2021). Quality control of herbal medicines: From traditional techniques to state-of-the-art approaches. *Planta Medica, 87*(12/13), 964–988.

Kumar, R., Aadil, K. R., Mondal, K., Mishra, Y. K., Oupicky, D., Ramakrishna, S., & Kaushik, A. (2022). Neurodegenerative disorders management: State-of-art and prospects of nano-biotechnology. *Critical Reviews in Biotechnology, 42*(8), 1180–1212.

Kumari, V. C., Patil, S. M., Ramu, R., Shirahatti, P. S., Kumar, N., Sowmya, B. P., & Patrick-Iwuanyanwu, K. C. (2022). Chromatographic techniques: Types, principles, and applications. In C. Egbuna, K. C. Patrick-Iwuanyanwu, M. A. Shah, J. C. Ifemeje, & A. Rasul (Eds.), *Analytical techniques in biosciences* (pp. 73–101). Academic Press.

Lai, Y. H., & Wang, Y. S. (2023). Advances in high-resolution mass spectrometry techniques for analysis of high mass-to-charge ions. *Mass Spectrometry Reviews, 42*(6), 2426–2445.

Lemma, D. T., Banjaw, D. T., & Megersa, H. G. (2020). Micropropagation of medicinal plants. *International Journal of Plant Breeding and Crop Science, 7*(2), 796–802.

Li, S., Cong, Y., Liu, Y., Wang, T., Shuai, Q., Chen, N., & Li, Y. (2017). Optimization of agrobacterium-mediated transformation in soybean. *Frontiers in Plant Science, 8*, 246.

Liu, C., Zhang, L., Liu, H., & Cheng, K. (2017). Delivery strategies of the CRISPR-Cas9 gene-editing system for therapeutic applications. *Journal of Controlled Release, 266*, 17–26.

Liu, Y., Zhang, D., Yang, T., Chen, R., & Luo, X. (2024). Pathway engineering of plant-derived bioactive compounds in microbes. In Y. Wei, J. Xiao-Jun, & M. Cao (Eds.), *Engineering biology for microbial biosynthesis of plant-derived bioactive compounds* (pp. 73–87). Academic Press.

Mittal, K. R., Pharasi, N., Sarna, B., Singh, M., Rachana, Haider, S., & Jha, N. K. (2022). Nanotechnology-based drug delivery for the treatment of CNS disorders. *Translational Neuroscience, 13*(1), 527–546.

Mollah, M. S., & Nyeem, M. A. (2021). Quality control and evaluation of herbal drugs in modern era. *International Journal of Research in Pharmacy and Pharmaceutical Sciences, 6*, 9–13.

Moreira, R., Pereira, D. M., Valentão, P., & Andrade, P. B. (2018). Pyrrolizidine alkaloids: Chemistry, pharmacology, toxicology, and food safety. *International Journal of Molecular Sciences, 19*(6), 1668.

Mosihuzzaman, M., & Choudhary, M. I. (2008). Protocols on safety, efficacy, standardization, and documentation of herbal medicine (IUPAC Technical Report). *Pure and Applied Chemistry, 80*(10), 2195–230.

Mugale, M. N., Dev, K., More, B. S., Mishra, V. S., Washimkar, K. R., Singh, K., Maurya, R., Rath, S. K., Chattopadhyay, D., & Chattopadhyay, N. (2024). A comprehensive review on preclinical safety and toxicity of medicinal plants. *Clinical Complementary Medicine and Pharmacology, 13*, 100129.

Mukherjee, P. K. (2019). *Quality control and evaluation of herbal drugs: Evaluating natural products and traditional medicine*. Elsevier.

Mustafa, G., Arif, R., Atta, A., Sharif, S., & Jamil, A. (2017). Bioactive compounds from medicinal plants and their importance in drug discovery in Pakistan. *Matrix Science Pharma, 1*(1), 17–26.

Muyumba, N. W., Mutombo, S. C., Sheridan, H., Nachtergael, A., Duez, P. (2021). Quality control of herbal drugs and preparations: The methods of analysis, their relevance, and applications. *Talanta Open, 4*, 100070. https://doi.org/10.1016/j.talo.2021.100070

Pankaj, R., Phade, S., Inamdar, P., Kakade, S., Ingale, V., Magar, S., & Thavare, A. (2024). Overview of herbal drugs: Regulatory perspective with special emphasis on the global market. *Yemen Journal of Medicine, 3*(3), 207–217.

Parveen, R., Khare, N., Singh, S., & Johri, P. (2023). An in silico outlook for the detection and surveillance of evolving and persistent plant pathogens. In U. B. Singh, R. Kumar, & H. B. Singh (Eds.), *Detection, diagnosis and management of soil-borne phytopathogens* (pp. 17–40). Springer Nature Singapore.

Parvin, K., Srivastava, A., Hidangmayum, N., Bansal, S., Meher, R., & Awasthi, R. (2023). Exploring the evolving role of herbal and alternative medicine in modern healthcare. *Acta Traditional Medicine, 2*, 35–42.

Patani, A., Patel, A., Prajapati, D., Khare, N., & Singh, S. (2023). Various agriculture crop plant-based bioactive compounds and their use in nanomaterial synthesis and applications. In A. Husen (Ed.), *Nanomaterials from agricultural and horticultural products* (pp. 223–241). Springer Nature Singapore.

Patra, J. K., Das, G., Fraceto, L. F., Campos, E. V. R., Rodriguez-Torres, M. D. P., Acosta-Torres, L. S., ... & Shin, H. S. (2018). Nano based drug delivery systems: Recent developments and future prospects. *Journal of Nanobiotechnology, 16*, 1–33.

Pant, M., Sohrab, S. S., & Husen, A. (2025). *Genetic improvement and conservation practice of medicinal plants.* Taylor & Francis Group. https://doi.org/10.1201/9781003482659

Peschel, W., & Alvarez, B. M. (2018). Harmonised European standards as a basis for the safe use of herbal medicinal products and their marketing authorisation in European Union member states. *Pharmaceutical Medicine, 32*(4), 275–293.

Raju, S., & Das, M. (2024). Medicinal plants industry in India: Challenges, opportunities, and sustainability. *Medicinal Plants-International Journal of Phytomedicines and Related Industries, 16*(1), 1–14.

Sachan, A. K., Vishnoi, G., & Kumar, R. (2016). Need of standardization of herbal medicines in modern era. *International Journal of Phytomedicine, 8*(3), 300–307.

Sahu, A., Srivastava, N., Jena, D., Singh, A., Kumar, S., & Chauhan, I. (2024). Important regulatory guidelines on natural products. *Current Traditional Medicine, 10*(6), 152–172.

Satish, L., Rency, A. S., Muthubharathi, B. C., Shamili, S., Rameshkumar, R., Swamy, M. K., & Ramesh, M. (2019). Transgenic plant cell cultures: A promising approach for secondary metabolite production. In M. Akhtar & M. Swamy (Eds.), *Natural bio-active compounds: Biotechnology, bioengineering, and molecular approaches* (Vol. 3, pp. 79–122). Springer.

Shaban, N. S., Abdou, K. A., & Hassan, N. E. H. Y. (2016). Impact of toxic heavy metals and pesticide residues in herbal products. *Beni-Suef University Journal of Basic and Applied Sciences, 5*(1), 102–106.

Sinha, S., Jeyaseelan, C., Singh, G., Munjal, T., & Paul, D. (2023). Spectroscopy: Principle, types, and applications. In A. K. Bhatt, R. K. Bhatia, & T. C. Bhatia (Eds.), *Basic biotechniques for bioprocess and bioentrepreneurship* (pp. 145–164). Academic Press.

Stanwood, P. C. (2024). Cryopreservation of seed germplasm for genetic conservation. In H. W. Pritchard (Ed.), *Cryopreservation of plant cells and organs* (pp. 199–226). CRC Press.

Thakkar, S., Anklam, E., Xu, A., Ulberth, F., Li, J., Li, B., Hugas, M., Sarma, N., Crerar, S., Swift, S., & Hakamatsuka, T. (2020). Regulatory landscape of dietary supplements and herbal medicines from a global perspective. *Regulatory Toxicology and Pharmacology, 114*, 104647.

Van Wyk, A. S., & Prinsloo, G. (2020). Health, safety and quality concerns of plant-based traditional medicines and herbal remedies. *South African Journal of Botany, 133*, 54–62.

Wang, H., Chen, Y., Wang, L., Liu, Q., Yang, S., & Wang, C. (2023). Advancing herbal medicine: Enhancing product quality and safety through robust quality control practices. *Frontiers in Pharmacology, 14*, 1265178.

Xuan, L., Ju, Z., Skonieczna, M., Zhou, P. K., & Huang, R. (2023). Nanoparticles-induced potential toxicity on human health: Applications, toxicity mechanisms, and evaluation models. *MedComm, 4*(4), e327.

Zhang, J., Wider, B., Shang, H., Li, X., & Ernst, E. (2012). Quality of herbal medicines: Challenges and solutions. *Complementary Therapies in Medicine, 20*(1–2), 100–106.

3 Scope of Molecular Markers against Other Conventional Markers in the Genetic Improvement of Medicinal Plants

Noopur Khare, Pragati Khare, Tanvi Jain, Sachidanand Singh, Megha Barot, and Azamal Husen

3.1 INTRODUCTION

3.1.1 Overview of Genetic Improvement in Medicinal Plants

Medicinal plants have been an essential part of traditional and modern medicine due to their bioactive compounds with therapeutic properties. With the increasing demand for high-quality medicinal plants in the pharmaceutical, nutraceutical, and cosmetic industries, genetic improvement has become a crucial aspect of their cultivation (Mickymaray, 2019). The genetic improvement of medicinal plants focuses on enhancing desirable traits such as high yield, disease resistance, stress tolerance, and the accumulation of bioactive compounds. Traditional breeding techniques, including selection and hybridization, have been used for centuries to improve plant characteristics. However, these methods have limitations, such as long breeding cycles, environmental influences, and the difficulty of selecting specific genetic traits (Tran et al., 2020).

Recent progress in plant biotechnology has enabled the use of molecular techniques that allow for precise and efficient genetic enhancement (Singha et al., 2022; Husen and Pant, 2024; Pant et al., 2025). Combining molecular markers with traditional breeding methods makes it possible to identify and select desirable genetic traits with improved accuracy. This integration accelerates breeding programmes by shortening the time needed to develop superior plant varieties. For medicinal plants, where the composition and concentration of bioactive compounds are crucial, molecular marker-assisted genetic improvement serves as an effective strategy to enhance their pharmacological potential (Nadeem et al., 2018).

3.1.2 Importance of Genetic Markers in Plant Breeding

Genetic markers are crucial tools in plant breeding, as they enable the detection, selection, and manipulation of specific genetic traits. Acting as identifiable reference points within the genome, these markers allow researchers to trace genes linked to desirable characteristics. Broadly, genetic markers are categorized into traditional types—such as morphological, biochemical, and cytological markers—and molecular markers, which rely on variations in DNA sequences. Incorporating genetic markers into breeding programmes facilitates the selection of superior plant varieties while effectively reducing the presence of undesirable traits (Kumar et al., 2024).

DOI: 10.1201/9781003542681-3

One of the primary advantages of genetic markers is their ability to facilitate marker-assisted selection (MAS). This technique allows breeders to screen plants at the seedling stage based on genetic markers linked to beneficial traits, thereby reducing the reliance on phenotypic selection, which is often influenced by environmental factors. In medicinal plants, genetic markers help in identifying plant varieties with high concentrations of bioactive compounds, ensuring consistent quality in commercial production. Furthermore, genetic markers are invaluable in germplasm conservation, genetic diversity assessment, and the authentication of medicinal plant species, which is crucial in preventing adulteration in herbal products (Salgotra and Stewart, 2020).

3.1.3 Comparison of Molecular and Conventional Markers

Conventional markers, such as morphological and biochemical markers, have historically been used in plant breeding. Morphological markers involve the identification of genetic variations based on observable traits such as plant height, leaf shape, flower colour, and fruit characteristics (Chesnokov et al., 2020). Although these markers are easy to use and do not require advanced technology, they are highly influenced by environmental conditions and often exhibit low heritability. Similarly, biochemical markers rely on variations in proteins or enzyme activity within plants, providing a more direct link to genetic differences. However, biochemical markers are also susceptible to environmental fluctuations, making them less reliable for precise genetic selection (Amiteye, 2021). In contrast, molecular markers provide a more robust and precise method for genetic analysis. Molecular markers are based on DNA sequence variations and are independent of environmental influences, allowing for more accurate selection of genetic traits. Several types of molecular markers, including restriction fragment length polymorphism (RFLP), random amplified polymorphic DNA (RAPD), simple sequence repeats (SSR), amplified fragment length polymorphism (AFLP), and single-nucleotide polymorphisms (SNPs), have been widely used in plant breeding programmes. These markers enable high-resolution mapping of genetic loci associated with desirable traits, facilitating the efficient selection of superior plant varieties (Grover and Sharma, 2016).

One of the key advantages of molecular markers over conventional markers is their ability to detect genetic variations at the DNA level. This capability allows for early selection in breeding programmes, reducing the time and resources required for the development of improved plant varieties. Additionally, molecular markers enable the identification of genetic linkages, assisting in the precise introgression of beneficial traits through marker-assisted backcrossing. In medicinal plants, molecular markers are particularly useful in identifying genes responsible for the biosynthesis of secondary metabolites, which determine the therapeutic properties of plant-based medicines (Avise, 2012). Despite their advantages, molecular markers require specialized equipment, technical expertise, and higher initial investment costs compared to conventional markers. However, with advancements in molecular biology and the decreasing costs of DNA sequencing technologies, the application of molecular markers in plant breeding is becoming more accessible. The integration of molecular markers with conventional breeding techniques offers a comprehensive approach to genetic improvement, ensuring that medicinal plants with enhanced traits can be developed efficiently and sustainably (Amiteye, 2021).

3.2 CONVENTIONAL MARKERS IN GENETIC IMPROVEMENT

3.2.1 Morphological Markers: Characteristics and Limitations

Morphological markers are the oldest and most widely used conventional markers in plant breeding. These markers are based on observable traits such as plant height, leaf shape, flower colour, seed texture, and fruit characteristics. They have been instrumental in the selection and classification of plant varieties, as they provide an easy and cost-effective method of distinguishing between genotypes (Chesnokov et al., 2020). Morphological markers have played a significant role in the domestication and improvement of medicinal plants by enabling farmers and breeders to select plants with desirable characteristics over generations. Despite their advantages, morphological markers have several limitations. One of the primary drawbacks is their high susceptibility to environmental influences (Khare et al., 2025). Factors such as soil quality, temperature, water availability, and nutrient levels can significantly alter the expression of morphological traits, making it challenging to differentiate between genetic and environmental variations. Additionally, many morphological markers are polygenic, meaning they are controlled by multiple genes, making it difficult to establish direct genetic correlations. Another limitation is the time-consuming nature of phenotypic selection, as some traits can only be assessed at specific developmental stages, delaying breeding programmes. These challenges highlight the need for more precise and reliable markers in genetic improvement (Jiang, 2013).

3.2.2 Biochemical Markers: Enzyme and Protein-Based Approaches

Biochemical markers provide an alternative approach to morphological markers by analysing variations in protein and enzyme expression in plants. These markers are based on the detection of specific biochemical compounds, such as secondary metabolites, isoenzymes, and storage proteins, which serve as indicators of genetic differences. One of the most commonly used biochemical markers in plant breeding is isozyme analysis, which involves the study of different enzyme variants present in plant tissues (Adhikari et al., 2017). Isozymes are multiple forms of an enzyme that catalyse the same reaction but differ in their amino acid composition due to genetic variation (Khare et al., 2025). The advantage of biochemical markers lies in their ability to provide a more direct link to genetic differences compared to morphological markers. As proteins and enzymes are the functional products of gene expression, their variations often reflect underlying genetic polymorphisms. This makes biochemical markers particularly useful in identifying plant varieties with high concentrations of bioactive compounds, an essential trait in medicinal plant breeding. Additionally, biochemical markers are less influenced by environmental factors compared to morphological traits, making them more reliable for genetic selection (Idrees and Irshad, 2014).

However, biochemical markers also have limitations. One of the key challenges is that they do not always provide comprehensive genome coverage, as they focus only on protein or enzyme variations rather than the entire genetic sequence (Singh and Singh, 2015). Furthermore, enzyme expression can still be influenced by external factors such as plant age, physiological conditions, and environmental stress, potentially affecting the accuracy of selection. The labour-intensive nature of protein and enzyme extraction, coupled with the requirement for specialized laboratory techniques, also limits the widespread application of biochemical markers in large-scale breeding programmes (Ghatak et al., 2017).

3.2.3 Cytological Markers: Chromosomal Studies in Breeding

Cytological markers involve the study of chromosome structure, number, and behaviour in plant cells to identify genetic variations. These markers are particularly useful in studying chromosomal aberrations, polyploidy, and hybridization in medicinal plant breeding. Techniques such as karyotyping and fluorescence in situ hybridization enable researchers to visualize chromosomal differences and identify genetic relationships between different plant species (Heslop-Harrison, 2000). Cytological markers have been widely used in cytogenetics to assess genome organization and detect chromosomal mutations that influence plant traits. One of the key applications of cytological markers in medicinal plant breeding is their role in polyploidization studies. Polyploidy, a condition in which plants possess more than two sets of chromosomes, is a common phenomenon in many medicinal plants and is often associated with increased biomass, enhanced secondary metabolite production, and improved stress tolerance. Cytological analysis allows breeders to select and stabilize polyploid varieties with superior medicinal properties, making it a valuable tool in genetic improvement (Talukdar and Sinjushin, 2015).

Despite their importance, cytological markers also have certain limitations. Chromosome analysis is often labour-intensive and requires advanced microscopy techniques and expertise. Additionally, while cytological markers provide valuable information about chromosomal variations, they do not offer the same level of genetic resolution as molecular markers. This limitation makes them less effective for fine-scale genetic mapping and precise selection of desirable traits. Furthermore, chromosomal changes identified through cytological markers do not always correlate directly with phenotypic variations, making it challenging to use them as standalone selection tools in breeding programmes (Doležel et al., 2012).

3.3 MOLECULAR MARKERS: ADVANCEMENTS IN GENETIC ANALYSIS

3.3.1 Importance of Molecular Markers

Molecular markers are DNA-based tools used to identify genetic variations within and between plant species (Patani et al., 2023). Unlike conventional markers, which rely on observable traits, molecular markers provide direct insights into the genetic makeup of an organism, allowing precise selection for desirable traits. These markers are widely used in plant breeding programmes to enhance the efficiency and accuracy of genetic improvement, particularly in medicinal plants where the quality and concentration of bioactive compounds are of prime importance (Hussain and Nisar, 2020).

The significance of molecular markers in genetic analysis stems from their capacity to detect DNA-level genetic polymorphisms that are unaffected by environmental factors. They enable the identification of genes responsible for important traits such as disease resistance, secondary metabolite biosynthesis, and stress tolerance. Moreover, molecular markers support MAS, allowing breeders to identify desirable traits at early developmental stages, thereby shortening breeding cycles and enhancing genetic improvement. In addition, these markers are essential for germplasm conservation, genetic fingerprinting, and plant authentication, ensuring the purity and quality of medicinal plants used in commercial cultivation (Amom and Nongdam, 2017).

3.3.2 Types of Molecular Markers Used in Medicinal Plants

Various types of molecular markers have been established for use in genetic analysis and plant breeding, each possessing unique features, benefits, and drawbacks. These differences determine their suitability for specific applications in medicinal plant research and breeding initiatives (Nam et al., 2020).

3.3.2.1 Restriction Fragment Length Polymorphism

RFLP was among the earliest molecular marker techniques developed for genetic analysis. The method involves digesting genomic DNA with restriction enzymes that cleave at specific recognition sites, producing DNA fragments of varying lengths. These fragments are then separated through gel electrophoresis, transferred onto a membrane, and hybridized with labelled DNA probes to identify target sequences. Due to their high specificity and reproducibility, RFLP markers have been effectively utilized in genetic diversity assessment, linkage mapping, and MAS (Grover and Sharma, 2016).

Despite its accuracy, RFLP has several limitations. It requires large amounts of high-quality DNA, making it labour-intensive and time-consuming. The need for radioactive or fluorescent probes and Southern blotting techniques further complicates its application. Due to these drawbacks, RFLP has been largely replaced by more advanced molecular marker technologies that offer higher throughput and efficiency (Al-Samarai and Al-Kazaz, 2015).

3.3.2.2 Random Amplified Polymorphic DNA

RAPD is a polymerase chain reaction (PCR)-based molecular marker technique that uses short, arbitrary primers to amplify random DNA sequences across the genome. This method does not require prior knowledge of the DNA sequence, making it a cost-effective and rapid tool for genetic analysis. RAPD has been widely used in medicinal plant research to assess genetic diversity, identify cultivars, and study population genetics (Babu et al., 2021). However, RAPD markers have limitations, including low reproducibility and sensitivity to experimental conditions such as primer selection, PCR parameters, and DNA quality. The dominant nature of RAPD markers (inability to distinguish between homozygous and heterozygous loci) also limits their utility in detailed genetic studies. Despite these drawbacks, RAPD remains a useful technique for preliminary genetic screening and diversity analysis (Kumari and Thakur, 2014).

3.3.2.3 SSR or Microsatellites

SSR markers, or microsatellites, are composed of short, tandemly repeated DNA sequences that exhibit length variations among individuals. These markers are highly polymorphic, co-dominant, and uniformly distributed across the genome, making them particularly useful for genetic fingerprinting, parentage analysis, and trait mapping in medicinal plants. Their ability to detect genetic diversity makes them valuable for breeding improved plant varieties with enhanced production of bioactive compounds (Ahmad et al., 2018). A key advantage of SSR markers lies in their high reproducibility and transferability across related species. They provide comprehensive genetic information, making them well suited for fine-scale genetic mapping and genome-wide association studies (GWAS). However, developing SSR markers requires prior sequence information and the synthesis of multiple primers, which can be labour-intensive and costly. Despite these challenges, SSR markers continue to be a powerful and reliable tool in plant genetics and breeding research (Mishra et al., 2022).

3.3.2.4 Amplified Fragment Length Polymorphism

AFLP integrates the principles of RFLP and PCR to produce highly informative molecular markers. The technique involves digesting genomic DNA with restriction enzymes, followed by the selective amplification of resulting DNA fragments using specific primers. Due to their high level of polymorphism, AFLP markers are valuable tools for assessing genetic diversity, conducting phylogenetic studies, and facilitating MAS in medicinal plants (Amom and Nongdam, 2017).

One of the key advantages of AFLP is its ability to generate a large number of markers without requiring prior sequence information. This makes it an excellent choice for studying species with limited genomic resources. Additionally, AFLP provides high-resolution genetic data, which is beneficial for mapping complex traits in medicinal plants. However, AFLP techniques are relatively labour-intensive and require advanced electrophoresis or capillary sequencing systems for fragment analysis, which may limit their accessibility in resource-constrained laboratories (Jajoriya et al., 2025).

3.3.2.5 Single Nucleotide Polymorphisms

SNPs are the most prevalent and stable form of genetic variation found in plant genomes. They arise from single base-pair substitutions in DNA sequences and serve as highly informative markers for genetic analysis. Owing to their abundance, genome-wide distribution, and suitability for automation, SNP markers have become increasingly valuable in medicinal plant research (Bragg et al., 2015). Their application has transformed genetic improvement programmes by facilitating detailed GWAS, quantitative trait loci (QTL) mapping, and genomic selection. Advanced SNP-based platforms, such as SNP chips and next-generation sequencing (NGS), enable the identification of genes linked to essential traits, including alkaloid biosynthesis, disease resistance, and drought tolerance. The high-throughput capacity and precision of SNP genotyping make it an indispensable tool for contemporary plant breeding programmes (Gabur et al., 2019).

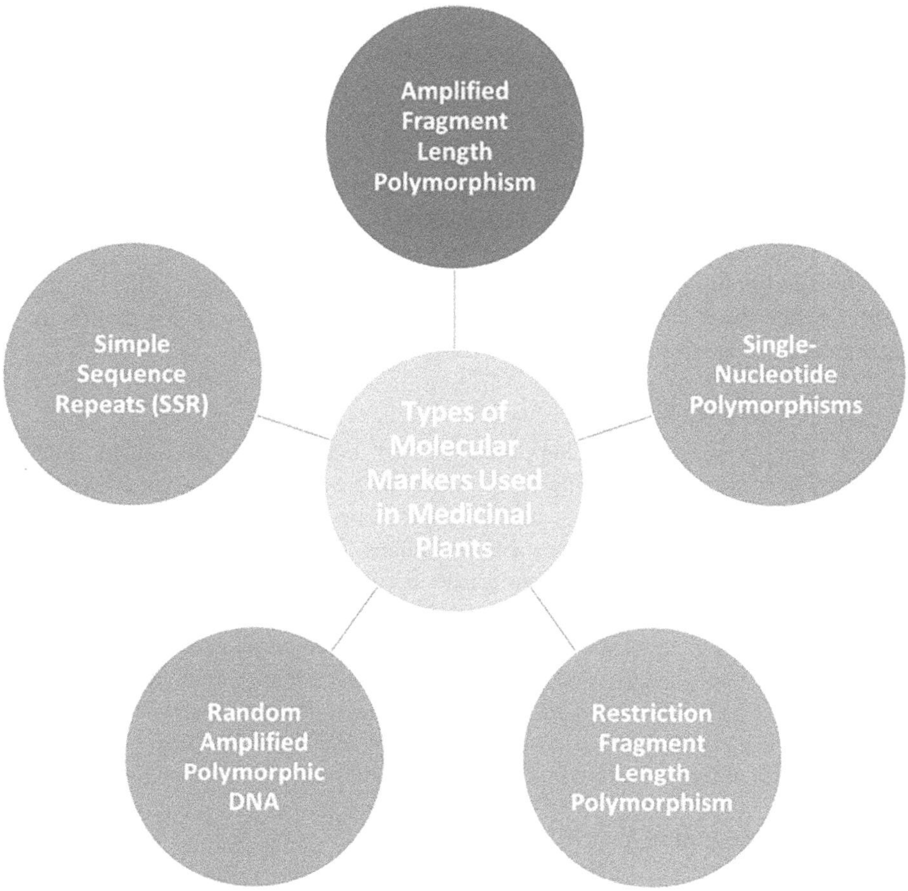

FIGURE 3.1 Types of molecular markers used in medicinal plants.

Despite their advantages, SNP markers require advanced bioinformatics tools and sequencing technologies, which may not be readily available in all research settings. The initial costs of SNP marker development and data analysis can be high, but the long-term benefits outweigh these challenges. As sequencing costs continue to decrease, SNP markers are expected to become the standard for precision breeding in medicinal plants (Figure 3.1) (Fischer et al., 2017).

3.4 COMPARATIVE ANALYSIS OF MOLECULAR AND CONVENTIONAL MARKERS

The use of genetic markers in plant breeding has significantly evolved, transitioning from conventional methods to molecular techniques that provide greater accuracy and efficiency. Conventional markers, such as morphological, biochemical, and cytological markers, have been instrumental in traditional breeding programmes. However, molecular markers have revolutionized genetic analysis by enabling precise identification of genetic variability at the DNA level. This chapter provides a comparative analysis of molecular and conventional markers in terms of accuracy, efficiency, reproducibility, and cost considerations in medicinal plant breeding (Kumar et al., 2016).

3.4.1 Accuracy and Sensitivity in Genetic Differentiation

One of the key advantages of molecular markers over conventional markers is their superior accuracy and sensitivity in genetic differentiation. Conventional markers, such as morphological traits, often exhibit variability due to environmental influences, which can obscure true genetic differences. For example, plant height, leaf shape, and flower colour can be affected by soil quality, climate conditions, and cultivation practices, making them unreliable for precise genetic identification (Ramesh et al., 2020).

In contrast, molecular markers operate at the DNA level, ensuring high specificity in detecting genetic polymorphisms. Techniques such as SNPs, SSRs, and AFLP provide high-resolution genetic profiles, allowing breeders to distinguish closely related plant varieties with unparalleled precision. SNP markers, in particular, offer single-base resolution, making them highly effective in differentiating genetic variations that are not detectable using conventional markers. This accuracy is critical in medicinal plant breeding, where the genetic composition directly influences the biosynthesis of therapeutic compounds (Platten et al., 2019).

3.4.2 Efficiency in Identifying Genetic Variability

Genetic variability is the foundation of successful breeding programmes, as it allows for the selection of superior plant varieties with desirable traits. Conventional markers have played a significant role in assessing genetic variability through phenotypic and biochemical evaluations. However, these methods are often time-consuming and require extensive field trials over multiple seasons to confirm genetic stability (Swarup et al., 2021).

Molecular markers, on the other hand, offer a more efficient and rapid means of identifying genetic diversity. Techniques such as RAPD and AFLP can scan the entire genome within a short period, enabling breeders to assess genetic relationships among plant populations without waiting for plants to reach maturity. SSR markers are particularly valuable in genetic variability studies due to their high polymorphism and co-dominance, which provide detailed insights into allelic diversity (Fu, 2015). Another major advantage of molecular markers is their ability to detect recessive alleles and mutations that are not observable through morphological assessments. This capability is crucial in medicinal plant breeding, where genetic modifications can enhance the production of bioactive compounds. For example, molecular markers have been used to identify genes responsible for alkaloid and flavonoid biosynthesis in medicinal plants, leading to the development of high-yielding and pharmacologically potent cultivars (Smith et al., 2015).

3.4.3 Reproducibility and Reliability in Breeding Programmes

Reproducibility is a critical factor in any genetic analysis method. Conventional markers, particularly morphological traits, often suffer from inconsistencies due to environmental interactions. Even biochemical markers, which rely on enzyme activity and secondary metabolite production, can be influenced by external factors such as nutrient availability and disease stress, making them less reliable for breeding programmes (Chesnokov et al., 2020).

Molecular markers, in contrast, exhibit high reproducibility due to their independence from environmental conditions. DNA-based techniques provide consistent results across different laboratories and research settings, ensuring reliability in genetic studies. For instance, RFLP and SSR markers produce highly repeatable results that can be used for genetic fingerprinting, variety authentication, and conservation of medicinal plant germplasm. The application of molecular markers greatly improves the accuracy and reliability of hybrid breeding and MAS. Incorporating DNA marker information into breeding strategies enables researchers to precisely monitor the inheritance of desirable traits and reduce uncertainties inherent in traditional selection approaches. This level of precision is especially valuable in medicinal plant cultivation, where maintaining consistency in phytochemical composition is critical for ensuring quality and efficacy in pharmaceutical use (Yoshiki et al., 2022).

3.4.4 Cost and Technical Expertise Considerations

Despite the numerous advantages of molecular markers, their widespread adoption is often limited by cost and technical expertise requirements. Conventional markers, such as morphological and cytological techniques, are relatively inexpensive and require minimal laboratory equipment. These methods have been widely used in traditional breeding programmes due to their simplicity and accessibility, especially in developing regions where resources for advanced genomic research may be limited (Mahender et al., 2015).

Despite their low initial cost, conventional markers often prove less efficient and reliable in practice. Morphological marker-based field evaluations demand substantial land, labour, and time, ultimately increasing the overall cost of breeding programmes. Conversely, molecular markers enable faster and more precise genetic analysis, significantly reducing both time and expenses compared to traditional selection methods (Akhtar et al., 2024). However, the adoption of molecular markers necessitates specialized laboratory equipment, including PCR machines, gel electrophoresis systems, and sequencing platforms. Although advanced technologies such as SNP genotyping and NGS require considerable initial investment, they provide long-term advantages by enhancing breeding efficiency and expediting the development of superior plant varieties (Tardieu et al., 2017).

Technical expertise is another crucial factor influencing the adoption of molecular markers. While conventional markers can be assessed through simple observational methods, molecular techniques require knowledge of molecular biology, bioinformatics, and statistical analysis. Researchers and breeders must be trained in DNA extraction, PCR amplification, and data interpretation to effectively utilize molecular markers. However, with the advancement of automated genotyping platforms and bioinformatics tools, the accessibility of molecular markers is gradually increasing, making them more viable for large-scale breeding programmes (Delrot et al., 2020).

3.5 APPLICATIONS OF MOLECULAR MARKERS IN MEDICINAL PLANT IMPROVEMENT

Molecular markers have transformed the landscape of medicinal plant breeding by enabling precise genetic analysis and targeted trait selection. Unlike conventional markers, which rely on observable phenotypic characteristics, molecular markers provide insights at the DNA level, allowing breeders to enhance genetic diversity, improve plant quality, and optimize secondary metabolite production. This chapter explores key applications of molecular markers in medicinal plant improvement, focusing on their role in germplasm conservation, MAS, genetic fingerprinting, and molecular breeding for enhanced phytochemical synthesis (Chaturvedi et al., 2022).

3.5.1 Identification and Conservation of Elite Germplasm

The conservation of elite germplasm is fundamental to the sustainable utilization of medicinal plants. Traditional germplasm conservation strategies, such as field gene banks and botanical gardens, have limitations due to environmental influences, genetic drift, and space constraints. Molecular markers offer a more effective approach by enabling the precise identification and characterization of genetic resources. DNA-based markers such as SSRs and SNPs allow researchers to assess genetic diversity and establish core collections of medicinal plants. These markers help distinguish between closely related species and detect genetic redundancies in germplasm collections. For example, molecular markers have been successfully used to identify distinct chemotypes in medicinal plants like *Withania somnifera* (Ashwagandha) and *Rauwolfia serpentina* (Indian snakeroot), ensuring the preservation of high-value strains with superior therapeutic properties (Sun et al., 2024).

Moreover, molecular markers aid in ex situ and in situ conservation programmes by identifying genetically resilient plant populations. This is particularly crucial for endangered medicinal species, as precise genetic characterization allows conservationists to implement targeted breeding and restoration efforts. Through molecular marker-assisted germplasm conservation, breeders can maintain the genetic integrity of medicinal plant species and prevent the loss of valuable bioactive compounds (Sharma et al., 2024).

3.5.2 MAS for Desirable Traits

MAS is an advanced breeding approach that enables the precise identification and selection of desirable traits using genetic markers associated with specific genes. Unlike conventional breeding, which depends on phenotypic evaluation and requires several generations to stabilize traits, MAS significantly accelerates the breeding process by facilitating early selection at the seedling stage. Molecular markers have been widely applied in MAS to enhance both agronomic and medicinal characteristics in plants. For example, SSR and SNP markers have been utilized to improve drought tolerance, disease resistance, and yield potential in medicinal crops such as *Curcuma longa* (turmeric) and *Panax ginseng* (Ginseng). Through the identification of marker-trait linkages, breeders can efficiently develop high-yielding, stress-tolerant cultivars with stable and consistent medicinal qualities (Arabzai and Gul, 2021). One of the most significant advantages of MAS in medicinal plants is its role in standardizing phytochemical content. The production of secondary metabolites, such as alkaloids, flavonoids, and glycosides, varies across plant populations due to genetic and environmental factors. Molecular markers enable the selection of genotypes with high and stable metabolite concentrations, ensuring uniformity in herbal medicine production. Consequently, MAS contributes to the pharmaceutical industry's demand for high-quality, standardized medicinal plant extracts (Cobb et al., 2019).

3.5.3 Genetic Fingerprinting and Authentication of Medicinal Plant Species

The authentication of medicinal plants is critical for ensuring product quality, safety, and efficacy in herbal medicine markets. Adulteration and misidentification of medicinal plants are common challenges that can compromise therapeutic efficacy and pose health risks to consumers. Molecular markers provide a reliable solution by enabling genetic fingerprinting, a technique used to identify and authenticate plant species based on their unique genetic profiles (Khoobchandani, 2024).

DNA barcoding, a molecular technique that utilizes specific gene sequences (e.g., rbcL, matK, ITS), has been widely applied for medicinal plant authentication. This method allows researchers to distinguish between genuine medicinal plants and their adulterants with high accuracy. For example, genetic fingerprinting using SSR and RAPD markers has been employed to differentiate *Tinospora cordifolia* from its closely related but less effective species, preventing fraudulent substitution in herbal formulations. Molecular markers also play a crucial role in maintaining the genetic purity of commercially cultivated medicinal plants. By developing DNA-based reference databases, regulatory agencies and herbal industries can implement quality control measures to verify the authenticity of raw materials. This ensures consumer safety and enhances the credibility of herbal medicine products in global markets (Ibrahim et al., 2024).

3.5.4 Enhancing Secondary Metabolite Production through Molecular Breeding

Secondary metabolites are the principal bioactive compounds that confer medicinal properties to plants. Enhancing their production through conventional breeding is often difficult because of the intricate nature of their biosynthetic pathways. Molecular markers serve as effective tools for unravelling these pathways and improving phytochemical yields through molecular breeding approaches. The integration of quantitative trait locus (QTL) mapping with molecular markers has been pivotal in pinpointing genes involved in secondary metabolite biosynthesis. For example, research on *Artemisia annua* has utilized SNP markers to identify genes associated with elevated artemisinin

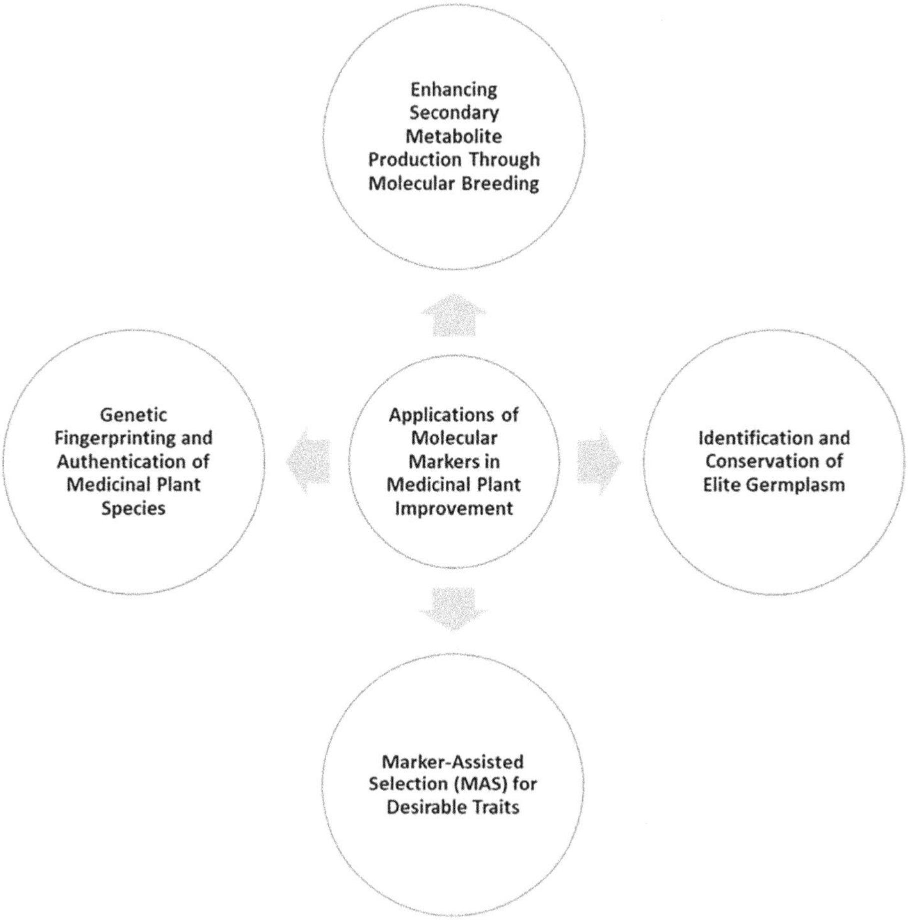

FIGURE 3.2 Applications of molecular markers in medicinal plant improvement.

production, resulting in the development of high-yielding cultivars for malaria therapy. Likewise, SSR markers have been employed in *Ocimum sanctum* (Holy Basil) to enhance essential oil content with potent anti-inflammatory and antimicrobial activities (González-Rodríguez et al., 2025).

Molecular breeding strategies, such as genomic selection and metabolic engineering, further leverage marker-based approaches to optimize phytochemical synthesis. By integrating transcriptomic and metabolomic data with molecular markers, researchers can manipulate key biosynthetic pathways and enhance the accumulation of therapeutic compounds. This targeted approach accelerates the development of medicinal plant varieties with superior pharmacological profiles, benefiting both traditional medicine practitioners and modern pharmaceutical industries (Figure 3.2) (Abdelrahman et al., 2018).

3.6 CHALLENGES AND LIMITATIONS OF MOLECULAR MARKERS

While molecular markers have significantly advanced genetic research and breeding programmes in medicinal plants, several challenges and limitations hinder their widespread adoption. These challenges include high initial costs, the need for specialized technology and expertise, and ethical and regulatory concerns surrounding genetic modification. Addressing these limitations is crucial for ensuring the effective and sustainable application of molecular markers in medicinal plant improvement (Piasecka et al., 2017).

3.6.1 High Initial Costs and Technological Requirements

One of the primary barriers to the widespread adoption of molecular markers in medicinal plant research is the high initial cost associated with laboratory infrastructure, sequencing technologies, and molecular analysis tools. Unlike conventional breeding methods, which rely on visual or biochemical assessments, molecular marker techniques require sophisticated equipment such as PCR machines, gel electrophoresis systems, DNA sequencers, and computational resources for data analysis. The cost of reagents and consumables, such as DNA extraction kits, primers, and enzymes, further adds to the financial burden, making these technologies less accessible, especially in developing countries (Barut et al., 2022).

Additionally, the maintenance of molecular biology laboratories requires substantial investments in quality control, calibration, and technical training. Many research institutions and small-scale breeding programmes struggle to allocate resources for these requirements, limiting their ability to conduct large-scale molecular marker studies. Unlike morphological markers, which can be assessed through field observations, molecular techniques necessitate controlled laboratory environments, further increasing operational costs. To overcome these financial constraints, researchers are exploring cost-effective alternatives such as high-throughput sequencing and multiplex PCR, which allow simultaneous analysis of multiple genetic markers. Government funding, international collaborations, and public–private partnerships are also essential to support research initiatives in molecular breeding. By reducing the cost barriers, the application of molecular markers in medicinal plant breeding can become more feasible and widespread (Kumar et al., 2025).

3.6.2 Need for Advanced Bioinformatics Tools and Expertise

The integration of molecular markers into plant breeding programmes requires expertise in molecular genetics, bioinformatics, and statistical analysis. The interpretation of genetic data involves complex computational processes, including genome mapping, sequence alignment, and marker-trait association studies. Bioinformatics tools such as Basic Local Alignment Search Tool (BLAST), Molecular Evolutionary Genetics Analysis (MEGA), and software for QTL mapping play a crucial role in analyzing large datasets generated from molecular marker studies. However, a significant limitation is the shortage of trained professionals with expertise in bioinformatics and molecular biology. Many institutions lack access to advanced computational tools and skilled personnel to handle the vast amount of genetic data produced through molecular marker research. This gap in expertise can lead to errors in data interpretation, affecting the reliability of breeding programmes (Thriveni et al., 2024).

Moreover, the rapid evolution of molecular marker technologies requires continuous learning and adaptation. New advancements in NGS, GWAS, and CRISPR-based MAS demand that researchers stay updated with the latest techniques and methodologies. Training programmes, workshops, and interdisciplinary collaborations are essential to build capacity in bioinformatics and molecular breeding. Cloud-based bioinformatics platforms and open-access databases have emerged as potential solutions to bridge the expertise gap. These resources allow researchers to analyse genetic data without the need for expensive in-house computational infrastructure. Additionally, collaborative research networks can facilitate knowledge exchange and capacity building among scientists working in medicinal plant genomics (Chavhan et al., 2024).

3.6.3 Ethical and Regulatory Concerns in Genetic Modification

The use of molecular markers in genetic improvement often intersects with ethical and regulatory considerations, particularly when genetic modification is involved. While MAS itself does not involve direct genetic modification, some breeding programmes integrate genetic engineering techniques to introduce desirable traits in medicinal plants. The modification of plant genomes raises concerns related to biosafety, biodiversity conservation, and consumer acceptance. One of the key ethical issues is the potential impact of genetically modified (GM) medicinal plants on traditional herbal medicine practices. Many medicinal plants have been used for centuries in their natural form, and alterations to their genetic composition may raise scepticism among consumers and traditional healers. The introduction of genetically enhanced medicinal plants into herbal markets requires transparency, regulatory approvals, and clear labelling to ensure consumer awareness and trust (Moyo, 2023).

Furthermore, regulatory frameworks governing the use of molecular markers and genetic engineering vary across countries. Some nations have stringent regulations on GM organisms (GMOs), while others have more flexible policies. The approval process for GM medicinal plants can be time-consuming and costly, delaying their commercial application. Regulatory bodies such as the Food and Drug Administration (FDA), European Food Safety Authority (EFSA), and national biosafety committees play a crucial role in assessing the safety and efficacy of GM medicinal plants. Intellectual property rights (IPR) and patent issues also pose challenges in the application of molecular markers. Many molecular marker technologies are patented by private companies, restricting their use by public research institutions and small-scale farmers. The cost of licensing these technologies can be prohibitive, limiting access to advanced breeding tools. Open-source molecular marker databases and collaborative research initiatives can help mitigate these barriers by promoting knowledge sharing and accessibility (Kanwal, 2024).

3.7 FUTURE PROSPECTS

The continuous advancements in molecular marker technologies have opened new avenues for the genetic improvement of medicinal plants. The integration of molecular markers with emerging genomic and transcriptomic studies, the expansion of NGS applications, and the role of CRISPR-based gene editing offer significant opportunities for enhancing breeding efficiency and precision. These developments will shape the future of medicinal plant research, ensuring better conservation, improved therapeutic efficacy, and sustainable agricultural practices (Shahid et al., 2021).

3.7.1 Integration of Molecular Markers with Genomic and Transcriptomic Studies

Molecular markers, when combined with genomic and transcriptomic studies, can provide a deeper understanding of plant genetics and gene expression patterns. Genomics involves the study of an organism's entire genetic material, while transcriptomics focuses on the expression levels of genes under different environmental and developmental conditions. The integration of these approaches with molecular markers allows for the identification of key genes associated with desirable traits such as enhanced secondary metabolite production, disease resistance, and stress tolerance. High-throughput sequencing technologies have facilitated whole-genome sequencing (WGS) and GWAS, enabling researchers to discover novel genetic variations linked to specific traits in medicinal plants. Molecular markers, particularly SNPs and SSRs, play a crucial role in associating these genetic variations with phenotypic characteristics. This integration helps breeders develop superior medicinal plant varieties with enhanced bioactive compound profiles and adaptability to different climatic conditions (Wolf, 2013).

Moreover, transcriptomic studies, using RNA sequencing (RNA-Seq), allow researchers to analyse gene expression profiles in medicinal plants under varying conditions. By mapping molecular markers to transcriptomic data, scientists can identify regulatory networks controlling important metabolic pathways. For instance, key genes responsible for the biosynthesis of medicinal alkaloids, flavonoids, and terpenoids can be pinpointed, allowing targeted breeding strategies to enhance their production. Despite the advantages, challenges such as data complexity, computational requirements, and the need for interdisciplinary expertise must be addressed. Advances in bioinformatics and artificial intelligence (AI)-based analytics are helping overcome these hurdles by providing powerful tools for integrating and interpreting large datasets (Dar et al., 2019).

3.7.2 Advances in NGS for Plant Breeding

NGS technologies have revolutionized the field of plant genomics by enabling rapid and cost-effective sequencing of entire genomes. In medicinal plant breeding, NGS facilitates the discovery of novel molecular markers, accelerates MAS, and enhances genetic diversity studies. Compared to traditional marker systems like RFLP and AFLP, NGS-based approaches offer higher resolution and accuracy in detecting genetic variations. One of the most promising applications of NGS in molecular marker research is genotyping-by-sequencing (GBS). GBS allows high-throughput SNP discovery and genotyping in large plant populations, making it a valuable tool for improving medicinal plant breeding efficiency. Another approach, RNA-Seq, provides insights into gene expression patterns, enabling researchers to link molecular markers with functional traits (Abdi et al., 2024).

In addition to identifying genetic variations, NGS contributes to de novo genome assembly of medicinal plants, many of which have complex genomes. The sequencing of medicinal plants such as *W. somnifera* (Ashwagandha) and *P. ginseng* (Ginseng) has provided valuable genomic resources for breeding programmes. These studies help breeders select high-yielding and disease-resistant cultivars with optimal phytochemical composition. Another breakthrough in NGS is the development of pan genomics, which examines the genetic diversity across multiple plant accessions rather than a single reference genome. This approach enhances the discovery of novel alleles that contribute to key traits, thus improving the adaptability and resilience of medicinal plants in changing environmental conditions. Despite its transformative impact, the implementation of NGS in medicinal plant research faces challenges such as high data processing costs, complex bioinformatics analysis, and storage requirements for large genomic datasets. To address these limitations, cloud computing and big data analytics are being increasingly integrated into genomic research workflows (Satam et al., 2023).

3.7.3 Role of CRISPR and Gene Editing in Complementing Molecular Markers

Clustered Regularly Interspaced Short Palindromic Repeats (CRISPR) and related gene-editing technologies have revolutionized plant breeding by enabling precise and targeted genome modifications. These tools complement molecular markers by offering greater accuracy and efficiency compared to traditional breeding methods that depend on random genetic recombination. CRISPR facilitates the direct modification of genes controlling desirable traits, thereby shortening breeding cycles and enhancing precision. In medicinal plant research, CRISPR-based genome editing has been successfully utilized to boost the biosynthesis of pharmacologically active compounds. For instance, CRISPR has been employed to modify genes involved in the synthesis of key secondary metabolites such as alkaloids, flavonoids, and terpenoids. Through targeted activation or knockout of specific biosynthetic genes, researchers can generate high-yield medicinal plant varieties with enhanced therapeutic potential (Ghosh and Chatterjee, 2024).

Another promising application of CRISPR is in disease resistance breeding. Many medicinal plants suffer from fungal, bacterial, and viral infections, which reduce their yield and bioactive compound production. By using CRISPR to introduce disease-resistant alleles, breeders can develop robust plant varieties with enhanced resistance to pathogens, thereby reducing the need for chemical pesticides. CRISPR can also be used in metabolic pathway engineering to optimize the accumulation of medicinal compounds. For example, by modifying transcription factors that regulate metabolic pathways, researchers can increase the yield of valuable plant-based pharmaceuticals. Such advancements have the potential to revolutionize the herbal medicine industry by ensuring a sustainable and consistent supply of high-quality medicinal plants. Despite its advantages, gene editing in medicinal plants faces regulatory and ethical challenges. Some countries have strict regulations on genetically edited organisms, making commercialization difficult. Additionally, concerns over unintended genetic modifications and off-target effects require rigorous validation and safety assessments before field deployment (Gao et al., 2025).

3.8 CONCLUSION

The use of molecular markers has revolutionized medicinal plant breeding by offering precise, efficient, and reliable tools for genetic improvement. Unlike conventional methods that rely on morphological, biochemical, and cytological markers, molecular markers provide a deeper understanding of genetic variability, allowing targeted breeding strategies to enhance desired traits. This chapter summarizes the advantages of molecular markers over traditional approaches, highlights future research priorities, and emphasizes their role in sustainable medicinal plant cultivation. Molecular markers have several advantages over conventional markers in genetic improvement programmes. One of the key advantages of molecular markers is their ability to detect polymorphisms at the genomic level, even in closely related plant varieties. Techniques such as SNPs, SSRs, and AFLP provide high-resolution genetic fingerprints, enabling breeders to identify elite germplasm with superior traits. Moreover, molecular markers facilitate MAS, which enhances breeding efficiency by enabling the precise transfer of desirable genes without the need for extensive phenotypic screening. This approach is particularly beneficial for medicinal plants where traits such as secondary metabolite production, disease resistance, and environmental adaptability are critical for their therapeutic and commercial value. Additionally, molecular markers contribute to genetic conservation and authentication of medicinal plant species, ensuring the purity and consistency of herbal products. By using DNA barcoding and genetic fingerprinting techniques, researchers can verify plant identity, prevent adulteration, and protect endangered medicinal plant species from extinction. To fully leverage the potential of molecular markers, future research should focus on expanding genomic databases for medicinal plants. Many medicinal species have complex genomes, and sequencing efforts must be intensified to generate comprehensive reference genomes. Another crucial area for research is the integration of molecular markers with advanced genomic tools such as NGS, transcriptomics, and metabolomics. By combining these technologies, breeders can gain a more holistic understanding of gene expression patterns and metabolic pathways, leading to the development of high-yielding and therapeutically potent medicinal plant varieties. Furthermore, efforts should be directed towards making molecular breeding technologies more cost-effective and accessible to researchers and farmers in developing regions. Establishing collaborative networks between academic institutions, research organizations, and industry stakeholders can accelerate the adoption of molecular marker-assisted breeding programmes. Molecular breeding plays a vital role in ensuring the sustainability and productivity of medicinal plant cultivation. As global demand for medicinal plants continues to rise, conventional cultivation methods alone may not be sufficient to meet the growing needs of the pharmaceutical and herbal industries. Molecular breeding offers solutions to improve yield, enhance disease resistance, and increase the bioactive compound content of medicinal plants without relying on excessive chemical inputs. Additionally, molecular marker-assisted breeding helps mitigate the effects of climate change by developing stress-tolerant plant varieties that can thrive under adverse environmental conditions. Drought-resistant and disease-resistant medicinal plants can be cultivated in diverse agro-climatic zones, ensuring a stable supply of high-quality raw materials for herbal medicine production. Moreover, the application of molecular markers in conservation programmes aids in preserving genetic diversity and protecting wild medicinal plant populations from overexploitation. By identifying genetically diverse accessions, conservationists can establish seed banks and ex situ collections to safeguard valuable genetic resources for future breeding efforts.

REFERENCES

Abdelrahman, M., Burritt, D. J., & Tran, L. S. P. (2018). The use of metabolomic quantitative trait locus mapping and osmotic adjustment traits for the improvement of crop yields under environmental stresses. *Seminars in Cell & Developmental Biology*, *83*, 86–94.

Abdi, G., Tarighat, M. A., Jain, M., Tendulkar, R., Tendulkar, M., & Barwant, M. (2024). Revolutionizing genomics: exploring the potential of next-generation sequencing. In V. Singh & A. Kumar (Eds.), *Advances in Bioinformatics* (pp. 1–33). Springer Nature Singapore.

Adhikari, S., Saha, S., Biswas, A., Rana, T. S., Bandyopadhyay, T. K., & Ghosh, P. (2017). Application of molecular markers in plant genome analysis: a review. *The Nucleus*, *60*, 283–297.

Ahmad, A., Wang, J. D., Pan, Y. B., Sharif, R., & Gao, S. J. (2018). Development and use of simple sequence repeats (SSRs) markers for sugarcane breeding and genetic studies. *Agronomy*, *8*(11), 260.

Akhtar, M. S., Zafar, Z., Nawaz, R., & Fraz, M. M. (2024). Unlocking plant secrets: a systematic review of 3D imaging in plant phenotyping techniques. *Computers and Electronics in Agriculture*, *222*, 109033.

Al-Samarai, F. R., & Al-Kazaz, A. A. (2015). Molecular markers: an introduction and applications. *European Journal of Molecular Biotechnology*, *9*(3), 118–130.

Amiteye, S. (2021). Basic concepts and methodologies of DNA marker systems in plant molecular breeding. *Heliyon*, *7*(10).

Amom, T., & Nongdam, P. (2017). The use of molecular marker methods in plants: a review. *International Journal of Current Research and Review*, *9*(17), 1–7.

Arabzai, M. G., & Gul, H. (2021). Application techniques of molecular marker and achievement of marker assisted selection (MAS) in three major crops rice, wheat and maize. *International Journal for Research in Applied Sciences and Biotechnology*, *8*(1), 82–93.

Avise, J. C. (2012). *Molecular markers, natural history and evolution*. Springer Science & Business Media.

Babu, K. N., Sheeja, T. E., Minoo, D., Rajesh, M. K., Samsudeen, K., Suraby, E. J., & Kumar, I. P. V. (2021). Random amplified polymorphic DNA (RAPD) and derived techniques. In P. Besse (Eds.), *Molecular plant taxonomy: Methods and protocols* (pp. 219–247). Humana.

Barut, M., Nadeem, M. A., Akgür, Ö., Tansi, L. S., Aasim, M., Altaf, M. T., & Baloch, F. S. (2022). Medicinal and aromatic plants in the omics era: application of plant breeding and biotechnology for plant secondary metabolite production. *Turkish Journal of Agriculture and Forestry*, *46*(2), 182–203.

Bragg, J. G., Supple, M. A., Andrew, R. L., & Borevitz, J. O. (2015). Genomic variation across landscapes: insights and applications. *New Phytologist*, *207*(4), 953–967.

Chaturvedi, T., Gupta, A. K., Lal, R. K., & Tiwari, G. (2022). March of molecular breeding techniques in the genetic enhancement of herbal medicinal plants: present and future prospects. *The Nucleus*, *65*(3), 413–436.

Chavhan, R. L., Hinge, V. R., Wankhade, D. J., Deshmukh, A. S., Mahajan, N., & Kadam, U. S. (2024). Bioinformatics for molecular breeding and enhanced crop performance: applications and perspectives. In J. Tsung Chen (Ed.), *Bioinformatics for Plant Research and Crop Breeding* (pp. 21–74). Wiley Online Library.

Chesnokov, Y. V., Kosolapov, V. M., & Savchenko, I. V. (2020). Morphological genetic markers in plants. *Russian Journal of Genetics*, *56*, 1406–1415.

Cobb, J. N., Biswas, P. S., & Platten, J. D. (2019). Back to the future: revisiting MAS as a tool for modern plant breeding. *Theoretical and Applied Genetics*, *132*, 647–667.

Dar, A. A., Mahajan, R., & Sharma, S. (2019). Molecular markers for characterization and conservation of plant genetic resources. *Indian Journal of Agricultural Sciences*, *89*(11), 1755–1763.

Delrot, S., Grimplet, J., Carbonell-Bejerano, P., Schwandner, A., Bert, P. F., Bavaresco, L., & Vezzulli, S. (2020). Genetic and genomic approaches for adaptation of grapevine to climate change. In C. Kelo (Ed.), *Genomic Designing of Climate-Smart Fruit Crops* (pp. 157–270). Springer.

Doležel, J., Vrána, J., Šafář, J., Bartoš, J., Kubaláková, M., & Šimková, H. (2012). Chromosomes in the flow to simplify genome analysis. *Functional & Integrative Genomics*, *12*, 397–416.

Fischer, M. C., Rellstab, C., Leuzinger, M., Roumet, M., Gugerli, F., Shimizu, K. K., & Widmer, A. (2017). Estimating genomic diversity and population differentiation: An empirical comparison of microsatellite and SNP variation in *Arabidopsis halleri*. *BMC genomics*, *18*, 1–15.

Fu, Y. B. (2015). Understanding crop genetic diversity under modern plant breeding. *Theoretical and Applied Genetics*, *128*, 2131–2142.

Gabur, I., Chawla, H. S., Snowdon, R. J., & Parkin, I. A. (2019). Connecting genome structural variation with complex traits in crop plants. *Theoretical and Applied Genetics*, *132*, 733–750.

Gao, H., Pei, X., Song, X., Wang, S., Yang, Z., Zhu, J., & Yang, X. (2025). Application and development of CRISPR technology in the secondary metabolic pathway of the active ingredients of phytopharmaceuticals. *Frontiers in Plant Science*, *15*, 1477894.

Ghatak, A., Chaturvedi, P., & Weckwerth, W. (2017). Cereal crop proteomics: systemic analysis of crop drought stress responses towards marker-assisted selection breeding. *Frontiers in Plant Science*, *8*, 757.

Ghosh, S. K., & Chatterjee, T. (2024). Clustered regularly interspaced short palindromic repeats (CRISPR)-associated proteins (Cas)[CRISPR–Cas]: an emerging technique in plant disease detection and management. In A. Kumar, S. Arora, S. Ogita, Y.-Y. Yau, & K. Mukherjee (Eds.), *Gene Editing in Plants: CRISPR-Cas and Its Applications* (pp. 589–645). Springer Nature Singapore.

González-Rodríguez, T., Montero-Vargas, J. M., Gamboa-Becerra, R., García-Lara, S., & Winkler, R. (2025). Metabolic markers and metabolite quantitative trait loci (mQTLs) for the breeding of crop quality and resistance. In J. Tsung Chen (Ed.), *Plant Stress Tolerance* (pp. 70–91). CRC Press.

Grover, A., & Sharma, P. C. (2016). Development and use of molecular markers: past and present. *Critical Reviews in Biotechnology*, *36*(2), 290–302.

Heslop-Harrison, J. S. (2000). Comparative genome organization in plants: from sequence and markers to chromatin and chromosomes. *The Plant Cell*, *12*(5), 617–635.

Hussain, H., & Nisar, M. (2020). Assessment of plant genetic variations using molecular markers: a review. *Journal of Applied Biology and Biotechnology*, *8*(5), 99–109.

Husen, H., & Pant, M. (2024). *Tissue Culture Techniques and Medicinal Plants: Enhancing Propagation and Production*. Taylor & Francis Group. https://doi.org/10.1201/b23374

Ibrahim, M., Detroja, A., Sheth, B. P., Bhadja, P., Sanghvi, G., & Bishoyi, A. K. (2024). Existing status and future advancements of adulteration detection techniques in herbal products. *Molecular Biology Reports*, *51*(1), 151.

Idrees, M., & Irshad, M. (2014). Molecular markers in plants for analysis of genetic diversity: a review. *European Academic Research*, *2*(1), 1513–1540.

Jajoriya, R., Sharma, R., Meena, V. K., Vadithya, A. S., Singh, K., Kumar, P., ... & Nathawat, T. S. (2025). Emergence of molecular marker from RFLP to SNP: a review. *Plant Archives*, *25*(1), 2062–2074.

Jiang, G. L. (2013). Molecular markers and marker-assisted breeding. In G.-L. Jiang (Ed.), *Plant Breeding from Laboratories to Fields* (pp. 45–85). IntechOpen.

Kanwal, N. (2024). Ethical considerations in molecular research: balancing innovation with societal responsibility. *Multidisciplinary Journal of Biochemistry*, *1*(2), 84–93.

Khare, N., Khare, P., & Singh, S. (2025). Molecular and physiological concepts: macronutrients in crop plant growth and development. In A. Husen (Ed.), *Agricultural Crop Improvement* (pp. 148–164). CRC Press.

Khoobchandani, M. (2024). Unveiling the complexity of herbal medicine: safety, toxicity, and regulatory challenges. In M. Khoobchandani & S. Ghosh (Eds.), *Medicinal Applications of Phytopharmaceuticals* (pp. 269–282). Springer Nature Switzerland.

Kumar, N., Sindhu, S., Gawaria, B., Ansari, A., Dhariwal, A., Sharma, P., & Kumar, R. (2025). GWAS approaches in medicinal plants for enhanced secondary metabolite production. In J.-T. Chen (Ed.), *Biotechnology, Multiple Omics, and Precision Breeding in Medicinal Plants* (pp. 119–134). CRC Press.

Kumar, R., Das, S. P., Choudhury, B. U., Kumar, A., Prakash, N. R., Verma, R., & Mishra, V. K. (2024). Advances in genomic tools for plant breeding: harnessing DNA molecular markers, genomic selection, and genome editing. *Biological Research*, *57*(1), 80.

Kumar, S., Banks, T. W., & Cloutier, S. (2016). SNP discovery through next-generation sequencing and its applications. *Crop Breeding*, *1*, 209–244.

Kumari, N., & Thakur, S. K. (2014). Randomly amplified polymorphic DNA: a brief review. *American Journal of Animal and Veterinary Sciences*, *9*(1), 6–13.

Mahender, A., Anandan, A., & Pradhan, S. K. (2015). Early seedling vigour, an imperative trait for direct-seeded rice: an overview on physio-morphological parameters and molecular markers. *Planta*, *241*, 1027–1050.

Mickymaray, S. (2019). Efficacy and mechanism of traditional medicinal plants and bioactive compounds against clinically important pathogens. *Antibiotics*, *8*(4), 257.

Mishra, A., Singh, P. K., Bhandawat, A., Sharma, V., Sharma, V., Singh, P., & Sharma, H. (2022). Analysis of SSR and SNP markers. In D. B Singh & R. K. Pathak (Eds.), *Bioinformatics* (pp. 131–144). Academic Press.

Moyo, T. (2023). Navigating the ethical landscape: challenges and opportunities in biotechnology for human enhancement and genetic modification. *Scientific Academia Journal*, *6*(1), 1–8.

Nadeem, M. A., Nawaz, M. A., Shahid, M. Q., Doğan, Y., Comertpay, G., Yıldız, M., & Baloch, F. S. (2018). DNA molecular markers in plant breeding: current status and recent advancements in genomic selection and genome editing. *Biotechnology & Biotechnological Equipment*, *32*(2), 261–285.

Nam, V. T., Hang, P. L. B., Linh, N. N., Ly, L. H., Hue, H. T. T., Ha, N. H., & Hien, L. T. T. (2020). Molecular markers for analysis of plant genetic diversity. *Vietnam Journal of Biotechnology*, *18*, 589–608.

Pant, M., Sohrab, S. S., & Husen, A. (2025). *Genetic Improvement and Conservation Practice of Medicinal Plants*. Taylor & Francis Group. https://doi.org/10.1201/9781003482659

Patani, A., Patel, A., Prajapati, D., Khare, N., & Singh, S. (2023). Various agriculture crop plant-based bioactive compounds and their use in nanomaterial synthesis and applications. In A. Husen (Ed.), *Nanomaterials from Agricultural and Horticultural Products* (pp. 223–241). Springer Nature Singapore.

Piasecka, A., Sawikowska, A., Kuczyńska, A., Ogrodowicz, P., Mikołajczak, K., Krystkowiak, K., & Kachlicki, P. (2017). Drought-related secondary metabolites of barley (*Hordeum vulgare* L.) leaves and their metabolomic quantitative trait loci. *The Plant Journal*, *89*(5), 898–913.

Platten, J. D., Cobb, J. N., & Zantua, R. E. (2019). Criteria for evaluating molecular markers: comprehensive quality metrics to improve marker-assisted selection. *PLoS ONE*, *14*(1), e0210529.

Ramesh, P., Mallikarjuna, G., Sameena, S., Kumar, A., Gurulakshmi, K., Reddy, B. V., & Sekhar, A. C. (2020). Advancements in molecular marker technologies and their applications in diversity studies. *Journal of biosciences*, *45*, 1–15.

Salgotra, R. K., & Stewart Jr, C. N. (2020). Functional markers for precision plant breeding. *International journal of molecular sciences*, *21*(13), 4792.

Satam, H., Joshi, K., Mangrolia, U., Waghoo, S., Zaidi, G., Rawool, S., & Malonia, S. K. (2023). Next-generation sequencing technology: current trends and advancements. *Biology*, *12*(7), 997.

Shahid, M. S., Sattar, M. N., Iqbal, Z., Raza, A., & Al-Sadi, A. M. (2021). Next-generation sequencing and the CRISPR-Cas nexus: a molecular plant virology perspective. *Frontiers in Microbiology*, *11*, 609376.

Sharma, P., Doultani, S., Hadiya, K. K., George, L. B., & Highland, H. N. (2024). Overview of marker-assisted selection in animal breeding. *History*, *9*, 11.

Singh, B. D., & Singh, A. K. (2015). *Marker-Assisted Plant Breeding: Principles and Practices*. Springer.

Singha, D. L., Das, D., Paswan, R. R., Chikkaputtaiah, C., & Kumar, S. (2022). Novel approaches and advanced molecular techniques for crop improvement. In B. B. Biswas & H. K. Das (Eds.), *Plant-Microbe Interactions* (pp. 1–27). CRC Press.

Smith, S., Bubeck, D., Nelson, B., Stanek, J., & Gerke, J. (2015). Genetic diversity and modern plant breeding. In M. Ahuja & S. Jain (Eds.), *Genetic Diversity and Erosion in Plants: Indicators and Prevention* (pp. 55–88). Springer.

Sun, L., Lai, M., Ghouri, F., Nawaz, M. A., Ali, F., Baloch, F. S., & Shahid, M. Q. (2024). Modern plant breeding techniques in crop improvement and genetic diversity: from molecular markers and gene editing to artificial intelligence—a critical review. *Plants*, *13*(19), 2676.

Swarup, S., Cargill, E. J., Crosby, K., Flagel, L., Kniskern, J., & Glenn, K. C. (2021). Genetic diversity is indispensable for plant breeding to improve crops. *Crop Science*, *61*(2), 839–852.

Talukdar, D., & Sinjushin, A. (2015). Cytogenomics and mutagenomics in plant functional biology and breeding. In D. Barh, M. S. Khan, & E. Davies (Eds.), *PlantOmics: The Omics of Plant science* (pp. 113–156). Springer.

Tardieu, F., Cabrera-Bosquet, L., Pridmore, T., & Bennett, M. (2017). Plant phenomics, from sensors to knowledge. *Current Biology*, *27*(15), R770–R783.

Thriveni, V., Teotia, J., Hazra, S., Bharti, T., Kumar, M., Lallawmkimi, M. C., & Panwar, D. (2024). A review on integrating bioinformatics tools in modern plant breeding. *Archives of Current Research International*, *24*(9), 293–308.

Tran, N., Pham, B., & Le, L. (2020). Bioactive compounds in anti-diabetic plants: from herbal medicine to modern drug discovery. *Biology*, *9*(9), 252.

Wolf, J. B. (2013). Principles of transcriptome analysis and gene expression quantification: an RNA-seq tutorial. *Molecular ecology resources*, *13*(4), 559–572.

Yoshiki, A., Ballard, G., & Perez, A. V. (2022). Genetic quality: a complex issue for experimental study reproducibility. *Transgenic Research*, *31*(4), 413–430.

4 Species Identification, Diversity, and Conservation of Medicinal Plants Using Molecular Markers

Ravi Shah

4.1 INTRODUCTION

Due to their capacity to produce variety of chemicals, plants are now not only representing the source of food and feed, but also becoming an area of interests for various industries like pharmaceuticals, dyes, agrochemicals, paints, and textiles. Medicinal plants are being consumed since 2600 BCE and in the present condition more than 80% of the population consume traditional herbal medicines in one or the other way (Davis and Choisy, 2024). The consumption of medicinal plants is being done through direct plant or plant parts in a dried or fresh form, syrup, powder, tablet, capsule, or as a pharmaceutical formulation of active ingredients. From the crude extract to the extraction and formulation of particular secondary metabolites, it is reflected that a single plant species has multiple use in medical conditions. In different medicinal practices such as Ayurveda, Unani, Traditional Chinese Medicine, Kampo, Sowa Rigpa, Homoeopathy, Naturopathy, Osteopathy use of different herbal formulations is mentioned (Husen, 2022; 2023a; 2023b; 2024a; 2024b; Husen and Pant, 2024; Pant and Husen, 2025; Bhapkar et al., 2025). This fact depicts how the diversity in the medicinal plant species provides us the resource for valuable substances based on which the modern medicines are designed and utilized to treat common to fatal diseases. However, for the incorporation of traditional medicines in the form of modern formulations, World Health Assembly (WHA) has provided a detailed guideline in which emphasis has been given on regulatory mechanism and evidence-based research (Burki, 2025). Some of the medicinal plant species are rare and undefined or classified and therefore it is very important to generate a library of herbaria. There are also methods of preservation like tissue culture and cryopreservation for the future use of endangered species (Pandey et al., 2022; Kadam and Pawar, 2020; Gupta et al., 2022). Molecular tools have no direct relation with the conservation strategies of the endangered plant species. However, there is a need to know the genetic diversity of the plant species to be conserved, which requires the use of molecular markers. Molecular markers are useful when tissue culture techniques are employed for the conservation. This is done by assessing the genetic fidelity of the mother plants as the tissue culture technique may introduce mutations called somaclonal variations (Modi et al., 2012; Kumar et al., 2010; Kumar et al., 2013). Moreover, when reintroduction or restoration of such plant species is carried out, the diversified and most nearby plant species are required for the breeding purposes for which molecular markers are needed (Vu et al., 2020).

Molecular markers are certain DNA sequences that function as chromosomal genetic markers, allowing the association of desired traits or genes with recognizable genetic loci. Through their utilization, specific genomic areas linked to beneficial features that contribute to yield, nutritional value, disease resistance, and environmental adaptability can be identified. Simple sequence repeats (SSRs), single nucleotide polymorphisms (SNPs), amplified fragment length polymorphisms

 DOI: 10.1201/9781003542681-4

(AFLPs), restriction fragment length polymorphisms (RFLPs), random amplified polymorphic DNA (RAPD), and inter-simple sequence repeats (ISSRs) are frequently employed molecular markers for plant species identification, characterization, and diversity studies because of their high efficacy in identifying genetic variation and relationships among plant species. Identification of the genetic composition of these traits is made possible by molecular markers, which are known to speed up and improve breeding efforts. In certain situations, it is important to understand that a molecular marker can include one or more primers, one or more restriction enzymes, or a combination of the primers, restriction enzymes, and other pertinent elements, in addition to the marker's methods (Amiteye, 2021). In the 1980s, the first molecular marker at the DNA level was proposed. Molecular markers are not influenced by environmental influences, have a large number of accessible markers, demonstrate Mendelian inheritance, have high heterozygosity, identify variations at the DNA level, and are independent of their phenotypes (Al-Samarai and Al-Kazaz, 2015). Use of molecular markers has not only increased the breeding cycle's efficiency but also enabled precise gene modification using techniques like genome editing and marker-assisted breeding. In addition to improving plant types and related agricultural practices, these methods guarantee that only the targeted attribute is altered, reducing the unintentional alterations that come with plant breeding. Several characteristics define an ideal genetic marker. It should only exist in one copy, adhere to the co-dominance criterion, and have no effect on the phenotype. The perfect marker should also be inexpensive, highly polymorphic, simple to identify, and multifunctional. When working with polyploid species, the marker should be genome-specific, publicly accessible for study, and able to be analysed within a genome. Additionally, the marker ought to facilitate automation and multiplexing (Raza et al., 2019).

4.2 SPECIES IDENTIFICATION USING MOLECULAR MARKERS

The purpose of the use of medicinal plants in a healthcare setting is to apply specific metabolite(s) for the given disease condition. Genotypic variations in the given plant species may also lead to chemotypic variations, and therefore, it is very important that the authenticity of the plant species to be used should not be compromised. A database of various molecular markers for different medicinal plants is available in the form of literature and sequence database. Table 4.1 shows the information on how different molecular markers are employed to identify various medicinal plant species. Dominant and co-dominant markers give valuable and genotype-specific information, which can be further used for the development of the Sequence-Characterized Amplified Region (SCAR) marker. SCAR markers are generated from the specific polymorphic amplicon(s) and can be used to identify plant species, genotype, variety, and gender (Modi et al., 2018; Feng et al., 2018; Hon et al., 2021). Apart from the molecular markers listed in the introduction section, DNA barcoding is also done which is specific region-based and provides significant information on the plant species. Markers used in DNA barcoding are from large subunit of ribulose bisphosphate carboxylase oxygenase enzyme (*rbcL*), maturase K enzyme (*matK*), cytochrome *c* oxidase (*COI*), internal transcribed spacer (*ITS*), and 5S rRNA region.

4.3 GENETIC DIVERSITY

Genetic diversity is highly important in medicinal plants, especially when each genotype corresponds to different chemotypes. If we keep the biotechnological interventions like genetic engineering and genome editing aside, the traditional breeding methodologies are also highly useful for the generation of a new cultivar with a distinct phenotype and chemotype. However, in many cases, identification of two closely and distinctly related species becomes crucial before going for the breeding strategies. Some of the medicinal plants have longer life cycle and therefore marker-assisted selection (MAS) methodologies are required (Sharma et al., 2018). Table 4.2 represents the use of various molecular markers to assess the genetic diversity in different medicinal plants.

TABLE 4.1
Identification of Different Plant Species or Genotype Using Various Dominant, Co-Dominant, SCAR, PCR-Based or PCR-Free, and DNA Barcoding Markers

Plant Species	Marker Type	Result	References
Rhodiola crenulate	Cleaved amplified polymorphic sequence (CAPS)	• Single-nucleotide polymorphism (SNP) in closely related species could be detected. • Based on the SNPs, the fluorescence based loop mediated isothermal amplification system was designed.	Hao et al. (2025)
Cirsium setosum	PCR – lateral flow assay	• Authentication of the plant species was done with high sensitivity, and the presence of adulterant species *Cirsium japonicum* could be assessed.	Chien et al. (2025)
Litchi chinensis Sonn.	RAPD-SCAR	• Variety-specific three SCAR markers were developed. • One of the three markers showed 100% efficiency in the authentication of the particular variety.	Cheng et al. (2015)
Dracocephalum oldavica L.	DNA barcoding-RFLP	• Adulteration of surrogate species sold under the common name could be detected using DNA barcoding followed by RFLP	Horn et al. (2014)
Ginkgo biloba	DNA barcoding	• Authentication of the plant in dietary supplement was successfully done and 16.2% of the samples taken were found have filler materials instead of *G. biloba*	Little (2014)
Convolvulus pluricaulis	RAPD	• Four out of 25 RAPD primers showed the distinct amplicons that can be used to authenticate the Shankhpushpi (*C. pluricaulis*) and adulteration of other similar species could be detected.	Ganie et al. (2012)
Rheum officinale Baill., Rheum palmatum L., *Rheum tanguticum Maxim.*	ISSR	• These species are important for Chinese Pharmacopoeia and very difficult to identify using morphological markers for the use in different formulation. • Out of 15 ISSR primers employed, three (UBC807, UBC811, and UBC816) showed distinct amplicon which can be effectively used for the future work.	Wang (2011)
Curcuma comosa Roxb.	AFLP	• Identification of *C. comosa* among other species in a herbal formulation was successful. • The AFLP marker system can be utilized to monitor the adulterants.	Keeratinijakal et al. (2010)

4.4 MOLECULAR MARKERS AND CONSERVATION STRATEGIES

For the loss of genetic resources of medicinal plants, many factors are directly or indirectly responsible. These include anthropogenic activities, harvesting irregularities, pollution, climate degradation, biotic stresses, and slower growth of the plant species (Kumar et al., 2024). Many of these factors are unavoidable in the present situation. One of the promising mitigation strategies to tackle this challenge is to restore the germplasm through conservation. Conservation of medicinal plants consists of two pre-requisite steps before the actual procedure is carried out. These include identification and characterization of species to be stored or preserved (Rohini, 2020; Marakli, 2018). Molecular markers used for the DNA barcoding (and authentication of plant species) are first employed to precisely identify the medicinal plant species. Sequence variations in genes such as *rbcL*, *matK*, *COI*, and ITS region depict the molecular signature of any plant and its species (Wattoo et al., 2016). This data is first generated and then stored to be used for the future identification of the same or related species, as well as breeding programme when reintroduction or restoration is carried out. After successful identification and reporting of the data related to it on the concerned

TABLE 4.2
Plant Species Diversity Assessed by Various Molecular Markers and the Efficiencies or Outcome of Their Application

Sr. No.	Plant	Molecular Marker	Outcome	References
1	Aloe vera	AFLP	• Percentage polymorphism was found between 38% and 57%. • Balaghat variety was completely distinct from the other group	Tripathi et al. (2011)
2		Nuclear markers (Illumina Hi Seq)	• Accurate and reliable method of identification of the more than 300 species. • Provided promising protocol for the traded endangered plants.	Woudstra et al. (2025)
3	Ashwagandha	EST-SSR, ISSR and SSR	• Average number of alleles per primer is 5.32 based on all the three markers studied. • EST marker EMS-12 was found to be linked with withanolides.	Chawla et al. (2025)
4		RAPD	• AGB-053 and AGB-055 random primers were able to show maximum diversity among studied population.	Khanna et al. (2014)
5	Chamomile	SSR	• Among 118 genotypes studied for the diversity, 670 alleles were present with the average of 20.3 allele per loci. • Trait specialized populations are grouped together.	Sharan et al. (2024)
6		ISSR	• Out of 20 ISSR primers, 5 were found to show maximum polymorphism among population. • Average PIC value was found to be 0.340.	Okoń et al. (2013)
7		SNP	• From the sequencing data, 89 SNPs were generated which could be used to differentiate 19 mitochondrial haplotypes. • Genetic distance was ranged between 0.011 and 0.851.	Ruzicka and Novak (2020)
8	*Curcuma longa* L.	SSR	• Accessions with more than 1.5% curcumin content were assessed with 39 SSR markers. • 2–8 allele per loci were observed. • A total of 163 alleles were found to have 100% polymorphism.	Gogoi et al. (2023)
9		ISSR	• Thirty-five turmeric genotypes were assessed with ISSR markers. The result showed 94.82% polymorphic loci. • Seven ISSR markers were found to identify 100% polymorphism.	Gowda et al. (2025)
10	Garlic	SSR and EST-SSR	• Seventeen SSR markers along with 12 polymorphic EST-SSR markers were used to check the polymorphism in 127 garlic accessions. • Combined average PIC value was found to be 0.36. A total of 79 polymorphic loci were detected among the testing population.	Li et al. (2022)
11		Sequence-related amplified polymorphism (SRAP)	• High genetic variability was observed among 40 germplasm of garlic. • Twenty-three SRAP primer combinations produced 130 amplicons that were polymorphic.	Chen et al. (2013)

(continued)

TABLE 4.2 (Continued)
Plant Species Diversity Assessed by Various Molecular Markers and the Efficiencies or Outcome of Their Application

Sr. No.	Plant	Molecular Marker	Outcome	References
12	Ginger	SSR	• One hundred ginger accessions were checked for the genetic diversity using 12 SSR markers. • The result showed 139 polymorphic loci and suggested that similar markers could be used to assess the genetic diversity when high rhizome yield and disease tolerance characters are targeted.	Gezahegn et al. (2025)
13		AFLP	• An average number of 113.5 polymorphic loci were found with 13 primer combination in 6 accessions of ginger. • The study was very useful for the introduction of new varieties through breeding programme.	Blanco et al. (2016)
14	Peppermint	ISSR	• A total of 58 accession belonging to 7 different taxonomic group were assessed using 9 highly polymorphic ISSR markers. • Authors suggested that the blend of morphological and molecular data is very vital for the breeding programme in *Mentha* spp.	Soilhi et al. (2020)

repository, conservation practices such as cryopreservation, somatic embryogenesis, cell culture, tissue culture, synthetic seed production or germplasm conservation are followed (Rai, 2022; Dixit et al., 2022; Rihan et al., 2017). While reviving the mother plant after tissue culture operations, molecular markers again play important role to assess whether the genetic fidelity is maintained or not. Selection for in vivo propagation is carried out by comparing the genetic fingerprinting of the clones to original mother culture. This process is governed using molecular markers.

4.5 CHALLENGES AND FUTURE DIRECTIONS

The most common bottleneck of using molecular marker for the crop improvement strategy is the correlation between genotypic and phenotypic data. In case of medicinal plants, the bar still gets higher because the correlation of genotypic data is done with chemotypic data. Apart from this, tissue culture protocols for the conservation of the plant species are not available and many a times the medicinal plants are highly recalcitrant to tissue culture and transformation procedures. Conservation of wild resources in the ecosystem is another challenge as the anthropogenic activities and urbanization prevails in many developed and developing nations. Biotechnological interventions like genetic engineering and gene editing may prove as promising strategies for the conservation and improvement of the medicinal plant species and to produce novel medicinally important metabolites. On the other hand, due to land scarcity, pharmaceutically important metabolites can be synthesized in the plant cells and harvested in a continuous or semi-discontinuous mode.

4.6 CONCLUSION

In summary, medicinal plants' amazing diversity, which is fuelled by their distinct chemical compositions, highlights their importance in healthcare and research. Given that some of these priceless species are in danger of going extinct, it is becoming increasingly crucial to conserve them. In addition to providing precise identification and characterization of medicinal plants, molecular markers have become vital tools for assisting with breeding and conservation initiatives. The production of new plant varieties with desirable qualities is made easier by these markers, which enable precise genetic analysis. This ensures that medicinal plant resources are used sustainably and preserved for future generations.

REFERENCES

Al-Samarai, F.R. and Al-Kazaz, A.A., 2015. Molecular markers: An introduction and applications. *European Journal of Molecular Biotechnology*, *9*(3), pp.118–130.

Amiteye, S., 2021. Basic concepts and methodologies of DNA marker systems in plant molecular breeding. *Heliyon*, *7*(10), e08093.

Bhapkar, V., Nisargand, V., Godatwar, P. and Bhalerao, S., 2025. Research status of traditional & complementary medicine systems across the world. *Journal of Ayurveda and Integrative Medicine*, *16*(2), 101078.

Blanco, E.Z., Bajay, M.M., Siqueira, M.V.B.M., Zucchi, M.I. and Pinheiro, J.B., 2016. Genetic diversity and structure of Brazilian ginger germplasm (*Zingiber officinale*) revealed by AFLP markers. *Genetica*, *144*(6), pp.627–638.

Burki, T., 2025. WHA adopts new strategy on traditional medicine. *The Lancet*, *405*(10493), pp.1897–1898.

Chawla, R., Sharma, H., Dadheech, A., Jattan, M., Devi, S., Kumar, P. and Singh, T., 2025. Genetic insights into diversity and population structure of ashwagandha (*Withania somnifera* (L.) Dunal) using EST-SSR, ISSR and SSR markers: implications for enhancing agricultural and industrial value. *Industrial Crops and Products*, *224*, 120242.

Chen, S., Zhou, J., Chen, Q., Chang, Y., Du, J. and Meng, H., 2013. Analysis of the genetic diversity of garlic (*Allium sativum* L.) germplasm by SRAP. *Biochemical Systematics and Ecology*, *50*, pp.139–146.

Cheng, J., Long, Y., Khan, A., Wei, C., Fu, S. and Fu, J., 2015. Development and significance of RAPD-SCAR markers for the identification of *Litchi chinensis* Sonn: by improved RAPD amplification and molecular cloning. *Electronic Journal of Biotechnology*, *18*(1), pp.35–39.

Chien, L.H., Lin, C.H., Ling, J.A., Sheu, S.C., Wu, C.T., You, W.R., Xu, M.R., Cheng, J.H. and Lee, M.S., 2025. Rapid authentication of the medicinal crop *Cirsium setosum* and discrimination from its adulterants in herbal markets via polymerase chain reaction integrated with a lateral flow assay (PCR-LFA). *Industrial Crops and Products*, *228*, 120849.

Davis, C.C. and Choisy, P., 2024. Medicinal plants meet modern biodiversity science. *Current Biology*, *34*(4), pp.R158–R173.

Dixit, J., Saini, R., Mishra, A.K., Verma, P., Kumar, P. and Tiwari, K.N., 2022. Somatic embryogenesis in medicinal plants. In A.C. Rai, A. Kumar, A. Modi, & M. Singh (Eds.), *Advances in Plant Tissue Culture* (pp. 223–258). London, UK: Academic Press.

Feng, S., Zhu, Y., Yu, C., Jiao, K., Jiang, M., Lu, J., Shen, C., Ying, Q. and Wang, H., 2018. Development of species-specific SCAR markers, based on a SCoT analysis, to authenticate Physalis (Solanaceae) species. *Frontiers in Genetics*, *9*, p.192.

Ganie, S.H., Srivastava, P.S., Narula, A., Ali, Z. and Sharma, M.P., 2012. Authentication of shankhpushpi by RAPD markers. *EurAsian Journal of BioSciences*, *6*, pp.39–46.

Gezahegn, G., Rezene, Y. and Feyissa, T., 2025. Genetic diversity analysis of Ethiopian ginger (*Zingiber officinale* Roscoe) accessions using simple sequence repeat (SSR) markers. *Genetic Resources and Crop Evolution*, *72*(1), pp.141–152.

Gogoi, A., Munda, S., Paw, M., Begum, T., Siddiqui, M.H., Gaafar, A.R.Z., Kesawat, M.S. and Lal, M., 2023. Molecular genetic divergence analysis amongst high curcumin lines of Golden Crop (*Curcuma longa* L.) using SSR marker and use in trait-specific breeding. *Scientific Reports*, *13*(1), 19690.

Gowda, M.S., Soundarya, D., Hiremath, C. and Shetty, N.P., 2025. Assessing genetic diversity of indigenous turmeric (*Curcuma longa* L.) through inter-simple sequence repeat (ISSR) markers. *Genetic Resources and Crop Evolution*, *72*(2), pp.2413–2423.

Gupta, G., Paul, S., Singh, S., Pietramellara, G., Pathan, S.I., Danish, S., Jabborova, D., Datta, R. and Jha, P.N., 2022. Exploring functional diversity and community structure of diazotrophic endophytic bacteria associated with *Pennisetum glaucum* growing under field in a semi-arid region. *Land*, *11*(7), p.991.

Hao, L., Shi, X., Wen, S., Yang, C., Chen, Y., Yue, S., Chen, J., Luo, K., Liu, B., Sun, Y. and Zhang, Y., 2025. Single nucleotide polymorphism-based visual identification of *Rhodiola crenulata* using the loop-mediated isothermal amplification technique. *Frontiers in Plant Science*, *15*, 1492083.

Hon, Y.K., Yong, C.S.Y., Abdullah, J.O. and Go, R., 2021. Development of species-specific SCAR markers for identification and authentication of three rare Peninsular Malaysian endemic Coelogyne (Orchidaceae) orchids. *F1000Research*, *9*, p.1161.

Horn, T., Völker, J., Rühle, M., Häser, A., Jürges, G. and Nick, P., 2014. Genetic authentication by RFLP versus ARMS? The case of Moldavian dragonhead (*Dracocephalum moldavica* L.). *European Food Research and Technology*, *238*(1), pp.93–104.

Husen, A., 2022. *Herbs, Shrubs and Trees of Potential Medicinal Benefits*. Boca Raton, FL: CRC Press. https://doi.org/10.1201/9781003205067

Husen, A., 2023a. *Exploring Poisonous Plants: Medicinal Values, Toxicity Response and Other Potential Uses.* Boca Raton, FL: CRC Press. https://doi.org/10.1201/b23017

Husen, A., 2023b. *Antidiabetic Medicinal Plants and Herbal Treatments.* Boca Raton, FL: CRC Press. https://doi.org/10.1201/b23347

Husen, A., 2024a. *Antimalarial Medicinal Plants.* Boca Raton, FL: Taylor & Francis Group. https://doi.org/10.1201/9781003378396

Husen, A., 2024b. *Medicinal Spices and Condiment Crops*. Boca Raton, FL: Taylor & Francis Group. https://doi.org/10.1201/9781003387046

Husen, H. and Pant, M., 2024. *Exploring Medicinal Orchids.* Boca Raton, FL: Taylor & Francis Group. https://doi.org/10.1201/b23375

Kadam, S.T. and Pawar, A.D., 2020. Conservation of medicinal plants: A review. *International Ayurvedic Medical Journal*, *8*, pp.3890–3895.

Keeratinijakal, V., Kladmook, M. and Laosatit, K., 2010. Identification and characterization of *Curcuma comosa* Roxb., phytoestrogens-producing plant, using AFLP markers and morphological characteristics. *Journal of Medicinal Plants Research*, *4*(24), pp.2651–2657.

Khanna, P.K., Chandra, R., Kumar, A., Dogra, N., Gupta, H., Gupta, G. and Verma, V., 2014. Correlation between morphological, chemical and RAPD markers for assessing genetic diversity in *Withania somnifera* (L.) Dunal. *Journal of Crop Science and Biotechnology*, *17*(1), pp.27–34.

Kumar, A., Prakash, K., Sinha, R.K. and Kumar, N., 2013. In vitro plant propagation of *Catharanthus roseus* and assessment of genetic fidelity of micropropagated plants by RAPD marker assay. *Applied biochemistry and biotechnology*, *169*(3), pp.894–900.

Kumar, N., Modi, A.R., Singh, A.S., Gajera, B.B., Patel, A.R., Patel, M.P. and Subhash, N., 2010. Assessment of genetic fidelity of micropropagated date palm (*Phoenix dactylifera* L.) plants by RAPD and ISSR markers assay. *Physiology and Molecular Biology of Plants*, *16*(2), pp.207–213.

Kumar, R., Sharma, V. and Kumar, S., 2024. Biotechnological tools for conservation and sustainable utilization of medicinal plants. In R.C. Sobti (Ed.), *Role of Science and Technology for Sustainable Future: Sustainable Development: A Primary Goal* (Vol. 1, pp. 303–319). Singapore: Springer Nature Singapore.

Li, X., Qiao, L., Chen, B., Zheng, Y., Zhi, C., Zhang, S., Pan, Y. and Cheng, Z., 2022. SSR markers development and their application in genetic diversity evaluation of garlic (*Allium sativum*) germplasm. *Plant Diversity*, *44*(5), pp.481–491.

Little, D.P., 2014. Authentication of Ginkgo biloba herbal dietary supplements using DNA barcoding. *Genome*, *57*(9), pp.513–516.

Marakli, S., 2018. A brief review of molecular markers to analyse medically important plants. *International Journal of Life Sciences and Biotechnology*, *1*(1), pp.29–36.

Modi, A., Suthar, K., Thakkar, P., Mankad, M.C., Kumari, S., Narayanan, S., Singh, A.S. and Kumar, N., 2018. Evaluation of sex specific RAPD and SCAR markers linked to papaya (*Carica papaya* L.). *Biocatalysis and agricultural biotechnology*, *16*, pp.271–276.

Modi, A.R., Patil, G., Kumar, N., Singh, A.S. and Subhash, N., 2012. A simple and efficient in vitro mass multiplication procedure for *Stevia rebaudiana* Bertoni and analysis of genetic fidelity of in vitro raised plants through RAPD. *Sugar Tech*, *14*(4), pp.391–397.

Okoń, S., Surmacz-Magdziak, A. and Paczos-Grzęda, E., 2013. Genetic diversity among cultivated and wild chamomile germplasm based on ISSR analysis. *Acta Scientiarum Polonorum Hortorum Cultus*, *12*(2), pp.43–50.

Pandey, G., Patil, G., Modi, A., Desai, S., Desai, P. and Narayanan, S., 2022. Expression pattern of in vitro organogenesis associated genes as transcriptional marker in sandalwood (*Santalum album* L.) micropropagation. *Plant Tissue Culture and Biotechnology*, *32*(2), pp.103–113.

Pant, M. and Husen, A., 2025. *Exploring Medicinal Bamboos.* Boca Raton, FL: Taylor & Francis Group. https://doi.org/10.1201/9781003485933

Rai, M.K., 2022. Plant tissue culture targeting germplasm conservation. In A.C. Rai, A. Kumar, A. Modi, & M. Singh (Eds.), *Advances in Plant Tissue Culture* (pp. 205–221). London, UK: Academic Press.

Raza, A., Mehmood, S.S., Shah, T., Zou, X., Yan, L., Zhang, X. and Khan, R.S.A., 2019. Applications of molecular markers to develop resistance against abiotic stresses in wheat. In M. Hasanuzzaman, K. Nahar, & M. Hossain (Eds.), *Wheat Production in Changing Environments: Responses, Adaptation and Tolerance* (pp. 393–420). Singapore: Springer Singapore.

Rihan, H.Z., Kareem, F., El-Mahrouk, M.E. and Fuller, M.P., 2017. Artificial seeds (principle, aspects and applications). *Agronomy*, *7*(4), p.71.

Rohini, M.R., 2020. Genomics and molecular characterization of threatened medicinal plants. In P. Rajasekharan & S. Wani (Eds.), *Conservation and Utilization of Threatened Medicinal Plants* (pp. 317–346). Cham: Springer International Publishing.

Ruzicka, J. and Novak, J., 2020. Mitochondrial genome variation between different accessions of *Matricaria chamomilla* L. (Asteraceae) based on SNP mutation analysis. *Genetic Resources and Crop Evolution*, *67*(4), pp.853–864.

Sharan, H., Kumar, A. and Singh, S., 2024. Newly developed SSRs based genetic diversity patterns in trait specific populations of German chamomile (*Matricaria chamomilla* L.). *Industrial Crops and Products*, *210*, 118152.

Sharma, A., Kumar, N. and Mishra, I.G., 2018. Role of molecular marker in the genetic improvement of the medicinal and aromatic plants. In *Biotechnological Approaches for Medicinal and Aromatic Plants: Conservation, Genetic Improvement and Utilization* (pp. 557–567). Singapore: Springer Singapore.

Soilhi, Z., Trindade, H., Vicente, S., Gouiaa, S., Khoudi, H. and Mekki, M., 2020. Assessment of the genetic diversity and relationships of a collection of *Mentha spp.* in Tunisia using morphological traits and ISSR markers. *The Journal of Horticultural Science and Biotechnology*, *95*(4), pp.483–495.

Tripathi, N., Saini, N. and Tiwari, S., 2011. Assessment of genetic diversity among Aloe vera accessions using amplified fragment length polymorphism. *International Journal of Medicinal and Aromatic Plants*, *1*(2), pp.115–121.

Vu, D.D., Shah, S.N.M., Pham, M.P., Bui, V.T., Nguyen, M.T. and Nguyen, T.P.T., 2020. De novo assembly and Transcriptome characterization of an endemic species of Vietnam, *Panax vietnamensis* Ha et Grushv., including the development of EST-SSR markers for population genetics. *BMC Plant Biology*, *20*(1), p.358.

Wang, X.M., 2011. Inter-simple sequence repeats (ISSR) molecular fingerprinting markers for authenticating the genuine species of rhubarb. *Journal of Medicinal Plants Research*, *5*(5), pp.758–764.

Wattoo, J.I., Saleem, M.Z., Shahzad, M.S., Arif, A., Hameed, A. and Saleem, M.A., 2016. DNA barcoding: Amplification and sequence analysis of rbcL and matK genome regions in three divergent plant species. *Advancements in Life Sciences*, *4*(1), pp.3–7.

Woudstra, Y., Rees, P., Rakotoarisoa, S.E., Rønsted, N., Howard, C. and Grace, O.M., 2025. An updated DNA barcoding tool for aloe vera and related CITES-regulated species. *Conservation Letters*, *18*(4), e13127.

5 Advances in Techniques Used in Molecular Marker Development

Susan Maria Binu, Rebecca Jacey Kurian, and Anu Jacob

5.1 INTRODUCTION

Present-day plant science has undergone many significant changes through biotechnology, and the sector that covers plant breeding is not an exception. Currently, one of the major technological discoveries that has greatly contributed to increasing accuracy and effectiveness in crop improvement programmes is the identification of molecular markers. Once the scientists know what type of molecular markers occur, the farmers can obtain useful information about the features and aspects of plants, thus ensuring that only the selected, most desirable characters, such as increased yield or disease resistance, are expressed. Molecular markers are specific sequences of DNA that act as genetic markers on chromosomes, enabling the connection of traits or genes of interest to identifiable genetic loci. They are used to pinpoint certain genome regions associated with advantageous traits responsible for yield, nutritional value, disease resistance, and adaptability to environmental conditions. Molecular markers are known to increase the speed and effectiveness of breeding activities, making it possible to identify the genetic makeup of these features. It is essential to recognize that, in specific contexts, a molecular marker may consist of a single or multiple primers, one or more restriction enzymes, or a combination of both the primers, restriction enzymes, and other relevant components, alongside the techniques employed for the detection (Amiteye, 2021). The first DNA-level molecular marker was put forth in the 1980s. In addition to being independent of phenotype, molecular markers have high heterozygosity, detect differences at the DNA level, are not impacted by environmental factors, have a large number of accessible markers, and exhibit Mendelian inheritance (Karlik & Tombuloğlu, 2016). Along with improving the efficiency of the breeding cycle, the use of molecular markers has made it possible to accurately change the genes using methods such as genome editing and marker-assisted breeding. These techniques ensure that only the desired trait is modified and thereby minimize the unintended changes associated with plant breeding, along with the enhancement of plant varieties and associated agricultural practices.

An ideal genetic marker is characterized by a number of features. It should have no effect on the phenotype, follow the rule of co-dominance, and be present in a single copy. In addition, the ideal marker should be cheap and highly polymorphic, easy to determine, and perform multiple functions. The marker should be freely available for research and be examined within a genome and be genome-specific when it is necessary to work with polyploid species. The marker should also support the possibility of multiplexing and automation (Raza et al., 2019).

 DOI: 10.1201/9781003542681-5

5.2 PROPERTIES OF MOLECULAR MARKERS

5.2.1 Genomic Abundance

The frequency with which specific key sites appear in the genome can help to find potential markers. Restriction fragment length polymorphisms (RFLPs) and amplified fragment length polymorphisms (AFLPs) are widely known for their abundance due to the availability of restriction enzymes along with genome recognition sites. Microsatellites are very common in eukaryotic genomes. Furthermore, random amplified polymorphic DNA (RAPD) markers are abundant because primers can be generated from a large range of random sequences. When aiming for both genomic abundance and coverage, markers must be carefully chosen because some are uniformly dispersed over the genome while others, such as AFLP markers, cluster in specific places (Karlik & Tombuloğlu, 2016).

5.2.2 Level of Polymorphism

The efficacy of genetic markers is dictated by the level of polymorphism discovered, which is influenced by the rate of mutations at the relevant genomic sites. Mutations at the core sequence, resulting in changes in the number of repeat units, cause mutations at the locus of the minisatellite and microsatellites, which are relatively high. When examining closely related samples, increased resolving power is necessary; for example, fast-evolving markers such as microsatellites may be useful for study within or among closely related species. Conversely, AFLPs or RFLPs may be more suited for investigating genetic relationships at a broader taxonomic level as rapidly changing markers are less likely to be homologous. Marker selection is based on the idea that as evolutionary distances rise, more permanent markers with slower evolutionary rates are necessary, and vice versa (Teama, 2018).

5.2.3 Co-Dominance of Alleles

The concept of molecular markers' dominance or co-dominance is crucial for distinguishing them. Co-dominant markers permit both alleles to be expressed when they coexist in an individual. Consequently, co-dominant markers make it possible to differentiate between heterozygotes and homozygotes, and therefore, it becomes possible to calculate genotypes and allele frequencies at particular loci. Generally, such co-dominant markers are preferred for most applications. Most of these markers are of the single locus type, which usually gives less information per assay than multi-locus techniques (Karlik & Tombuloğlu, 2016). By a dominant marker, a homozygous one cannot be distinguished from a heterozygous one. However, the heterozygote of a co-dominant marker can be distinguished. Sometimes a molecular marker can be found within a transposon, which, however, does not obey Mendel's laws of inheritance but continues to function. Therefore, Mendelian inheritance is not always necessary. Figure 5.1 illustrates that dominant markers and co-dominant markers have different abilities to distinguish homozygotes from heterozygotes; co-dominant markers express both alleles, and dominant markers express the dominant allele only in heterozygotes.

5.3 CLASSIFICATION OF MOLECULAR MARKERS

The molecular markers indicate the presence of genetic variation, which can be revealed at the level of DNA or protein, and help in tracking the current status of a given gene or genomic region that is associated with a targeted trait (Dida, 2022). As shown in Figure 5.2, molecular markers based on polymerase chain reaction (PCR) include arbitrarily primed polymerase chain reaction (AP-PCR), start codon targeted (sCoT), inter-simple sequence repeats (ISSR), sequence characterized amplified

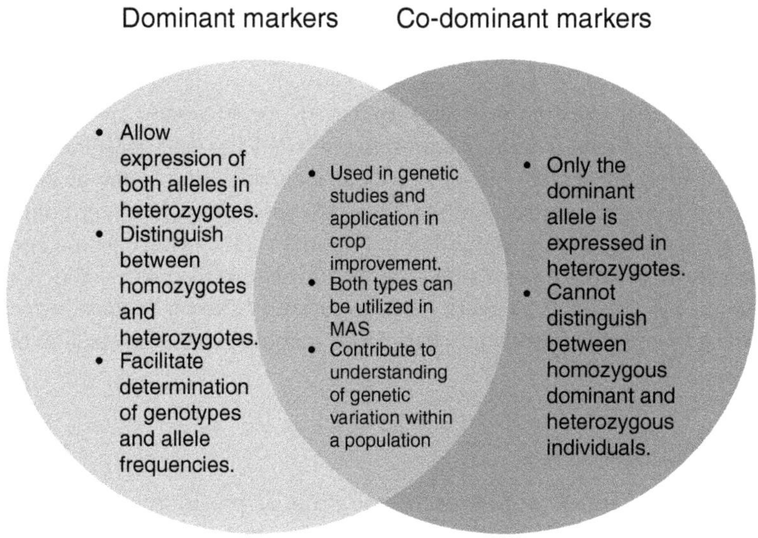

FIGURE 5.1 Comparison of dominant and co-dominant markers in plant breeding.

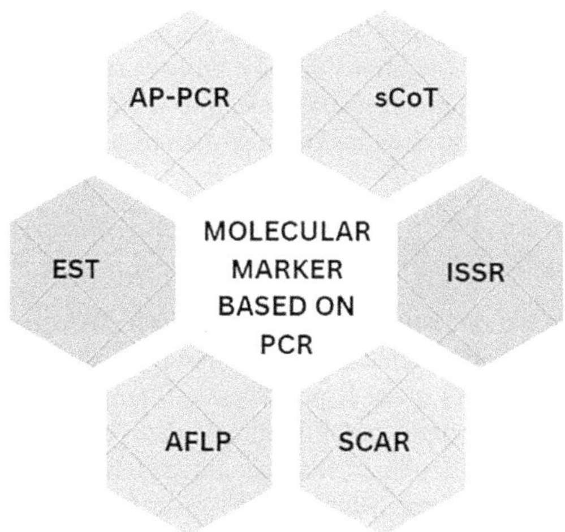

FIGURE 5.2 Molecular marker classification based on PCR.

region (SCAR), AFLP, and expressed sequence tags (EST), each serving distinct applications in genetic analysis and marker-assisted selection (MAS).

5.3.1 Restriction Fragment Length Polymorphisms

RFLPs are one of the first molecular markers developed. It is based on the existence of the DNA sequence variation in the target that is recognized by a certain restriction enzyme. The majority of RFLP markers are specific to the locus, and they are co-dominant. Its procedure is simple, with no need for sophisticated equipment, and the RFLP phenotype is highly reproducible. Variations in the sizes of digestion enzyme-generated fragments can be observed through agarose gel electrophoresis. Variabilities in individuals are exhibited by the appearance of the DNA band pattern found in RFLP. It is due to the variation that is caused by a mutation with deletions and/or insertions of a single-nucleotide base pair. The mutation either creates a new RE recognition sequence or deletes the existing enzyme recognition sequences, which cause a fragment of different sizes and polymorphism (Amiteye, 2021). It, however, yields sufficient information but requires an enormous amount of DNA and can be labour-intensive.

5.3.2 Amplified Fragment Length Polymorphisms

AFLPs can be used to amplify DNA fragments selectively by PCR. It is performed with the digestion of specific DNA by two restriction enzymes recognized by rare and frequent cutters, followed by ligating the adaptors to the DNA ends. AFLP generates a number of DNA fragments that are separated by gel electrophoresis to analyse the polymorphic markers. Following the description of modified dye-labelled primers, the fragments are detected and separated on a capillary or gel-based DNA sequencers either fluorescently or radioactively. The generation of the desired bands depends on the number of nucleotide motifs selected and the quantity of a specific subgroup (Vos et al., 1995).

5.3.3 Simple Sequence Repeats or Microsatellites

Simple sequence repeats (SSRs) are short DNA sequences made up of repeating units (e.g., ATATATAT), which vary in length depending on the number of repeats. They are easily identified by PAGE or AGE and can be analysed using PCR (Ahmad et al., 2018). Multiplexing, high-throughput genotyping, and automation are all conceivable with SSR markers. SSR assays require very small amounts of DNA (about 100 ng per person), and manual assay methods offer cheap initial costs. They find considerable usage in plant breeding due to their abundance in the genome as well as their highly variable nature.

5.3.4 Single Nucleotide Polymorphisms

Single-nucleotide polymorphisms (SNPs) consist of single base pair polymorphisms. These substitutions can be further classified as a transversion or a transition. They are the most common type of variation in the DNA sequence, and, therefore, many of these markers have been reliably detected, especially due to microarrays and next-generation sequencing (NGS).

5.3.5 Cleaved Amplified Polymorphic Sequences and Derived Cleaved Amplified Polymorphic Sequences

Cleaved amplified polymorphic sequences (CAPS) and derived cleaved amplified polymorphic sequences (dCAPS) markers are SNP-based and may modify or generate enzyme recognition sites. The resulting markers, following amplification by PCR, can be utilized to detect different genotypes and generate useful markers through digestion with the appropriate enzymes and separation by gel electrophoresis based on fragment size. The patterns are referred to as 'polymorphisms' and provide a foundation for marker development (Park et al., 2024).

5.3.6 Random Amplified Polymorphic DNA

Unlike RFLP analysis, which relies on the restriction enzyme fragment sizes, RAPDs result from the PCR amplification of DNA segments with short, random primers, which are then separated by gel electrophoresis. The differences in banding patterns among DNA samples stand for genetic diversity between them. One advantage of using RAPD markers is that they do not need any prior knowledge of the DNA sequences because of the random design of primers allowing them to bind to any region of the genome (Rafalski et al., 2020). This makes them simple to use and facilitates rapid production of markers in a wide range of species with low-quality sequence data. An ideal primer sequence designed for RAPD amplification must have a Guanine–Cytosine (GC) concentration of 40–60%, as it influences the primer's binding effectiveness during a PCR. A primer with less than 40% GC concentration does not bind strongly to the template, resulting in poor amplification. However, a GC content higher than that may interfere with the primer annealing temperature (Fadel et al., 2022). Furthermore, while RAPDs are a quick and inexpensive approach, the arbitrary nature of the primers may cause reproducibility issues, necessitating careful optimization of experimental conditions. While they are a quick and easy way to generate markers, they are not always repeatable and require ongoing optimization.

5.3.7 Inter-Simple Sequence Repeats

Inter-simple sequence repeats (ISSRS) amplify the microsatellites using primers that are anchored, and they can be used to determine genetic diversity by combining the benefits from SSRs and RAPDs (Zietkiewicz et al., 1994).

5.3.8 Single Nucleotide Polymorphism Arrays

High-density SNP arrays: SNP arrays represent one of the most crucial advances in genomics. These arrays have a high density of SNP markers, ranging from thousands to millions. Thus, powerful and expedient genotyping of vast populations might be achieved. Additionally, SNP arrays are widely used in genome-wide association studies (GWAS) and genomic selection (GS) (Gunderson et al., 2005).

5.3.9 Genotyping-by-Sequencing

Genotyping-by-sequencing (GBS) utilizes NGS that can target subsets of any particular genome. Thereby enabling the genotyping of many individuals simultaneously at a lower cost and making it suitable for both well-characterized and non-model species (Elshire et al., 2011).

5.3.10 Methylation-Based Markers

These markers identify epigenetic modifications like DNA methylation that can impact gene expression and phenotype. They are also significant for studying complex traits (Shaikh et al., 2022).

5.3.11 Gene-Specific Markers

These markers assist in the tracking and selection of distinct populations by targeting particular genes of interest (Swarup et al., 2021).

5.3.12 Transcriptome-Based Markers

Transcriptome sequencing provides crucial insights into gene function and regulation by uncovering variations in gene expression associated with traits (Fujito et al., 2021).

5.3.13 Exome Capture Markers

These markers offer cost-effective genotyping information by focusing on the sequencing of only protein-coding regions (exons) (Mu et al., 2022).

5.3.14 Insertion–Deletion Polymorphisms

Insertion–deletion polymorphisms (InDels) are also unique when used to distinguish genotypes of very closely related persons, most commonly used by SNP nucleotide varieties in a given location within a genotype (Pan et al., 2021).

5.3.15 Sequence-Tagged Sites

Sequence-tagged site (STS) markers are identifiable sequences from within the DNA text by using a pair of short primers and PCR to produce a faint band at the right measurement. Molecular markers may be selected depending on the objectives of the specific breeding programme, as well as the available resources and the species of interest. The fast growth of genomics and sequencing-specific technology has contributed to the development of an increasing range of molecular markers available for utilization by plant breeders to speed up the development of crop varieties that are superior to their predecessors (Olson et al., 1989).

5.4 MARKER-ASSISTED SELECTION

Plant breeding has enormous promise as DNA markers can identify allelic variability in the genes that underpin traits. DNA markers can increase the efficiency and precision of plant breeding. MAS refers to the use of DNA markers in plant breeding. Because of molecular MAS, new crop varieties are now offered to the market significantly faster (Hasan et al., 2021).

5.4.1 Comparison with Traditional Breeding

Traditional plant breeding has helped to generate improved crop types by selecting desired phenotypic features across generations. This method, while successful, is typically time-consuming and influenced by ambient circumstances, which might mask the manifestation of particular characteristics. Furthermore, traditional breeding has difficulties in enhancing complex qualities regulated by several genes, such as tolerance to drought and resistance to disease.

In comparison to traditional plant breeding, which mostly depends on the increased variability and mutation to alter the genetic composition of the parent plants, molecular breeding employs novel genomic tools to ensure the desired results are predictable and maximally efficient. Thus, by using such techniques as MAS and GS, breeders can rapidly and unequivocally pinpoint the genes or DNA sequences that are linked to the target phenotype to eliminate the natural effect and variability in the environment (Collard & Mackill, 2008). The comparison (Kushanov et al., 2021) is depicted in Table 5.1.

5.4.2 Quantitative Trait Locus Mapping and Marker Development

The concept of MAS is based on ideas formalized by Sax in 1923 as quantitative trait loci (QTLs) mapping. Sax observed the association of a monogenic and a polygenic trait located on the same chromosome: the seed coat colour gene indicated the seed size (Sax, 1923). In 1961, the more sophisticated approach was proposed by Thoday to map all QTLs to single gene markers and study their mapping function (Thoday, 1961). This approach overcame the existing limits on studying quantitative variation, as no high-resolution genetic map was available.

QTL mapping involves identifying associations between genetically determined phenotypes and genetic markers. Unlike conventional single-marker analysis techniques, which do not utilize the marker order on a genetic map, QTL mapping takes this into account. Populations are classified based on their genotype at each marker locus, and a QTL is identified if there is significant variation in the mean phenotypic scores between the groups (Powder, 2020).

Numerous DNA markers linked to genes of QTLs have been reported in earlier years. As shown in Table 5.2, it was assumed that the huge amounts of markers linked to QTLs from early mapping research were immediately applicable in MAS (Hasan et al., 2021). However, since then, it has been widely recognized in recent years that fine or high-resolution mapping and QTL validation and/or QTL confirmation may be required. While it is true that further QTL mapping research has identified instances of extremely correct initial QTL mapping data, a validation phase, if possible,

TABLE 5.1
Comparison of QTL Mapping and Marker Development in Plant Breeding

Aspect	Conventional Breeding	Molecular Marker-Assisted Backcross Selection (MABS)
Basis of selection	Rely on phenotypic traits, which can be influenced by the environment.	Utilizes genotypic information linked to traits, independent of environmental influences.
Efficiency	Time-intensive, requiring multiple generations to fix traits.	Accelerate selection using genotypic markers, reducing breeding cycles.
Complex traits	Improvement is challenging due to polygenic inheritance and environmental effects.	Facilitates precise improvement by identifying QTLs linked to complex traits.
Trait introgression	High risk of linkage drags when transferring traits from wild relatives.	Minimizes linkage drag by targeting specific genomic regions for introgression.
Disease resistance	Identification of resistant genotypes can be inconsistent due to latent infections or environmental masking.	Molecular markers enable accurate identification and pyramid of resistance genes.
Cost and resources	Requires extensive field trials and resources for phenotypic evaluations.	Reduces resource use by focusing on genotypic data and early-stage selection.
Environmental dependence	Strongly influenced by environmental factors, leading to inconsistent results.	Genotypic markers ensure stability and reproducibility regardless of environmental effects.

TABLE 5.2
Recent Examples of QTL Mapping and Marker Development in Crops

Crop	Trait	Identified QTLs	Markers Developed	References
Rice	Heading date	qHD8, qHD10	SNP markers for MAS	Zhang et al. (2024)
Barley	Yield and nitrogen traits	Qldr.sau-5H.1, Qnhi.sau-5H	SNP markers for yield improvement	Chen et al. (2024a)
Rice	Nitrogen content, root traits, and grain yield	MQTL3.3, MQTL3.5	Markers for improving Nitrogen-use efficiency in rice	Sandhu et al. (2021)
Rice	Nitrogen traits	G393-C922, RM232-C63	Markers linked to these QTLs for potential use in breeding programmes.	Wei et al. (2012)

is always recommended as sampling error and other factors can lead to QTL positions and effects being inaccurate.

5.4.3 Genetic Diversity and Selection

MAS is the major way to introduce genetic variation to the breeding programmes, excluding the accompanying genetic drag from the otherwise inferior genomes of donor types. Genetic diversity itself gives critical variance to the breeding of crop plants. However, in practice, many programmes have difficulty applying MAS. A lack of confidence in MAS as a process and uncertainty about how it integrates with traditional breeding programmes are among the contributors to this deficiency. Such challenges might be overcome through progress in genomic tools, locus validation techniques, and fresh methodologies for applying QTLs to breeding activity (Hasan et al., 2021). It is crucial to find reliable QTLs, to devise dependable marker technology to pick out these QTLs, and to incorporate these QTLs into elite genomic contexts, making these QTLs actionable without any genetic drag, to make MAS-forward breeding possible. Conversely, other breeding techniques involve the introduction of unwanted variations requiring screening and selection to characterize desired changes. Genome editing is the latest breeding technology that can be utilized to introduce genetic variation. Just like multiple applications integrated to create a specific smart device in the digital world, several traits are integrated into multiple varieties of germplasm. The end product of plant breeding is distinguished as a commercial cultivar with a novel genetic composition known as a novel phenotype regardless of the source of genetic diversity, for example, native gene divers, biotechnology, or gene editing (Paul & Serita Frey, 2023). Despite the potential of GS, it is still challenging to predict quantitative polygenic traits effortlessly. Some intragenic interactions exist, and it is important to include their models to improve prediction. The choice of the single-trait or multiple-trait model should also be made concisely; at a high correlation level, multiple-trait models are preferred.

5.4.4 Types of MAS

5.4.4.1 Marker-Assisted Backcrossing

The marker-assisted backcrossing (MABC) strategy enables the deployment of the molecular markers to facilitate the recovery and selection of the recipient parent genome, facilitating the transfer of genes of interest or QTLs from the donor parents, therefore, making it easier to eliminate the undesirable traits from the high-yielding and popular varieties, for example, the pest and disease susceptibility or the anti-nutritional factors, among others from popular varieties (Osei et al., 2022), as illustrated in Figure 5.3.

5.4.4.2 Marker-Assisted Gene Pyramiding

Crop breeding has advanced dramatically in recent years, and the advent of novel molecular techniques has made precision breeding possible in a short period of time. Plant breeders can introduce many genes from many sources into cultivars and increase the diversity of resistance genes by employing the gene pyramiding approach, which allows the breeder to insert several resistance genes into a single genotype. Multiple genes/QTLs have been pyramided from donor to recipient parent utilizing various approaches such as recurrent selection, backcrossing, and multiple-parent or complex crossing. Three to four favourable genes from other lines are adopted via convergent or stepwise backcrossing, whereas additional genes are usually incorporated via serial crossing or recurrent selection (Kumawat et al., 2020).

5.4.4.3 Marker-Assisted Recurrent Selection

To address MABC's inability to transfer numerous QTLs controlling complex traits like yield or wide-ranging disease resistance, marker-assisted recurrent selection (MARS) was implemented (Cholin & Kulkarni, 2023). This method works well for cross-pollinating species and utilizes the F2 population. In contrast to MABC, both parents may contribute advantageous alleles, resulting in a chosen improved genotype that is a chimaera of its parents. MARS is widely used for the selection of inheritable traits, such as those of agricultural productivity and abiotic and biotic stress tolerance. MARS is a process that is referred to as forward breeding, used to improve numerous genes or QTLs; it increases the effectiveness of selection through the combination of phenotypic data with marker scores, resulting in reduced breeding time.

5.4.4.4 Genomic Selection

In GS, breeding value (BV) is estimated using marker information that covers the entire genome, rather than relying on precise gene location data. NGS technology has made it faster to genotype SNPs, enabling the GS of superior parents based on genomic-estimated breeding value (GEBV). SNPs are genotyped by every individual in the reference population to represent the full genome, and both phenotypic and genotypic data from this population are used to create a prediction model for estimating GEBV. Increasing the reference population sample size enhances the model's precision,

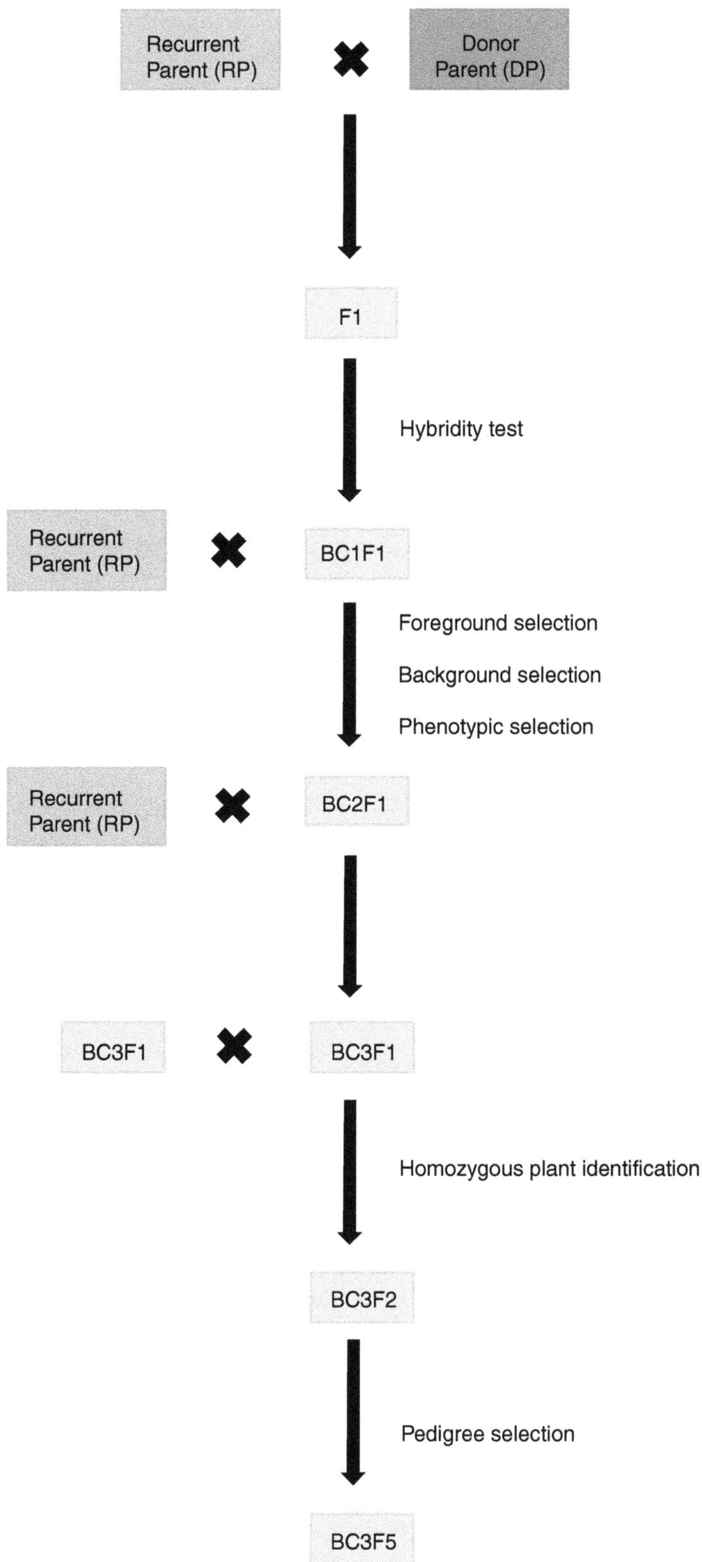

FIGURE 5.3 Backcross breeding methodology for introgression of desired traits.

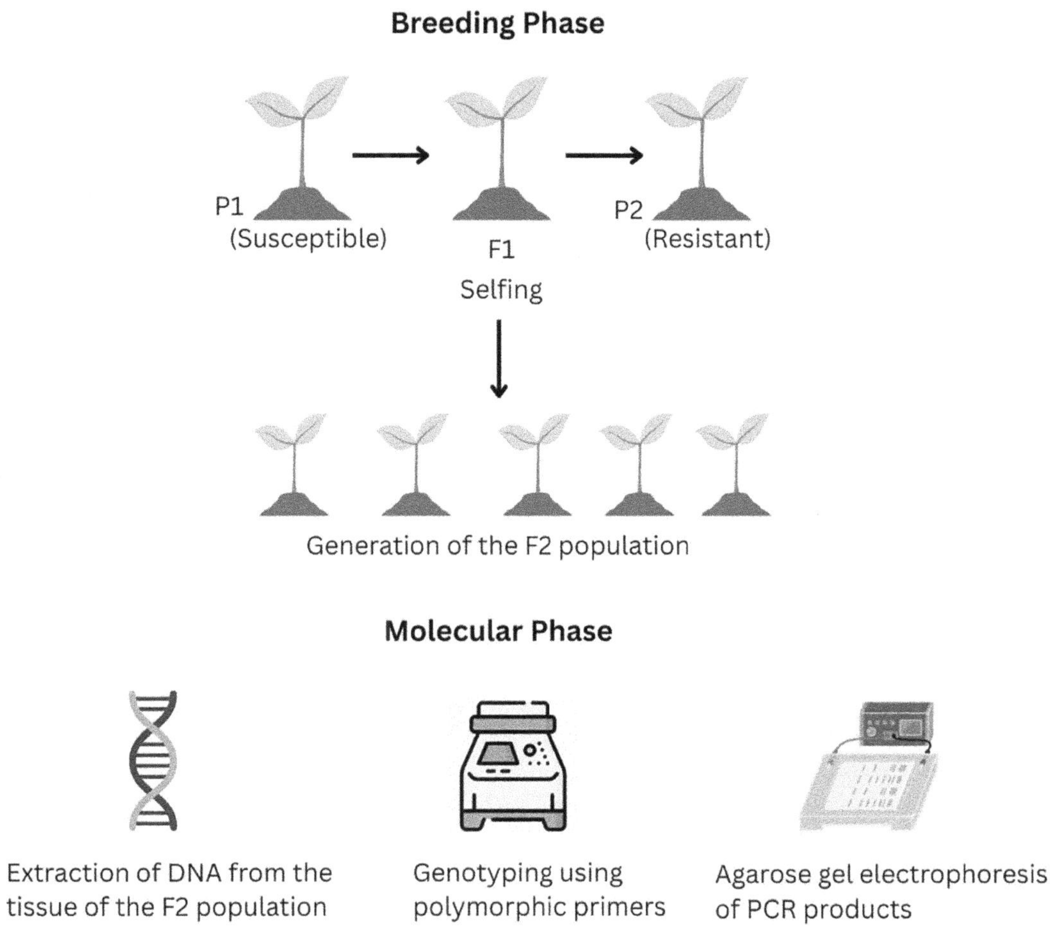

FIGURE 5.4 Overview of MAS.

and the model is trained using the collected genotype and phenotype data (Majumdar et al., 2020). These methods combine genetic information with observed traits to enable breeders to select for desired traits accurately and more rapidly. They include MAS as illustrated in Figure 5.4, shorter breeding cycles, more genetic gain, prediction of BV for instance, milk production in cows, integration of genes for environmental stress in crossbreeding, improved double haploid production for ensuring genetic variability, quick gene encoding from the wild populations, GS of multiple quantitative traits, evaluation of diversity, and germplasm preservation (Sun et al., 2024). Although the potential of GS appears promising, the prediction of multi-trait polygenic performance is challenging due to the trait's complexity. To increase the accuracy of the predictions, the models should be combined with gene-to-gene interactions in the gene.

5.5 MOLECULAR MARKERS IN PLANT STRESS TOLERANCE

5.5.1 MAS in Abiotic Stress Breeding

5.5.1.1 Drought Tolerance

Molecular markers have greatly contributed to the study of the tolerance mechanisms that have been coupled with the deployment of markers linked to specific attributes, such as root architecture and osmolyte production and water-use efficiency and targeted through identification of QTLs of drought-responsive genes, in the breeding programmes (Chugh et al., 2022). Furthermore, specific QTLs have been applied to develop drought-tolerant rice varieties. In the case of sorghum, MAS has allowed establishing the improvement of drought performance. Consequently, molecular markers were harnessed to identify the genes associated with the complex features of drought tolerance to develop the resilient sorghum lines (Mwamahonje et al., 2021).

Moreover, the introduction of high-throughput genotyping platforms has made it possible to rapidly discover the SNPs associated with drought tolerance and increase the rate of breeding new resistant cultivars (Farooqi et al., 2022). In addition, the use of molecular MAS has led to successful improvements in crop drought tolerance, such as complex trait phenotyping in sorghum, where molecular markers were applied to explore the gene source of complex drought tolerance (Mwamahonje et al., 2021). These developments emphasize the central role that molecular markers have the potential to play in improving drought tolerance, enabling the creation of more resilient crop varieties due to changes in the global climate.

5.5.1.2 Salt Tolerance

Salt tolerance is indispensable for providing cultivated crops that grow in soil with a high content of salt minerals, and molecular markers have facilitated discovering the genetic processes that provide this tolerance. It was determined that, in creating crops with improved salt tolerance, the markers related to ion homeostasis, osmolyte accumulation, and processes of stress signalling would prove efficient. In one example, the famous Saltol locus, which imparts salinity tolerance at the seedling stage in rice, has been effectively introgressed into temperate japonica cultivars by marker-assisted backcrossing. The practice was based on the selection of polymorphic competitive allele specific PCR (KASP) markers and enabled the salt tolerance of some of the newly obtained lines to be greatly enhanced (Marè et al., 2023). Moreover, the present approach of integrating several stress-tolerant genes via marker-assisted pyramiding has accelerated the progress of developing rice genotypes resistant to multiple abiotic stresses, including salinity. The method involves the combination of several target QTLs/genes, ultimately improving the overall stress resistance in the resultant cultivars (Mohanavel et al., 2024; Patani et al., 2023).

The subsequent rise of the CRISPR/Cas9 genome-editing technology has continued to transform the sector by making it possible to make targeted changes to plant genomes that can increase their tolerance of abiotic stresses. As shown by recent publications, CRISPR/Cas9 has already proven to be a valuable resource in developing crops that are less vulnerable to high salinity, lack of moisture, and other stressful factors in the environment (Erdoğan et al., 2023). Such progress underlines the importance of molecular markers and genome-editing approaches in the field of salt-tolerant crop breeding, opening up a variety of new and prospective possibilities for agriculture in arid saline areas.

5.5.2 MAS in Stress Tolerance Breeding

In combination with GS, the MAS shows great potential in developing crops with high tolerance to stress. This approach to integration has many advantages over conventional QTL-based selection processes, especially for modest features such as drought tolerance (Farooqi et al., 2022). For example, in maize, the combination of MAS and GS is presented as a new method to ameliorate functional traits. This approach allows the extracting of the best individuals using GEBVs, broadening the spectrum of the potential alleles to be selected, including the ones with a small effect that are indiscernible in conventional techniques (Cerrudo et al., 2018).

In addition, the combination of high-throughput phenotyping and genotyping, machine learning, and speed breeding into GS has exponentially increased genetic gains in stress tolerance breeding. This holistic approach enables the reduction in the time and resources for plant breeding and the enhanced genetic gain for complex traits (Sandhu et al., 2022).

5.5.3 MAS for Disease Resistance

It is apparent that the MAS has helped significantly in advancing the creation of disease-resistant plants, as it makes it easier to be identified through genetic markers that are associated with their resistance genes. Previously, breeding for resistance would involve phenotypic selection and field testing, but these methods were time-consuming and likely to be influenced by environmental factors. By using MAS, it is likely that breeders can accelerate the breeding of crops with desired characteristics, which can be quickly discovered at a molecular level (Zhou et al., 2022). Recently, the most significant achievements included identifying molecular markers that are in close proximity to the genetic loci that confer resistance to specific pathogens. Thus, for example, MAS was designed for targeting the *Pi-ta* gene, which provides resistance to *Magnaporthe oryzae* in rice (Jia et al., 2016). In a similar way, SNP markers that are near to the *Fhb1* gene in the wheat genome have significantly improved the development of wheat cultivars resistant to FHB. Using these markers, breeders can screen and select for resistance homozygotes, which creates a foundation for developing high-performing cereal varieties that are resistant to the most damaging diseases (Serajazari et al., 2023).

Various high-throughput genotyping technologies, such as SNP arrays and NGS, have been integrated into MAS to enhance its resolution and efficiency. With the ability to analyse thousands or millions of markers throughout the genome, these technologies can detect even the smallest variations in disease resistance, which might be impossible to pinpoint using conventional methods. Additionally, bioinformatics resources have been critical in processing and making sense of complex marker-trait associations, making it easier to identify genetic combinations that are most likely to express the desired trait. Finally, there is great potential to improve MAS further by combining it with GS and CRISPR technologies to improve the accuracy and efficiency of selecting for disease-resistance factors in crop breeding (Zhu et al., 2020).

5.5.4 MAS for Improving Yield Traits

MAS enhances the effectiveness of developing high-yielding crops. Through MAS, the fast-track process is utilized to select the ideal plants and develop new varieties with a high yield at a faster rate. This is made possible through MAS, which enables breeders to identify plants with a better genetic configuration based on appropriate molecular markers at the beginning of the breeding cycle (Hasan et al., 2021). MAS makes it possible to choose individual genes linked to yield-related traits, making the process more precise than conventional breeding. Therefore, MAS results in less time and resources being spent on phenotypic evaluations as human–QTL associations typically occur early in the process. Setting up MAS is more expensive initially, but the eventual cost of developing high-yielding material could be justified (Alminderej et al., 2020). MAS was also conducted for grain weight and yield traits in rice, and fine-mapped yield-related genes were promoted by high-throughput sequencing (HTS), assisting high-yield breeding efforts (Kumar et al., 2024). Thus, MAS was widely utilized to enhance the grain weight, kernel size, and the number of spikes per wheat plant. These made it possible to breed new wheat cultivars with an enhanced yield potential. All these studies mostly concentrated on the most important and significant QTLs associated with the above-mentioned complex traits (Singh et al., 2022).

5.6 ADVANCES IN MOLECULAR MARKER TECHNOLOGIES

5.6.1 HTS Technologies

In recent years, HTS technologies have been widely used in molecular genetics and breeding of plants. These technologies include large-scale DNA and RNA analysis performed quickly and precisely (Kalendar et al., 2022; Datta et al., 2024). During the past few years, it has revolutionized the process of marker discovery, presenting an attractive way to get the complete picture of a genome, transcriptome, or epigenome for a relatively low price and in a short time. More recently released sequencing platforms include Illumina, Oxford Nanopore, and PacBio. They have introduced a variety of novel features that enabled researchers to greatly widen the scope of genomic projects. One of the most important innovations is their higher accuracy and the longer nature of the reads, which has enabled simpler sequencing of whole genomes that otherwise were difficult to assemble due to size and repeat content, a frequent trait of plant species (Nizamani et al., 2023). It also includes the long-read sequencing platforms such as the PacBio and Oxford Nanopore, which have enhanced the detection of structural variations and sequence polymorphisms that were challenging to find using short-read technologies (Dohm et al., 2020). Moreover, the reduced cost of HTS has facilitated genome sequencing for investigators involved in breeding programmes; this has considerably fastened the marker discovery for crucial traits, such as resistance to diseases, yield, and tolerance to drought. Additionally, conducting high-throughput genotyping and large-scale GWAS has enabled obtaining numerous novel genetic markers that can be linked to separate agronomic characteristics. Further, other high-throughput processes such as transcriptome sequencing have become vital in revealing gene expression markers linked to phenotypic diversity, which offers an in-depth comprehension of gene regulation in crops (Dohm et al., 2020).

5.6.2 Automation in Molecular Marker Analysis

The molecular marker analysis process has revolutionized plant breeding in that it has accelerated the process of genotyping and thus enhanced accuracy and scalability. The use of such systems has made it easy to achieve high-throughput genotypic screening, a basic requirement for large-scale breeding programmes. Moreover, due to reduced human error, systems can be used to analyse hundreds of samples at once, making the process efficient and cost-effective (Kalendar et al., 2022).

Moreover, technology advancements, such as robot liquid handling systems and the creation of high-density microarrays, have made it possible to perform large-scale genotyping projects, such as GWAS and MAS, much more precisely (Liu et al., 2022). Consequently, novel genetic markers have been discovered and used to predict complex traits, including disease resistance and drought tolerance with higher accuracy. Additionally, automation in PCR amplification and gel electrophoresis has accelerated the analysis of SSRs, SNPs, and other molecular markers used in marker-assisted breeding.

In addition, the application of machine learning and artificial intelligence made it possible to create more effective automated systems in the form of predictive models. They can predict marker–trait associations quickly and with the greatest reliability (Peng et al., 2025). Using this system, the analysis of large genomic datasets helps to identify combinations of gene variations that fit a specific trait. Therefore, the molecular markers' analysis autonomy is not only a manner of producing better output but also a manner of ensuring its quality. These improvements lead to faster and more dependable breeding output, which is why they are instrumental in delivering it in contemporary plant breeding (Nadeem et al., 2023).

5.6.3 Innovations in Bioinformatics Tools for Data Analysis

The development of molecular marker technologies has been further sped up by new bioinformatics tools, which facilitate the analysis of the data from molecular markers. With an exponential increase in the volume of genetic data generated using high-throughput technology such as NGS, bioinformatics tools have also grown indispensable in organizing, analysing, and making sense of the large datasets (Guler & Imamoglu, 2023). One significant progress in bioinformatics has been made in the development of software tools that can process genomic data complexity, such as molecular markers, explicitly and with high precision. The advent of molecular marker software tools like

GATK, VCF tools, and PLINK has empowered the researchers to work efficiently while conducting SNP and other molecular marker data analysis; for example, it has offered them statistical analysis and robust and trustworthy genotype calling. Moreover, these software tools also allow users to intermix data with genotypic records of the same; this is specifically needed to identify which markers are associated with which traits (Song et al., 2023).

Furthermore, progress in the field of machine learning and artificial intelligence facilitated the development of prediction models to analyse marker data and more effectively predict the phenotypic outcomes of genetic traits. As a result, genomics-assisted breeding was advanced using the DeepGS and BayesB models. These and other tools aided in enhancing GS by predicting the performance of genotypes without the requirement for large field trials (Liu et al., 2022; Tong & Nikoloski, 2021).

Moreover, bioinformatics platforms like Geneious and Galaxy support the integration of multi-omics data, and as a result, scientists have been able to integrate data from genomics, transcriptomics, and proteomics to obtain a fuller view on marker-trait relationships. These pathways let breeders consider molecular markers within entire biological pathways, which simplifies the identification of genes underlying complex traits like disease resistance or drought tolerance (Hayes et al., 2024).

5.6.4 Next-Generation Sequencing and Its Impact on Marker Discovery

The emergence of NGS has fundamentally transformed molecular marker discovery, allowing researchers to identify a more comprehensive range of genetic variations. These are SNPs, InDels, and structural variations; they are crucial for enhancing MAS and GS in plant breeding. As the technology is high-throughput, millions of DNA fragments can be sequenced simultaneously, which makes the discovery of markers more comprehensive and faster compared to the use of previous methods (Fahida et al., 2024).

The NGS technologies have made substantial progress in the identification of molecular markers in both model and non-model species due to the limited extent of available genomic resources. One remarkable progress was achieved through whole-genome sequencing (WGS) and GBS, used as one of the most frequently implemented approaches for massive-scale marker discovery in a growing number of crop species in recent years. As these methods do not rely on a preliminary reference genome, they create added value for target non-model species with a poorly characterized genome (Kumar et al., 2024).

Additionally, due to their characteristic speed and capability to detect rare and low-frequency variations unnoticeable by conventional marker development methods, novel marker systems found on NGS procedures have already passed the discovery stage. Given the increasing number of genetic variations, it is a very logical step in acquiring comprehensive data on genetic variation and phenotyping in complex traits such as resistance to diseases, performance, and stress tolerance. In maize, for example, numerous SNPs discovered by NGS strategies were proven to be associated with beneficial agricultural characteristics, such as disease resistance and drought tolerance, and could thus be used to develop MASs for the species (Xu et al., 2024). Most importantly, the combination of NGS with other high-throughput platforms like SNP arrays and novel bioinformatics software solutions for data analysis has dramatically enhanced and popularized the discovery of markers and additional acceleration of GS model advancement. With pricing reduction as the main concept and goal of technological progress, in the future, NGS technologies are likely to be employed more attractively by breeder communities on a global level, allowing for extensive GS agendas and more accurate and rapid breeding for a broader range of crops and plant species (Burridge et al., 2024).

Due to the integration of NGS with multiple other high-throughput technologies, such as an SNP array and improved bioinformatics tools, marker discovery has reached an even higher pace over the past years, and the development of the GS model has been faster. Given the ever-decreasing costs of sequencing, NGS platforms are expected to be more and more broadly used by breeders in the next decade as well, allowing for large GS projects to be launched and, therefore, improving the accuracy of plant breeding for a larger number of crops (Kumar et al., 2024).

5.7 CRISPR/CAS SYSTEMS IN MAS AND GENOME EDITING

By facilitating precise genetic alterations, CRISPR/Cas technologies have transformed plant breeding and improved genome editing and MAS. These methods make it possible to specifically change the genes that produce characteristics like increased yield, stress tolerance, and disease resistance (Chen et al., 2024a). CRISPR technology speeds up breeding and lessens dependency on traditional techniques by enabling precise selection of desired alleles when paired with molecular markers. CRISPR enables precise modifications at specific loci, which can be tracked through molecular markers to ensure the correct alleles are selected. It promotes quick identification and integration of favourable traits, thus decreasing the many generations that traditional breeding methods usually require (Ahmad, 2023). By editing particular genes for things like drought resistance or protection against diseases, CRISPR/Cas9-based technology can target new genes expressed in the plant genome in conjunction with marker-based selection to improve plant performance (Chen et al., 2024b). Lately, multiple editing methods using CRISPR/Cas9 have been invented to allow editing of numerous genes simultaneously. This has also accelerated the development of complex trait plants through the above-mentioned accuracy, as the selection of the most optimal genetic changes is now achieved due to the extremely effective monitoring of gene edits. CRISPR is combined with high-throughput genotyping technologies like SNP arrays, and NGS produces accurate lists of genetic changes and gene editing hotspots. It can be claimed that these technologies transform plant breeding (Saini et al., 2023).

5.8 CASE STUDIES ON THE INCORPORATION OF MOLECULAR MARKERS IN PLANT BREEDING

5.8.1 MAS for Drought Tolerance in Sorghum

Sorghum (*Sorghum bicolor*) is one of the important drought-tolerant cereal crops grown in the semi-arid regions of the world with unpredictable rainfall. For instance, Mwamahonje et al. (2021) and Gupta et al. (2022) conducted a multi-locus GWAS to unravel the genetic basis and map novel QTLs associated with grain yield and variation in drought tolerance in sorghum. The researchers used 216 sorghum accessions grown under optimal and stressed conditions and successfully identified some of the novel genomic regions that play a key role in drought tolerance. Such identified regions can be used in conjunction with other sources of variation to develop new sorghum varieties through MAS.

5.8.2 MAS for Disease Resistance in Rice

Rice blast caused by *Magnaporthe oryzae* is a critical disease that significantly affects rice yield. Liu et al. (2024a) utilized MAS to introduce the *Pi9* gene that conditions the resistance to rice blast into one of the water-saving and drought-resistant rice, Hanhui 3. After multiple backcrosses and majority selections, a new restoration line, Hanhui 8200, of rice with improved resistance to rice blast and without drought avoidance was successfully bred (Liu et al., 2024a).

5.8.3 MAS for Disease Resistance in Wheat

Wheat is known to suffer from a plethora of diseases; one of the diseases that pose a severe threat to the world's wheat production is wheat blast. Resistance to the blast disease in rice has largely involved the use of the *Pi9* gene, while its use in wheat has been rarely reported. Alternatively, several studies have been carried out to identify and transfer various resistance genes using MAS technologies. In some cases, genes such as *Rmg8* and 2NS translocation have been transferred from the wild relative through saturated linkage of molecular markers to accelerate breeding for resistance to wheat blast (Islam et al., 2020). Thus, molecular marker technology allows for the identification of resistant plants at a very early stage in the breeding programme.

5.8.4 MAS for Salt Tolerance in Tomato

Tomato production faces cuts in yield productivity because of salt stress. Scientists have employed MAS to develop tomato varieties that can survive under saline conditions. For example, research by Liu et al. (2024b) exposed that genetically modified tomato plants that simulate S-nitrosylated Δ1-pyrroline-5-carboxylate reductase enzyme recordings increased growth and yield following salt and drought exposure. This methodology shows the putative implementation of MAS for the creation of new salt-tolerant tomatoes, so necessary for the most widespread scenario of cultivation in saline areas. Some of the represented case studies can be characterized by high relative efficiency of the use of molecular markers, which ensure the development of varieties and breeding lines with increased drought and disease resistance, quality properties, and stress tolerance in a number of cultures. Integration of MAS technology enables us to implement breeding processes towards improved varieties and lines quickly and efficiently, which leads to more sustainable agriculture.

5.9 CHALLENGES IN MOLECULAR MARKER UTILIZATION AND CONCLUSION

Plant breeding has been significantly transformed with molecular markers; however, their application is associated with several challenges. Despite the potential provided to select the best genotypes through molecular markers, the low level of development of suitable markers and high-throughput technologies and the complex nature of trait control can hinder efforts to obtain the characteristic varieties.

5.9.1 Marker Development and Validation

The development of molecular markers can be time-consuming, as they must be rigorously validated across many different genetic backgrounds and environmental contexts. Markers that seem promising in one study may work poorly in another, making it difficult to apply them more broadly in breeding programmes. Thus, the establishment of standardized validation procedures is critical to guarantee the reliability of a marker in multiple different breeding settings (Hasan et al., 2021).

5.9.2 Complexity of Polygenic Traits

Drought tolerance and disease resistance are examples of qualities regulated by several genes, having little impact. Traits that are determined by multiple genes have a polygenic inheritance pattern, which makes it difficult to find markers that are consistently close to the genes. It is thus only through more advanced genomic methods, such as QTL mapping and GWAS, that the genetic architecture of these traits can be determined (Yi et al., 2023).

5.9.3 Genomic Resources for Non-Model Crops

Most model species, such as rice and maize, have well-characterized genomes, whereas the genomic resources of many crops remain largely incomplete. The deficiency prevents the creation of molecular markers. It also hinders the molecular marker in the given species. Therefore, the work on WGS and annotation of poorly studied crops plays a vital role in filling this gap (Bohra et al., 2020).

5.9.4 High Costs and Technical Barriers in Genotyping

High-throughput genotyping technologies exploiting Illumina and Affymetrix platforms are expensive and require specialized equipment and personnel. Thereby, these technologies are out of reach for many breeders, particularly in less developed countries. The expansion of affordable and easy-to-use genotyping is critical for making the resulting breeding technologies more inclusive (Begna, 2020).

5.9.5 Integration with Traditional Breeding Practices

However, integration of molecular markers into conventional breeding programmes is often difficult for two reasons. First, breeders prefer phenotypic selection and are reluctant to use new technologies due to the high cost and complexity. Second, regional breeding programmes and strategies developed without molecular markers make it challenging to introduce new methods without special training and capacity building (Tripodi et al., 2023). Tackling these challenges is essential for applying molecular markers successfully in breeding plants. Further research and technical progress are projected to alleviate these issues, making the breeding procedures more accurate and effective in developing crops with enhanced characteristics.

REFERENCES

Ahmad, E., Badoni, S., Akhtar, M. & Jaiswal, J. P. (2018). Molecular diversity of wheat (*Triticum aestivum* L.) genotypes resistance to rice weevil (*Sitophilus oryzae* L.) revealed by SSR markers. *International Journal of Current Microbiology and Applied Sciences*, *7*(7), 761–772.

Ahmad, M. (2023). Plant breeding advancements with "CRISPR-Cas" genome editing technologies will assist future food security. *Frontiers in Plant Science*, *14*, 1133036.

Alminderej, F., Bakari, S., Almundarij, T.I., Snoussi, M., Aouadi, K. & Kadri, A. (2020). Antioxidant activities of a new chemotype of *Piper cubeba* L. fruit essential oil (Methyleugenol/Eugenol): In silico molecular docking and ADMET studies. *Plants*, *9*(11), 1534.

Amiteye, S. (2021). Basic concepts and methodologies of DNA marker systems in plant molecular breeding. *Heliyon*, *7*(10).

Begna, T. (2020). The role of genotype by environmental interaction in plant breeding. *International Journal of Agriculture and Biosciences*, *9*(5), 209–215.

Bohra, A., Jha, U. C., Godwin, I. D. & Varshney, R. K. (2020). Genomic interventions for sustainable agriculture. *Plant Biotechnology Journal*, *18*(12), 2388–2405.

Burridge, A. J., Winfield, M., Przewieslik-Allen, A., Edwards, K. J., Siddique, I., Barral-Arca, R. & Griffiths, S. (2024). Development of a next generation SNP genotyping array for wheat." (). *Plant Biotechnology Journal*, *22*(8), 2235–2247.

Cerrudo, D., Cao, S., Yuan, Y., Martinez, C., Suarez, E. A., Babu, R., Zhang, X. & Trachsel, S. (2018). Genomic selection outperforms marker assisted selection for grain yield and physiological traits in a maize doubled haploid population across water treatments. *Frontiers in Plant Science*, *9*, 366.

Chen, B., Hou, Y., Huo, Y., Zeng, Z., Hu, D., Mao, X. & Zhong, C. (2024a). QTL mapping of yield, agronomic, and nitrogen-related traits in Barley (*Hordeum vulgare* L.) under low nitrogen and normal nitrogen treatments. *Plants*, *13*(15), 2137.

Chen, F., Chen, L., Yan, Z., Xu, J., Feng, L., He, N., Guo, M. & . (2024b). Recent advances of CRISPR-based genome editing for enhancing staple crops. *Frontiers in Plant Science*, *15*, 1478398.

Cholin, S. S. & Kulkarni, C. C. (2023). Prospects of marker-assisted recurrent selection: Current insights and future implications. *Tropical Plant Biology*, *16*(4), 259–275.

Chugh, V., Kaur, D., Purwar, S., Kaushik, P., Sharma, V., Kumar, H., Rai, A., Singh, C. M., Kamaluddin & Dubey, R. B. (2022). Applications of molecular markers for developing abiotic-stress-resilient oilseed crops. *Life*, *13*(1), 88.

Collard, B. C. & Mackill, D. J. (2008). Marker-assisted selection: an approach for precision plant breeding in the twenty-first century. *Philosophical Transactions of the Royal Society B: Biological Sciences*, *363*(1491), 557–572.

Datta, R., Kapadia, C., Danish, S. and Singh, S. (Eds.). (2024). *Non-coding RNA in plants: Modulation and stress responses*. Elsevier.

Dida, G. (2022). Molecular markers in breeding of crops: Recent progress and advancements. *International Journal of Novel Research in Life Sciences*, *7*(4), 10–21.

Dohm, J. C., Peters, P., Stralis-Pavese, N. & Himmelbauer, H. (2020). Benchmarking of long-read correction methods. *NAR Genomics and Bioinformatics*, *2*(2), lqaa037.

Elshire, R. J., Glaubitz, J. C., Sun, Q., Poland, J. A., Kawamoto, K., Buckler, E. S. & Mitchell, S. E. (2011). A robust, simple genotyping-by-sequencing (GBS) approach for high diversity species. *PLoS ONE*, *6*(5), e19379.

Erdoğan, İ., Cevher-Keskin, B., Bilir, Ö., Hong, Y. & Tör, M. (2023). Recent developments in CRISPR/Cas9 genome-editing technology related to plant disease resistance and abiotic stress tolerance. *Biology*, *12*(7), 1037.

Fadel, A. A., Abdulhamed, Z. A. & Yousif, S. A. (2022). RAPD technique to determine the genetic divergence of barley genotypes. *IOP Conference Series: Earth and Environmental Science*, *1060*(1), 012123.

Fahida, P. K., Sivasankarreddy, K. & Mushid Muhammed., M. M. (2024). Next-generation sequencing and its application in crop improvement. In M. K. Tripathi & N. Tripathi (Eds.), *Advances in Plant Biotechnology* (Vol. 1, pp. 69–82). Cornous Publications LLP.

Farooqi, M. Q. U., Nawaz, G., Wani, S. H., Choudhary, J. R., Rana, M., Sah, R. P. & Afzal, M. (2022). Recent developments in multi-omics and breeding strategies for abiotic stress tolerance in maize (*Zea mays* L.). *Frontiers in Plant Science*, *13*, 965878.

Fujito, S., Akyol, T. Y., Mukae, T., Wako, T., Yamashita, K., Tsukazaki, H. & Hirakawa, H. (2021). Construction of a high-density linkage map and graphical representation of the arrangement of transcriptome-based unigene markers on the chromosomes of onion, *Allium cepa* L. *BMC Genomics*, *22*(1), 481.

Guler, B. A. & Imamoglu, E. (2023). Molecular marker technologies in food plant genetic diversity studies: An overview. *Foods and Raw Material*, *11*(2), 282–292.

Gunderson, K. L., Steemers, F. J., Lee, G., Mendoza, L. G. & Chee., M. S. (2005). A genome-wide scalable SNP genotyping assay using microarray technology. *Nature Genetics*, *37*(5), 549–554.

Gupta, G., Paul, S., Singh, S., Pietramellara, G., Pathan, S. I., Danish, S., Jabborova, D., Datta, R. & Jha, P. N. (2022). Exploring functional diversity and community structure of diazotrophic endophytic bacteria associated with Pennisetum glaucum growing under field in a semi-arid region. *Land*, *11*(7), p.991.

Hasan, N., Choudhary, S., Naaz, N., Sharma, N. & Laskar, R. A. (2021). Recent advancements in molecular marker-assisted selection and applications in plant breeding programmes. *Journal of Genetic Engineering and Biotechnology*, *19*(1), 128.

Hayes, C. N., Nakahara, H., Ono, A., Tsuge, M. & Oka, S. (2024). From omics to multi-omics: A review of advantages and tradeoffs. *Genes*, *15*(12), 1551.

Islam, M. T., Gupta, D. R., Hossain, A., Roy, K. K., He, X., Kabir, M. R., Singh, P. K., Khan, M. A. R., Rahman, M. & Wang, G.-L. (2020). Wheat blast: A new threat to food security. *Phytopathology Research*, *2*, 1–13.

Jia, Z., Dumont, M. J. & Orsat, V. (2016). Encapsulation of phenolic compounds present in plants using protein matrices. *Food Bioscience*, *15*, 87–104.

Kalendar, R., Hunter, C. & Orbovic, V. (2022). Innovative applications of sequencing technologies in plant science. *Frontiers in Plant Science*, *13*, 1058347.

Karlik, E. & Tombuloğlu, H. (2016). Molecular markers and their applications. In K. R. Hakeem, H. Tombuloğlu & G. Tombuloğlu (Eds.), *Plant Omics: Trends and Applications* (pp. 137–157). Springer.

Kumar, R., Das, S. P., Choudhury, B. U., Kumar, A., Prakash, N. R., Verma, R. & Chakrabortii, M. (2024). Advances in genomic tools for plant breeding: harnessing DNA molecular markers, genomic selection, and genome editing. *Biological Research*, *57*(1), 80.

Kumawat, G., Kumawat, C. K., Chandra, K., Pandey, S., Chand, S., Mishra, U. N., Lenka, D. & Sharma, R. (2020). Insights into marker assisted selection and its applications in plant breeding. In I. Y. Abdurakhmonov (Ed.), *Plant Breeding-Current and Future Views* (pp. 123–135). Intechopen.

Kushanov, F. N., Turaev, O. S., Ernazarova, D. K., Gapparov, B. M., Oripova, B. B., Kudratova, M. K. & Rafieva, F. U. (2021). Genetic diversity, QTL mapping, and marker-assisted selection technology in cotton (*Gossypium spp.*). *Frontiers in Plant Science*, *12*, 779386.

Liu, D., Kong, H., Luo, X., Liu, W. & Subramaniam, R. (2022). Bringing AI to edge: From deep learning's perspective. *Neurocomputing*, *485*, 297–320.

Liu, Q., Xue, J., Zhang, L., Jiang, L. & Li, C. (2024a). Unveiling the roles of LncRNA MOIRAs in rice blast disease resistance. *Genes*, *15*(1), 82.

Liu, W., Wei, J.-W., Shan, Q., Liu, M., Xu, J. & Gong, B. (2024b). Genetic engineering of drought-and salt-tolerant tomato via Δ1-pyrroline-5-carboxylate reductase S-nitrosylation. *Plant Physiology*, *195*(2), 1038–1052.

Majumdar, G., Sayanti, A. R. & Mishra, D. C. (2020). Integrated framework for selection of additive and nonadditive genetic markers for genomic selection. *Journal of Computational Biology*, *2*(6), 845–855.

Marè, C., Zampieri, E., Cavallaro, V., Frouin, J., Grenier, C., Courtois, B. & Brottier, L. (2023). Marker-assisted introgression of the salinity tolerance locus Saltol in temperate Japonica rice. *Rice*, *16*(1), 2.

Mohanavel, V., Muthu, V., Kambale, R., Palaniswamy, R., Seeli, P., Ayyenar, B. & Rajagopalan, V. (2024). Marker-assisted breeding accelerates the development of multiple-stress-tolerant rice genotypes adapted to wider environments. *Frontiers in Plant Science*, *15*, 1402368.

Mu, Y., Gong, W., Qie, Y., Liu, X., Li, L., Sun, N. & Liu, W. (2022). Identification of the powdery mildew resistance gene in wheat breeding line Yannong 99102-06188 via bulked segregant exome capture sequencing. *Frontiers in Plant Science*, *13*, 1005627.

Mwamahonje, A., Eleblu, J. S. Y., Ofori, K., Deshpande, S., Feyissa, T. & Tongoona, P. (2021). Drought tolerance and application of marker-assisted selection in sorghum. *Biology*, *10*(12), 1249.

Nadeem, F., Rehman, A. & Petropoulos, S. A. (2023). Increasing production of crops with bioavailable micronutrients: A solution to reduce global malnutrition. *Frontiers in Plant Science*, *14*, 1348246.

Nizamani, M. M., Zhang, Q., Muhae-Ud-Din, G. and Wang, Y. (2023). High-throughput sequencing in plant disease management: A comprehensive review of benefits, challenges, and future perspectives. *Phytopathology Research*, *5*(1), 44.

Olson, R., Wise, B., Conners, F., Rack, J. &and Fulker, D. (1989). Specific deficits in component reading and language skills: Genetic and environmental influences. *Journal of Learning Disabilities*, *22*(6), 339–348.

Osei, K., Michael, A. D., Adu-Dapaah, H., Danquah, E., Blay, E., Massoudi, M. & Maxwell, D. (2022). Marker assisted backcrossing of alcobaca gene into two elite tomato breeding lines. *Frontiers in Horticulture*, *1*, 1024042.

Pan, G., Li, Z., Huang, S., Tao, J., Shi, Y., Chen, A. & Li, J. (2021). Genome-wide development of insertion-deletion (InDel) markers for Cannabis and its uses in genetic structure analysis of Chinese germplasm and sex-linked marker identification. *BMC Genomics*, *22*, 1–12.

Park, J.-S., Choi, Y., Kim, J.-H., Lee, C., Jeong, M.-G., Jeong, Y.-I., Kang, Y. J., Chung, Y.-S. & Choi, H.-K. (2024). Development of a web-based high-throughput marker design program: CAPS (cleaved amplified polymorphic sequence) Maker. *Plant Methods*, *20*(1), 192.

Patani, A., Prajapati, D., Singh, S., Enakiev, Y., Bozhkov, S., Jabborova, D. & Joshi, C. (2023). Beneficial effects of biochar application on mitigating the drought and salinity stress implications on plants. *Plant Science Today*, *10*, 188–193.

Paul, E. & Frey, S. (Eds.). (2023). *Soil Microbiology, Ecology and Biochemistry*. Elsevier.

Peng, J., Lei, X., Liu, T., Xiong, Y., Wu, J., Xiong, Y., You, M., Zhao, J., Zhang, J. & Ma, X. (2025). Integration of machine learning and genome-wide association study to explore the genomic prediction accuracy of agronomic trait in oats (*Avena sativa* L.). *The Plant Genome*, *18*(1), e20549.

Powder, K. E. (2020). Quantitative trait loci (QTL) mapping. In X. M. Shi (Ed.), *eQTL Analysis: Methods and Protocols* (pp. 211–229). Springer.

Rafalski, J. A., Hanafey, M. K., Tingey, S. V. & Williams, J. G. (2020). Technology for molecular breeding: RAPD markers, microsatellites and machines. In P. M. Gresshoff (Ed.), *Plant Genome Analysis* (pp. 19–27). CRC Press.

Raza, A., Mehmood, S. S., Shah, T., Zou, X., Yan, L., Zhang, X. & Khan, R. S. A. (2019). Applications of molecular markers to develop resistance against abiotic stresses in wheat. In M. Hasanuzzaman, K. Nahar, & M. A. Hossain (Eds.), *Wheat Production in Changing Environments: Responses, Adaptation and Tolerance* (pp. 393–420). Springer.

Saini, H., Thakur, R., Gill, R., Tyagi, K. & Goswami, M. (2023). CRISPR/Cas9-gene editing approaches in plant breeding. *GM Crops & Food*, *14*(1), 1–17.

Sandhu, K. S., Shiv, A., Kaur, G., Meena, M. R., Raja, A. K., Vengavasi, K. & Mall, A. K. (2022). Integrated approach in genomic selection to accelerate genetic gain in sugarcane. *Plants*, *11*(16), 2139.

Sandhu, N., Pruthi, G., Raigar, O. P., Singh, M. P., Phagna, K., Kumar, A., Sethi, M., Singh, J., Ade, P. A. & Saini, D. K. (2021). Meta-QTL analysis in rice and cross-genome talk of the genomic regions controlling nitrogen use efficiency in cereal crops revealing phylogenetic relationship. *Frontiers in Genetics*, *12*, 807210.

Sax, K. (1923). The association of size differences with seed-coat pattern and pigmentation in *Phaseolus vulgaris*. *Genetics*, *8*(6), 552.

Serajazari, M., Torkamaneh, D., Gordon, E., Lee, E., Booker, H., Pauls, K. P. & Navabi, A. (2023). Identification of fusarium head blight resistance markers in a genome-wide association study of CIMMYT spring synthetic hexaploid derived wheat lines. *BMC Plant Biology*, *23*(1), 290.

Shaikh, A. A., Chachar, S., Chachar, M., Ahmed, N., Guan, C. & Zhang, P. (2022). Recent advances in DNA methylation and their potential breeding applications in plants. *Horticulturae*, *8*(7), 562.

Singh, L., Coronejo, S., Pruthi, R., Chapagain, S. & Subudhi, P. K. (2022). Integration of QTL mapping and whole genome sequencing identifies candidate genes for alkalinity tolerance in rice (*Oryza sativa*). *International Journal of Molecular Sciences*, *23*(19), 11791.

Song, B., Ning, W., Wei, D., Jiang, M., Zhu, K., Wang, X., Edwards, D., Odeny, D. A. & Cheng, S. (2023). Plant genome resequencing and population genomics: Current status and future prospects. *Molecular Plant*, *16*(8), 1252–1268.

Sun, L., Lai, M., Ghouri, F., Nawaz, M. A., Ali, F., Baloch, F. S., Nadeem, M. A., Aasim, M. & Shahid, M. Q. (2024). Modern plant breeding techniques in crop improvement and genetic diversity: From molecular markers and gene editing to artificial intelligence—A critical review. *Plants*, *13*(19), 2676.

Swarup, S., Cargill, E. J., Crosby, K., Flagel, L., Kniskern, J. & Glenn, K. C. (2021). Genetic diversity is indispensable for plant breeding to improve crops. *Crop Science*, *61*(2), 839–852.

Teama, S. (2018). DNA polymorphisms: DNA-based molecular markers and their application in medicine. In Y. Liu (Ed.), *Genetic Diversity and Disease Susceptibility* (pp. 25–40). IntechOpen.

Thoday, J. M. (1961). Location of polygenes. *Nature*, *191*, 368–370.

Tong, H. & Nikoloski, Z. (2021). Machine learning approaches for crop improvement: Leveraging phenotypic and genotypic big data. *Journal of Plant Physiology*, *257*, 153354.

Tripodi, P., D'Alessandro, A. & Francese, G. (2023). An integrated genomic and biochemical approach to investigate the potentiality of heirloom tomatoes: Breeding resources for food quality and sustainable agriculture. *Frontiers in Plant Science*, *13*, 1031776.

Vos, P., Hogers, R., Bleeker, M., Reijans, M., Lee, T., van de, Hornes, M. & Friters, A. (1995). AFLP: A new technique for DNA fingerprinting. *Nucleic Acids Research*, *23*(21), 4407–4414.

Wei, D., Cui, K., Ye, G., Pan, J., Xiang, J., Huang, J. & Nie, L. (2012). QTL mapping for nitrogen-use efficiency and nitrogen-deficiency tolerance traits in rice. *Plant and Soil*, *359*, 281–295.

Xu, X., Yang, X. & Chen, X. (2024). Genomic advances in cucurbitaceae: Implications for crop improvement and breeding. *International Journal of Horticulture*, *14*(3), 117–126.

Yi, Y., Hassan, M. A., Cheng, X., Li, Y., Liu, H., Fang, W., Zhu, Q. & Wang, S. (2023). QTL mapping and analysis for drought tolerance in rice by genome-wide association study. *Frontiers in Plant Science*, *14*, 1223782.

Zhang, H.-W., Wang, L.-Z., Xie, Y., Liu-Gen, H., Wang, Z.-Z., Yi, C.-F. & Guo, H. (2024). QTL mapping for heading date and plant height using a RIL population in rice in different photoperiod environments. *Czech Journal of Genetics and Plant Breeding*, *60*(3), 119–125.

Zhou, X., Li, X., Han, D., Yang, S. & Zhensheng Kang, R. R. (2022). Genome-wide QTL mapping for stripe rust resistance in winter wheat pindong 34 using a 90K SNP array. *Frontiers in Plant Science*, *13*, 932762.

Zhu, H., Li, C. & Gao, C. (2020). Applications of CRISPR–Cas in agriculture and plant biotechnology. *Nature Reviews Molecular Cell Biology*, *21*(11), 661–677.

Zietkiewicz, E., Rafalski, A. & Labuda, D. (1994). Genome fingerprinting by simple sequence repeat (SSR)-anchored polymerase chain reaction amplification. *Genomics*, *20*(2), 176–183.

6 Quantitative Trait Loci (QTLs) of Secondary Metabolism

Manu Yadav, Anjali Pathak, and Ananya Sharma

6.1 INTRODUCTION

The ability of plants to produce a wide range of secondary compounds, referred to as secondary metabolites or natural products, is astounding. These substances are essential for assisting plants in interacting with and adapting to their surroundings. Secondary metabolites have a variety of ecological functions, including as defence mechanisms, pollinator attraction, and stress management, in contrast to primary metabolites, which are necessary for basic biological activities like growth, reproduction, and development (Reshi et al., 2023). The production of these specialized compounds is driven by intricate networks of enzymatic reactions, tightly controlled gene expression, and transport systems that move substrates and products within plant cells. This results in a wide variety of structurally unique substances with distinct biological functions (Kochhar and Gujral, 2020; Seca and Trendafilova, 2021). The diverse pharmacological properties of these metabolites have sparked significant interest in their potential for drug discovery and development across biological, chemical, and medical research fields (Şimşek Sezer and Uysal, 2021).

Secondary metabolites are typically produced in specific plant cell types and at particular stages of development, often accumulating in much smaller amounts than primary metabolites (Kinghorn, 1994). The genetic framework behind their biosynthesis is usually complex, involving numerous genes, detailed regulatory systems, and environmental factors. Quantitative trait loci (QTLs) are genomic regions that influence variation in complex traits, including the types and amounts of secondary metabolites present (Niu and Tan, 2013). Studying QTLs linked to secondary metabolite production is a valuable strategy for uncovering the genetic mechanisms behind metabolic diversity and identifying key genes involved in these biosynthetic pathways (Thirumurugan et al., 2018). Successfully mapping and understanding QTLs requires a combination of quantitative genetics, genomics, and metabolomics, allowing for a more detailed picture of how these important compounds are regulated and produced (Dwivedi et al., 2018). Continued research in this field offers potential benefits for sustainable agriculture, human health, and environmental conservation (Seca and Trendafilova, 2021; Srivastav et al., 2020).

The process begins with creating mapping populations from parent plants that differ significantly in the levels of specific secondary metabolites. These populations often include recombinant inbred lines (RILs), doubled haploids, or F2 generations, all of which show genetic variation related to the metabolic traits of interest. Once the population is established, detailed phenotypic assessments are carried out to measure the concentration of the target secondary metabolites in each plant. This is typically done using sophisticated analytical methods such as high-performance liquid chromatography (HPLC), gas chromatography-mass spectrometry (GC-MS), or nuclear magnetic resonance (NMR) spectroscopy, which provide accurate and reliable metabolite data (Khoshakhlagh

DOI: 10.1201/9781003542681-6

et al., 2022). To create a high-resolution genetic map throughout the genome, the population is then genotyped using molecular markers such as amplified fragment length polymorphisms (AFLPs), simple sequence repeats (SSRs), or single nucleotide polymorphisms (SNPs). Researchers now have access to extensive genomic and transcriptome data because of advancements in sequencing technology. This improves the capacity to identify potential genes involved in metabolite production and fine-map QTLs. By statistically connecting phenotypic and genotypic data using methods like interval mapping, composite interval mapping, or mixed-model approaches, QTL mapping is accomplished. By identifying particular genomic areas that are linked to variations in secondary metabolite levels, these methods assist in identifying QTLs. The results of QTL mapping provide crucial information about the quantity, location, and significance of genetic loci involved in these metabolites' biosynthesis. Identifying molecular markers closely associated with key genes and determining their positions on chromosomes is a key goal in plant breeding, as it supports gene cloning efforts and marker-assisted selection (Shiri and Akhavan, 2014). Once QTLs are located, fine-mapping techniques are used to narrow down the genomic regions that contain the candidate genes responsible for the observed traits. This typically involves using high-resolution genetic maps in combination with methods such as association mapping or positional cloning to accurately identify the specific genes within the QTL region (Borevitz and Chory, 2004; Kumar et al., 2017).

To produce a comprehensive genetic map that encompasses the entire genome, the mapping population is subsequently genotyped using a variety of molecular markers, including AFLPs, SSRs, and SNPs. With the advancement of modern sequencing technologies, researchers are now able to obtain extensive genomic data, including whole-genome sequences and transcriptome profiles, which significantly enhance the accuracy of QTL fine-mapping and the identification of potential candidate genes. QTL mapping is then conducted by linking phenotypic data with genotypic data through statistical techniques such as interval mapping, composite interval mapping, or mixed-model approaches. These methods help identify genomic regions significantly associated with variations in secondary metabolite levels, thus revealing the presence of QTLs. The outcomes of QTL mapping provide valuable information about the number, location, and effect size of genetic loci that impact the biosynthesis of secondary metabolites. Moreover, identifying molecular markers closely associated with target genes and mapping their chromosomal locations is a key goal in plant breeding, aiding in gene cloning and marker-assisted selection (Shiri and Akhavan, 2014). After identifying QTLs, fine-mapping techniques are applied to further narrow down the genomic region containing the candidate genes responsible for the variation in secondary metabolite production. This typically involves utilizing high-resolution genetic maps alongside association mapping or positional cloning methods to locate the causative genes within the QTL interval (Borevitz and Chory, 2004; Kumar et al., 2017).

A critical first step in comprehending the molecular mechanisms underlying the QTL effects on secondary metabolite production is the identification of candidate genes. By utilizing data from various gene markers, new genes can be discovered (Grisel and Crabbe, 1995). This process involves combining multiple sources of information, such as gene annotation databases, expression profiling, and functional genomics resources, to pinpoint genes within the QTL region that may be involved in the biosynthetic pathway or regulation of secondary metabolite production. After identifying candidate genes, they undergo functional validation through methods like gene silencing, overexpression, or enzyme activity assays to confirm their involvement in secondary metabolite biosynthesis. These strategies can be enhanced by using suitable genetic materials and the latest genomics technologies (Salvi and Tuberosa, 2005). In the end, these methods will make plant breeding for complicated quantitative traits more feasible and effective (Grisel and Crabbe, 1995; Lan et al., 2020; Sun et al., 2019).

Plant breeding benefits from the strategic application of genetic markers that are tightly linked to genes of interest (Devi et al., 2017). Plant breeders can choose desired plants early on based on their genotype rather than waiting for phenotypic expression, because of the useful information that DNA markers offer (Meena et al., 2017). By finding individuals with advantageous alleles at QTLs linked to the synthesis of secondary metabolites, marker-assisted selection improves the effectiveness and precision of conventional breeding programmes (Lema, 2018; Vagndorf et al., 2018).

These markers can be used to enhance plant cultivars (Bernardo, 2008; Lema, 2018). The combination of multiple beneficial QTLs through marker-assisted selection or genomic selection methods shows enormous potential for improving both the levels and consistency of secondary metabolite production in crops (Pilet-Nayel et al., 2017). QTL analysis is a valuable tool for identifying genes that control specific traits and phenotypes, providing important insights into the functional roles of certain genetic regions (McMullen et al., 2003). Discovering new genes and gaining a deeper understanding of the genetic and biochemical processes involved in secondary metabolite synthesis will drive progress in metabolic engineering and synthetic biology. By clarifying the genetic foundation of secondary metabolite production, these studies open the door to developing new breeding techniques aimed at improving the nutritional content, medicinal benefits, and stress resilience of crops. Molecular markers can be used to identify high-performing genotypes, evaluate genetic variation, and speed up the breeding process (Yali, 2022). Additionally, a comprehensive approach to breeding is crucial for tackling future challenges in food production (Dreisigacker et al., 2016). This is accomplished by leveraging existing genomic resources (Dreisigacker et al., 2016). In the end, a thorough integration of genomic technologies and information speeds up the development of superior crop varieties and improves our understanding of complicated trait inheritance. These cultivars are especially made to meet the many demands of contemporary farming, which enhances human health and food security (Grover and Sharma, 2014; Heffner et al., 2010; Liu et al., 2019). To create commercially viable cultivars, breeding and field testing are essential (Stuthman et al., 2007).

6.2 OVERVIEW OF QTL MAPPING IN SECONDARY METABOLISM

A key method for determining the genetic foundation of complex traits is QTL mapping, which offers a strong framework for examining the complicated interactions between genotype and phenotype. This approach is especially crucial when studying secondary metabolism, as the generation of specific metabolites is frequently impacted by several genes, each of which makes a minor but cumulative contribution to the total metabolic output (Sun et al., 2019). Finding genomic areas that contain genes involved in the production, accumulation, and control of secondary metabolites is made easier by QTL mapping's capacity to connect genetic variation to phenotypic variability (Kumar et al., 2011). By statistically associating phenotypic differences, represented as quantitative trait variations, with genotypic markers spread across the genome, QTL mapping allows researchers to identify specific genomic regions, or QTLs, that are strongly linked to the traits under investigation, providing key insights into the genetic framework controlling these complex metabolic processes (Huang et al., 2021).

Numerous QTLs connected to a broad variety of secondary metabolites, including medicines and plant defence chemicals, have been found as a result of the application of QTL mapping in secondary metabolism research (Sahu and Sharma, 2017). QTL mapping in conjunction with conventional methods helps expedite the comprehension of the genetic underpinnings of complex traits such as disease resistance, resulting in more rapid and accurate population improvements (Pereira et al., 2000). Genes encoding enzymes in metabolic pathways, transcription factors that control gene expression, and structural proteins that impact metabolite transport and storage are frequently among the QTLs found. Additionally, QTL mapping can uncover epistatic interactions between different genomic regions, offering a deeper insight into the intricate regulatory networks that control secondary metabolite production. These identified QTLs serve as valuable targets for further research, such as fine-mapping, gene cloning, and functional characterisation, which can facilitate the genetic modification of secondary metabolite production in plants and other organisms.

Combining complementary resistance QTLs is also emerging as an effective strategy for breeding durable resistance (Pilet-Nayel et al., 2017; Moury et al., 2017).

An improved version of conventional QTL mapping is multi-trait analysis, which enables the simultaneous investigation of several linked traits. In secondary metabolism research, where intricate connections between several metabolic pathways and their by-products are common, this method is very helpful (Sukhwinder-Singh et al., 2012; Tong et al., 2025). By analysing multiple secondary metabolites as correlated traits, multi-trait QTL mapping can uncover pleiotropic QTLs that impact the production of several metabolites at the same time.

6.3 PRINCIPLES OF QTL MAPPING

Step 1: Mapping Population

- Establishing appropriate strategies for biparental or multiparent populations is essential. Common examples include F2, RILs, Near Isogenic Lines (NILs), and Multi-parent Advanced Generation Inter-cross (MAGIC) populations.

Step 2: Phenotyping of Secondary Metabolites

- Accurate metabolite quantification is typically performed using techniques such as liquid chromatography-mass spectrometry (LC-MS), GC-MS, or NMR to assess biochemical variation.

Step 3: Genotyping

- High-density genotyping through tools like SNP arrays, genotyping-by-sequencing (GBS), or whole genome sequencing (WGS) ensures precise mapping at a fine scale.

Step 4: QTL Analysis

- QTL detection utilizes statistical models such as composite interval mapping (CIM), multiple interval mapping (MIM), or genome-wide association studies (GWAS), depending on the population and intended outcomes.

Step 5: Candidate Gene Identification

- Combining data from transcriptomics (eQTL), metabolomics (mQTL), and co-expression networks helps identify the genes responsible for the observed QTLs. Different methods used for QTL mapping of secondary metabolites in Table 6.1.

The overall workflow of QTL mapping in secondary metabolism is illustrated in Figure 6.1.

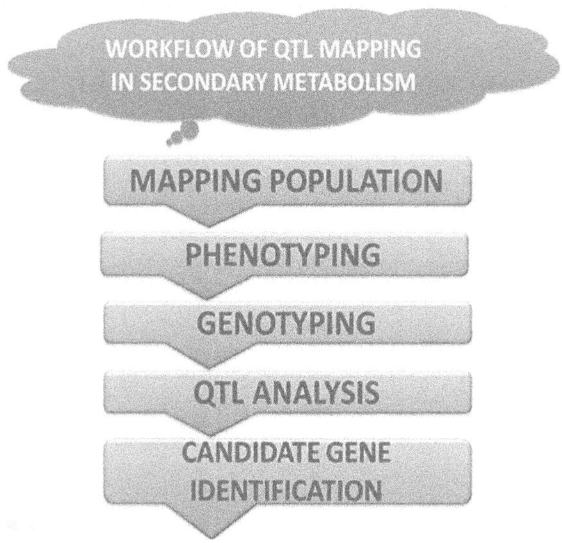

FIGURE 6.1 Workflow of QTL mapping in secondary metabolism.

6.4 CHALLENGES IN QTL MAPPING

A summary of QTLs associated with secondary metabolites in various crops is presented in Table 6.1. QTL mapping being a crucial technique for analysing the genetic foundation of secondary metabolites in plants, provides information on the biosynthesis, regulation, and evolutionary adaptability of these compounds. Although there has been a lot of progress recently, especially after 2015, there are still a number of challenges that prevent the precise identification and functional validation of QTLs associated with secondary metabolites. These restrictions result from tissue-specific expression, the intrinsic complexity of these features' genetic architecture, significant environmental effects, and technology limits in both genomic analysis and metabolite profiling.

Secondary metabolites are frequently regulated by polygenic networks, in which several loci have small and cumulative effects. It is challenging to discover significant associations because of

TABLE 6.1
Different Methods Used for QTL Mapping of Secondary Metabolites

Method	Description	Advantages	Limitations	Applications
Single Marker Analysis (SMA)	Tests association between markers and traits one at a time	Simple and easy to implement	Ignores marker order and linkage	Initial QTL screening
Interval Mapping (IM)	Uses likelihood ratio to detect QTL between two flanking markers	Estimates QTL position and effect	Can detect false positives if multiple QTLs are nearby	Locating QTLs in a defined region
Composite Interval Mapping (CIM)	Combines interval mapping with covariates to control background noise	Improves precision and power	Model complexity increases with many markers	Common in high-resolution mapping of metabolic QTLs
Multiple Interval Mapping (MIM)	Simultaneously fits multiple QTLs in a model	Accounts for epistasis and multiple QTLs	Requires large population size	Detailed dissection of complex traits
Association Mapping (GWAS)	Uses natural populations with high recombination	High resolution, uses historical recombination	Sensitive to population structure	Identifying candidate genes for secondary metabolism
Meta-QTL (MQTL) Analysis	Combines results from multiple QTL studies	Increases reliability and precision	Requires standardized datasets	Consensus QTL identification across environments and studies
eQTL Mapping	Maps loci controlling gene expression (expression traits)	Links genotype to gene regulation	Requires transcriptomic data	Understanding regulation of metabolic genes
mQTL Mapping	Maps QTLs using metabolite levels as phenotypes	Direct insight into metabolic traits	Needs high-throughput metabolomics data	Terpenoids, flavonoids, alkaloids mapping
G×E Interaction QTL Mapping	Considers genotype × environment interactions	Detects stable and environment-specific QTLs	Needs multi-location or multi-year trials	Selecting QTLs for stable metabolite production under stress
Bayesian QTL Mapping	Uses Bayesian statistics to estimate QTL parameters	Handles complex models and small datasets	Computationally intensive	Modelling complex traits like secondary metabolism

Source: Lee et al. (2025).

this intricacy, especially for metabolites with poor heritability. The accumulation of stilbenoids in grapevines, for instance, is controlled by a number of minor-effect loci, and the genetic variance is distributed among various genomic regions, as reported by Massonnet et al. (2020). QTLs for oxalic acid and L-tyrosine were found to explain a considerable amount of phenotypic variation in wheat; however, the underlying genetic structure was still mostly unknown (Toubiana et al., 2016). Due to small population sizes and restricted recombination events in biparental populations, traditional QTL mapping frequently has poor resolution. This hinders accurate gene identification by producing large confidence intervals with several candidate genes (Chen et al., 2024). An additional level of complexity is introduced by the impact of environmental factors. The accumulation of secondary metabolites is extremely malleable and frequently changes in response to environmental factors such as light, temperature, water availability, and nutritional conditions. Noting that even genetically identical plants can have distinct metabolite profiles under varying environmental conditions, Wen et al. (2015) highlighted the frequency of genotype-by-environment interactions (G×E interactions) in maize leaf metabolite levels. One of the biggest challenges in metabolite-QTL research is repeatability across studies and locales due to this environmental modification. The timing and spatial specificity of secondary metabolite production present another difficulty. As many chemicals are produced in particular tissues or stages of development, highly targeted sampling techniques are required. Sasaki et al. (2019) emphasized the function of epigenetic QTLs in the generation of metabolites specific to flowers in *Arabidopsis*, implying that neglecting tissue-specific expression can mask pertinent QTL signals. Furthermore, metabolite content is influenced by developmental stage, necessitating synchronized sampling to prevent the introduction of variance unrelated to genetic control. The constraints of traditional QTL mapping have been partially overcome by GWAS, which allows for higher resolution mapping across a variety of populations. However, population structure and low linkage disequilibrium in certain species also cause confounding in GWAS. Local Arabidopsis populations showed stronger connections at biosynthetic loci than geographically dispersed panels, according to Katz et al. (2021), indicating that population stratification could weaken genuine genetic signals.

Furthermore, the interpretation of both QTL and GWAS results is complicated by the high occurrence of pleiotropy and epistasis in metabolic pathways. Riedelsheimer et al. (2015) have shown in a maize study that numerous carbon and nitrogen metabolites shared common genetic determinants, suggesting extensive pleiotropy that can mask the precise role of particular loci. On a technical level, accurate metabolite profiling is still quite difficult. Accurately identifying and measuring a wide range of secondary metabolites requires sophisticated techniques like LC-MS. However, these platforms need standardized protocols and expensive infrastructure. Sauvage et al. (2017) used high-resolution LC-MS to identify 79 metabolite-trait relationships in tomatoes, although they emphasized the significance of replication and analytical consistency to lower false positives.

Lastly, even though identifying QTLs is an important beginning step, functional characterization and validation are still difficult undertakings. The underlying genes and their mechanisms are still unknown without integrative techniques like transcriptomics and Clustered Regularly Interspaced Short Palindromic Repeats (CRISPR)-based gene editing, despite mapping studies frequently finding genomic areas linked to metabolite characteristics. Although Massonnet et al. (2020) identified a significant mQTL hotspot in grapevine linked to stilbenoid production, they also pointed out that more research is required to identify the causative genes and their biological functions. In conclusion, QTL mapping of secondary metabolites has advanced significantly since 2015, although there are still many challenges. These include tissue- and stage-specific expression, robust environmental interactions, intricate genetic regulation, and technical constraints in metabolomics and genotyping.

6.5 INTEGRATION OF MULTI-OMICS APPROACHES OF SECONDARY METABOLITES

A thorough investigation of biological systems requires integrated approaches that go beyond the limitations of single-omics analyses, particularly in the complex area of secondary metabolites, where intricate regulatory networks and varying environmental factors play a key role in their biosynthesis and accumulation (Subramanian et al., 2020). Multi-omics integration, which includes genomics, transcriptomics, proteomics, and metabolomics, offers a pathway to uncover the complex interactions between genetic potential and phenotypic expression. This approach leads to a deeper understanding of the regulatory mechanisms, metabolic pathways, and environmental factors that influence secondary metabolite production (Twaij et al., 2022). By applying multi-omics strategies, researchers can develop predictive models that link genotypes to phenotypes, identifying potential traits for crop improvement and gaining valuable insights into the molecular regulatory networks involved (Yang et al., 2021). Additionally, these integrated analyses can reveal novel enzymes, regulatory genes, and previously unidentified metabolic pathways essential for the synthesis of valuable secondary metabolites (Zhu et al., 2023).

To integrate multi-omics data, advanced computational and bioinformatics tools are required to manage the complexity and diversity of the datasets generated (Haas et al., 2017). This process involves normalizing, transforming, and integrating data through various statistical and machine learning methods, which must account for the distinct characteristics of each omics layer while reducing biases (Bersanelli et al., 2016; Picard et al., 2021).

A comprehensive understanding of the variables affecting the generation of secondary metabolites in crops is made possible by the integration of genomics, transcriptomics, metabolomics, and epigenomics. Developing strategies to increase the synthesis of desirable metabolites is enabled by the discovery of important regulatory networks and pathways made possible by multi-omics approaches. For example, gene clusters in rice and tomato that are in charge of the manufacture of beneficial chemicals have been identified using the combination of transcriptomic and metabolomic data (Singh et al., 2022; Yu, 2023; Shahidi et al., 2022; Cai et al., 2023).

Genomics: The genome is the basis of any multi-omics strategy. The basic genetic blueprint of a plant is provided by genomics, and genome mining allows to identify biosynthetic gene clusters (BGCs), which are co-located and frequently co-expressed genes involved in the production of secondary metabolites (Nützmann et al., 2016). The discovery of genes involved in the manufacture of secondary metabolites is made possible by WGS. For example, potential genes encoding enzymes in the monoterpenoid indole alkaloid pathway were identified through sequencing of *Camptotheca acuminata*, a plant known for producing camptothecin (Liu et al., 2020). The biosynthesis of thebaine has been clarified by the genome sequencing of the opium poppy, providing information on the manufacture of alkaloids (Singh et al., 2022). The identification of genes regulating the accumulation of secondary metabolites has also been aided by GWAS, which have linked particular gene variations to secondary metabolite profiles (Zhou et al., 2019).

Transcriptomics: The regulatory mechanisms controlling the manufacture of secondary metabolites under various environmental or developmental situations might be better understood through transcriptomic analysis (using RNA-seq). Genes with unknown functions can be connected to established metabolic pathways by co-expression network analysis. Key transcription factors and co-expression networks involved in the synthesis of metabolites can be found by researchers by combining transcriptome data with metabolomic profiles. *Salvia miltiorrhiza* transcriptome profiling, for instance, has produced new pathway components by identifying genes that are co-expressed with known diterpene synthases (Xu et al., 2016). MYB transcription factors have been shown to have a part in controlling the production of flavonoids in crops such as grapes and tomatoes (Singh et al., 2022).

Metabolomics: The connection between genotype and phenotype can be established by metabolomics. To identify chemicals of interest and provide a picture of the metabolic state, it entails the thorough investigation of metabolites inside plant tissues. Based on this approach, the metabolic profiling of different crops have also led to the discovery of new secondary metabolites that may have uses in agriculture and medicine (Singh et al., 2022). Metabolomics enables the identification and quantification of secondary metabolites in plant tissues using techniques such as LC-MS, GC-MS, and NMR. Unknown connections between metabolic flux and regulatory genes, for instance, have been discovered using the integrated metabolomic and transcriptome profiling of *Arabidopsis thaliana* mutants (Sumner et al., 2015). Furthermore, metabolomic data can be combined with transcriptomic and genomic data to clarify metabolic pathways and their regulation (Zhou et al., 2019).

Epigenomics: Heritable variations in gene expression without changes to the underlying DNA sequence are studied via epigenomics. Histone modification and DNA methylation are two examples of epigenetic processes that affect the formation of secondary metabolites. According to Desgagné-Penix (2015), the tissue-specific synthesis of benzylisoquinoline alkaloids, such as morphine, in *Papaver somniferum* has been connected to epigenetic control. The dynamic character of plant metabolism is shown by studies demonstrating how environmental factors and stress conditions can cause epigenetic modifications that modify metabolite profiles (Singh et al., 2022). The influence of chromatin shape on gene expression in SM pathways can be revealed by integrating transcriptomics and epigenomic data. The integrated framework of multi-omics for studying secondary metabolite biosynthesis is illustrated in Figure 6.2.

6.6 QTLS FOR MAJOR CLASSES OF SECONDARY METABOLITES

6.6.1 Alkaloids

QTLs for alkaloid content have been mapped in species such as *Nicotiana tabacum* (nicotine), *Papaver somniferum* (morphine), and *Lupinus* species (quinolizidine alkaloids). Many of these QTLs co-localize with known biosynthetic genes or transcriptional regulators.

6.6.2 Terpenoids

In *Arabidopsis*, QTLs associated with glucosinolate-derived terpenoids have been identified. In crops like tomato and rice, QTLs for volatile terpenes and diterpenoid phytoalexins have been mapped, often linked to pathogen resistance.

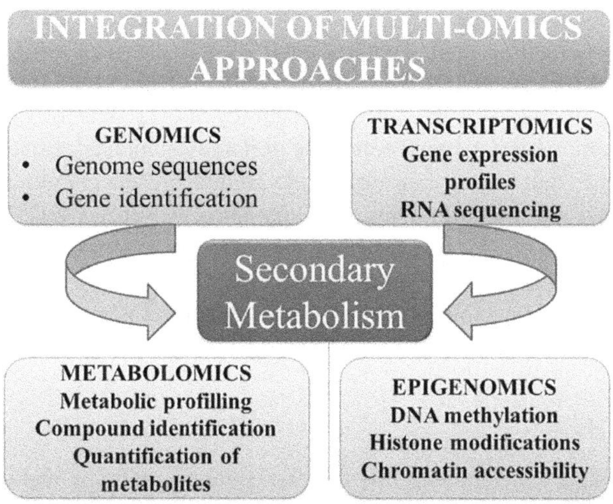

FIGURE 6.2 Integration of multi-omics approaches.

6.6.3 Phenolics and Flavonoids

Extensive work in *Zea mays*, *Vitis vinifera*, and *Medicago truncatula* has identified QTLs for flavonoid biosynthesis, anthocyanin pigmentation, and stilbenoid content. These QTLs are often regulated by MYB and bHLH transcription factors.

6.6.4 Glucosinolates

In *Brassica* species, glucosinolate content and profile have been dissected using QTL mapping and association studies, revealing clusters of loci with major effects and environment-dependent expression.

6.7 CHALLENGES AND FUTURE PROSPECTS

With contextual interactions being crucial to trait expression, QTL study is becoming increasingly acknowledged as a context-sensitive endeavour. To fully capture the range of genetic effects, multi-environment trials are required. Pleiotropy and epistasis, in which a single locus may influence several features and interact with other loci, further complicate the situation, and make simple interpretation difficult. The combination of high-throughput omics platforms—genomics, transcriptomics, metabolomics, and phenomics—offers previously unheard-of resolution despite these challenges, but at a considerable computational cost. To advance precision breeding and functional genomics, future directions in QTL research are expected to make use of pan-genome mapping, MAGIC populations for enhanced recombination, machine learning algorithms for predictive modelling, and genome-wide approaches specifically designed for metabolic trait prediction.

6.8 FUTURE PROSPECTS

Multi-omics approaches can be integrated to decipher intricate secondary metabolism pathways, a systems biology viewpoint is provided by combining QTL mapping with transcriptomics (eQTL), metabolomics (mQTL), and proteomics. For instance, metabolites and gene expression patterns can be directly linked by mQTL mapping and transcriptomics, improving the identification of candidate genes (Zhang et al., 2024). High-density genotyping and GWAS may facilitate the production of high-density markers by the introduction of GBS and next-generation sequencing technology. In contrast to traditional QTL mapping, GWAS across a variety of natural populations offers greater resolution mapping, which aids in identifying causative SNPs linked to secondary metabolites.

Complex genetic interactions and environmental effects are increasingly being captured by machine learning and Bayesian models. Particularly in big datasets with different settings and omics layers, these techniques increase the strength and accuracy of QTL detection (Wang et al., 2024). To incorporate more genetic variety beyond reference genomes, which is crucial for secondary metabolic traits, pan-genomes that incorporate structural variants, presence/absence variations, and gene copy number variations are revolutionizing QTL mapping (Li et al., 2023). To improve desired secondary metabolites in crops, identified QTLs and underlying genes provide targets for genome editing (e.g., CRISPR/Cas9) and marker-assisted selection. QTL insights can direct metabolic engineering utilizing synthetic biology techniques to generate useful chemicals in heterologous systems or plants (Chen et al., 2024).

6.9 CONCLUSION

QTL mapping of secondary metabolism fills the gap between genotype and metabolite phenotype, offering key insights into the genetic control of complex traits. The integration of mQTL data with other 'omics' technologies paves the way for next-generation crop improvement strategies that can be tailored to meet both predictive breeding and metabolic design in plants.

REFERENCES

Bernardo, R. (2008). Molecular markers and selection for complex traits in plants: Learning from the last 20 years. *Crop Science*, 48(5): 1649–1664. https://doi.org/10.2135/cropsci2008.03.0131

Bersanelli, M., Mosca, E., Remondini, D., Giampieri, E., Sala, C., Castellani, G., and Milanesi, L. (2016). Methods for the integration of multi-omics data: Mathematical aspects. *BMC Bioinformatics*, 17(Suppl 2): 15. https://doi.org/10.1186/s12859-015-0857-9

Borevitz, J.O., and Chory, J. (2004). Genetics of quantitative traits in *Arabidopsis thaliana. Nature Reviews Genetics*, 5(5): 353–364. https://doi.org/10.1038/nrg1342

Cai, Q., Jiao, F., Wang, Q., Li, X., Zhang, Y., and Liu, Z. (2023). Multiomics comparative analysis of the maize large grain mutant tc19 identified pathways related to kernel development. *BMC Genomics,* 24: 537. https://doi.org/10.1186/s12864-023-09567-z

Chen, Y., Liu, H., Zhao, X., and Xu, J. (2024). Synthetic biology approaches in metabolic engineering for plant secondary metabolites. *Plant Biotechnology Journal*, 22(3): 456–468. https://doi.org/10.1111/pbi.14045

Desgagné-Penix, I. (2015). Biosynthesis and regulation of alkaloids in *Papaver somniferum*: An overview of recent advances. *Botany*, 93(5): 311–323. https://doi.org/10.1139/cjb-2014-0243

Devi, E.L., Punita, D.L., and Sharma, R. (2017). QTL mapping for yield and drought tolerance traits in rice: A review. *Agricultural Reviews,* 38(1): 1–11. https://doi.org/10.20546/ijcmas.2017.606.169

Dreisigacker, S., Crossa, J., Sehgal, D., Herrera-Foessel, S., Luna, D., Reyes, J.A., and Singh, R.P. (2016). Genomic selection in plant breeding: A review of recent advances. *Plant Breeding,* 135(4): 429–440. https://doi.org/10.1111/pbr.12371

Dwivedi, G. R., & Sisodia, B. S. (2018). Secondary metabolites: Metabolomics for secondary metabolites. In V.J. Gupta (Ed.), *New and Future Developments in Microbial Biotechnology and Bioengineering* (pp. 333–344). Elsevier. https://doi.org/10.1016/B978-0-444-63504-4.00022-0

Grisel, J.E., and Crabbe, J.C. (1995). Quantitative trait loci mapping. *Alcohol Health and Research World,* 19(3): 220–227.

Grover, A., and Sharma, P.C. (2014). Development and use of molecular markers: Past and present. *Critical Reviews in Biotechnology*, 36(2): 290–302. https://doi.org/10.3109/07388551.2014.959891

Haas, R., Zelezniak, A., Iacovacci, J., Kamrad, S., Townsend, S., and Ralser, M. (2017). Designing and interpreting 'multi-omic' experiments that may change our understanding of biology. *Current Opinion in Systems Biology*, 6: 37–45. https://doi.org/10.1016/j.coisb.2017.08.009

Heffner, E.L., Lorenz, A.J., Jannink, L., and Sorrells, M.E. (2010). Plant breeding with genomic selection: Gain per unit time and cost. *Crop Science*, 50(5): 1681–1690. https://doi.org/10.2135/cropsci2009.11.0662

Huang, L., Li, M., Cao, D., and Yang, P. (2021). Genetic dissection of rhizome yield-related traits in *Nelumbo nucifera* through genetic linkage map construction and QTL mapping. *Plant Physiology and Biochemistry*, 160: 155–165. https://doi.org/10.1016/j.plaphy.2021.01.020

Kumar, A., Kumar, S., Pal, A.K., and Singh, G. P. (2017). Quantitative trait loci from identification to exploitation for crop improvement. *Plant Cell Reports*, 36(8): 1187–1213. https://doi.org/10.1007/s00299-017-2117-8

Kumar, J., and Gupta, D.S. (2017). Applications of molecular markers and QTL mapping in crop improvement: A review. *Plant Breeding and Biotechnology*, 5(4): 367–384. https://doi.org/10.1186/s12870-023-04257-0

Kumar, N., Sindhu, S., Gawaria, B., Ansari, A., Dhariwal, A., Sharma, P., and Kumar, R. (2025). GWAS approaches in medicinal plants for enhanced secondary metabolite production. In *Biotechnology, Multiple Omics, and Precision Breeding in Medicinal Plants* (pp. 119–134). CRC Press.

Katz, E., Li, J., Jaeger, C., Waldherr, S., and Sulpice, R. (2021). Local population genome-wide association study reveals candidate genes for *Arabidopsis* metabolic traits. *Philosophical Transactions of the Royal Society B: Biological Sciences,* 376(1817): 20200512. https://doi.org/10.1098/rstb.2020.0512

Khoshakhlagh, F., Darvishzadeh, R., Bernousi, I., and Maleki, H.H. (2022). Application of NMR-based metabolomics in plant breeding and functional genomics research: A review. *Metabolomics,* 18(3): 21. https://doi.org/10.1007/s11306-022-01880-0

Kinghorn, A.D. (1994). The discovery of drugs from higher plants. *Discovery of Novel Natural Products with Therapeutic Potential* (pp. 81–108). Elsevier: Ciba Foundation Symposium. https://doi.org/10.1016/B978-0-7506-9003-4.50010-1

Kochhar, S.L., and Gujral, S.K. (2020). Secondary plant metabolites. In *Plant Physiology: Theory and Applications* (pp. 590–610). Cambridge University Press.

Kumar, A., Dixit, S., Ram, T., Yadaw, R.B., Mishra, K.K., and Mandal, N.P. (2011). Breeding high-yielding drought-tolerant rice: Genetic variations and conventional and molecular approaches. *Journal of Experimental Botany,* 62(1): 199–214. https://doi.org/10.1093/jxb/erq300

Lan, L., Chen, G., Meng, S., and Zhang, J. (2020). Advances in QTL mapping and molecular breeding for complex traits in crop plants. *Frontiers in Plant Science*, 11: 943. https://doi.org/10.3389/fpls.2020.00943

Lee, S., Kelly, R.S., Mendez, K.M., Prokopenko, D., Hahn, G., Lutz, S.M., Celedón, J.C., Clish, C.B., Weiss, S.T., Lange, C., and Lasky-Su, J.A. (2025). On the analysis of metabolite quantitative trait loci: Impact of different data transformations and study designs. *Science Advances*, 11(15): eadp4532.

Lema, M. (2018). Marker-assisted selection in crop improvement: A review of recent developments. *Journal of Plant Breeding and Crop Science*, 10(8): 204–212. https://doi.org/10.5897/JPBCS2018.0750

Li, Q., Zhang, Y., Huang, L., and Zhou, W. (2023). Pan-genomics advances in understanding structural variations in crop species. *Trends in Genetics*, 39(9): 789–802. https://doi.org/10.1016/j.tig.2023.06.004

Liu, X., Wang, H., Hu, X., Li, K., Liu, Z., Wu, Y., and Huang, C. (2019). Improving genomic selection with quantitative trait loci and nonadditive effects revealed by empirical evidence in maize. *Frontiers in Plant Science*, 10: 476921. https://doi.org/10.3389/fpls.2019.01129

Liu, X., Wang, Y., Wang, H., Guo, Z., Xu, J., and Gao, W. (2020). The genome of *Camptotheca acuminata* provides insights into the biosynthesis of camptothecin, a plant-derived anticancer drug. *GigaScience*, 9(1): giz161. https://doi.org/10.1093/gigascience/giz161

Massonnet, M., Figueroa-Balderas, R., Galarneau, E.R.A., Miki, S., Lawrence, J., Sun, Q., and Cantu, D. (2020). Grapevine genome-wide association study reveals major mQTLs for stilbenoid accumulation. *Horticulture Research*, 7: 1–13. https://doi.org/10.1038/s41438-020-0294-3

McMullen, M.D., Byrne, P.F., Snook, M.E., Musket, T.A., Houchins, K.E., Coe, E.H., and Widstrom, N.W. (2003). Quantitative trait loci and metabolic pathways: Genetic control of flavone glucoside accumulation in maize silks. *Proceedings of the National Academy of Sciences of the United States of America,* 100(13): 8267–8272. https://doi.org/doi:10. 1073/pnas.1508681112

Meena, R.K., Singh, A., and Jha, S.K. (2017). Molecular approaches in plant breeding for crop improvement. *Journal of Applied and Natural Science,* 9(1): 632–644. https://doi.org/10.31018/jans.v9i1.1216

Moury, B., Caffier, V., Montarry, J., Kerlan, M., Fournet, S., Durel, C., and Delourme, R. (2017). Quantitative resistance to plant pathogens in pyramiding strategies for durable crop protection. *Frontiers in Plant Science*, 8: 295209. https://doi.org/10.3389/fpls.2017.01838

Niu, G., and Tan, H. (2013). Biosynthesis and regulation of secondary metabolites in microorganisms. *Science China Life Sciences,* 56: 581–583. https://doi.org/10.1007/s11427-013-4501-5

Nützmann, H.-W., Scazzocchio, C., and Osbourn, A. (2016). Metabolic gene clusters in eukaryotes. *Annual Review of Genetics*, 50, 305–330. https://doi.org/10.1146/annurev-genet-120215-035529

Pereira, M.G., Orlando de Oliveira, L., and Lee, M. (2000). QTL mapping and disease resistance in cereals. *Journal of New Seeds,* 2(2): 1–21. https://doi.org/10.1300/J153v02n02_01

Picard, M., Scott-Boyer, M.P., Bodein, A., Périn, O., Droit, A., and Lavoie, J. (2021). Integration strategies of multi-omics data for machine learning analysis. *Computational and Structural Biotechnology Journal*, 19: 3735–3746. https://doi.org/10.1016/j.csbj.2021.06.030

Pilet-Nayel, M.L., Moury, B., Caffier, V., Montarry, J., Kerlan, M.C., Fournet, S., Durel, C.E., and Delourme, R. (2017). Quantitative resistance to plant pathogens in pyramiding strategies to improve resistance durability. *Frontiers in Plant Science,* 8: 1838. https://doi.org/10.3389/fpls.2017.01838

Sun, Q., Wang, P., Li, W., Li, W., Lu, S., Yu, Y., Zhao, M., and Meng, Z. (2019). Genomic selection on shelling percentage and other traits for maize. *Breeding Science,* 69(2): 266–271. https://doi.org/10.1270/jsbbs.18141

Reshi, Z.A., Ahmad, W., Lukatkin, A.S., and Javed, S.B. (2023). From nature to lab: A review of secondary metabolite biosynthetic pathways, environmental influences, and *in vitro* approaches. *Metabolites,* 13(8): 895. https://doi.org/10.3390/metabo13080895

Riedelsheimer, C., Czedik-Eysenberg, A., Grieder, C., Lisec, J., Technow, F., Sulpice, R., Altmann, T., Stitt, M., and Melchinger, A.E. (2015). Genetic architecture of maize metabolic traits reveals pattern of pleiotropy and epistasis. *Genetics,* 199(2): 389–400. https://doi.org/10.1534/genetics.114.173765

Salvi, S., and Tuberosa, R. (2005). To clone or not to clone plant QTLs: Present and future challenges. *Trends in Plant Science,* 10(6): 297–304. https://doi.org/10.1016/j.tplants.2005.04.008

Sasaki, E., Zhang, P., Atwell, S., Meng, D., Nordborg, M., and Shimizu, K.K. (2019). Epigenetic QTLs for flower metabolic traits in *Arabidopsis. Genome Research,* 29(1): 96–107. https://doi.org/10.1101/gr.232371.117

Sauvage, C., Segura, V., Bauchet, G., Stevens, R., Do, P.T., Nikoloski, Z., Fernie, A.R., Causse, M., and Rothan, C. (2017). Genome-wide association in tomato reveals metabolic QTL hotspots and candidate genes for fruit quality traits. *New Phytologist,* 213(1): 526–539. https://doi.org/10.1111/nph.14283

Seca, A.M.L., and Trendafilova, A. (2021). Secondary metabolites in edible species: Looking beyond nutritional value. *Foods,* 10(5): 1131. https://doi.org/10.3390/foods10051131

Shahidi, F., Danielski, R., Rhein, S. O., Meisel, L. A., Fuentes, J., Speisky, H., ... & de Camargo, A. C. (2022). Wheat and rice beyond phenolic acids: Genetics, identification database, antioxidant properties, and potential health effects. *Plants*, 11(23): 3283.

Shiri, M., and Akhavan, A. (2014). QTL mapping: An effective approach for discovering genes controlling complex traits in plants. *International Journal of Advanced Biological and Biomedical Research,* 2(12): 2811–2821.

Sahu, J., and Sharma, D. (2017). Recent advances in marker-assisted selection and crop improvement: A review. *International Journal of Current Microbiology and Applied Sciences,* 6(7): 1977–1991. https://doi.org/10.20546/ijcmas.2017.607.234

Singh, S., Singh, R., Singh, G.P., and Gupta, R.K. (2012). Marker-assisted selection: A paradigm shift in wheat breeding for rust resistance and stress tolerance. *Indian Journal of Genetics and Plant Breeding,* 72(4): 481–493.

Simsek Sezer, E.N., and Uysal, T. (2021). Phytochemical analysis, antioxidant and anticancer potential of *Sideritis niveotomentosa:* Endemic wild species of Turkey. *Molecules,* 26(9): 2420. https://doi.org/10.3390/molecules26092420

Singh, D.P., Bisen, M.S., Shukla, R., Prabha, R., Maurya, S., Reddy, Y.S., Singh, P.M., Rai, N., Chaubey, T., Chaturvedi, K.K., Srivastava, S., Farooqi, M.S., Gupta, V.K., Sarma, B.K., Rai, A., and Behera, T.K. (2022). Metabolomics-driven mining of metabolite resources: Applications and prospects for improving vegetable crops. *International Journal of Molecular Sciences,* 23(20): 12062. https://doi.org/10.3390/ijms232012062

Srivastava, V.K., Egbuna, C., and Tiwari, M. (2020). Plant secondary metabolites as lead compounds for the production of potent drugs. In C. Egbuna and S. Kumar (Eds.), *Phytochemicals as Lead Compounds for New Drug Discovery* (pp. 3–14). Elsevier. https://doi.org/10.1016/B978-0-12-817890-4.00001-9

Stuthman, D., Leonard, K., and Miller-Garvin, J. (2007). Breeding crops for durable resistance to disease. *Advances in Agronomy,* 95: 319–367. https://doi.org/10.1016/S0065-2113(07)95004-X

Subramanian, I., Verma, S., Kumar, S., Jere, A., & Anamika, K. (2020). Multi-omics Data Integration, Interpretation, and Its Application. *Bioinformatics and Biology Insights.* Advance online publication. https://doi.org/10.1177/1177932219899051

Sumner, L.W., Amberg, A., Barrett, D., Beale, M.H., Beger, R., Daykin, C.A., Fan, T.W.M., Fiehn, O., Goodacre, R., Griffin, J.L., Hankemeier, T., Hardy, N., Harnly, J., Higashi, R., Kopka, J., Lane, A.N., Lindon, J.C., Marriott, P., Nicholls, A.W., Reily, M.D., Viant, M.R., and Wilson, I.D. (2015). Proposed minimum reporting standards for chemical analysis. *Metabolomics,* 11: 872–894. https://doi.org/10.1007/s11306-015-0821-3

Thirumurugan, D., Cholarajan, A., Raja, S.S.S., and Vijayakumar, R. (2018). An introductory chapter: Secondary metabolites. In *Secondary Metabolites – Sources and Applications.* IntechOpen. https://doi.org/10.5772/intechopen.79766

Tong, H., Ullah, A., Kamran, M., Chen, X., Lin, F., Zhu, T., Xu, H., & Xiao, B. (2025). QTL mapping and association analysis revealed complicate genetic architecture of 17 smoke and chemical traits in tobacco (*Nicotiana tabacum L.*). *Industrial Crops and Products*, 225: 120599. https://doi.org/10.1016/j.indcrop.2025.120599

Toubiana, D., Semel, Y., Tohge, T., Beleggia, R., Cattivelli, L., Rosental, L., Nikoloski, Z., Zamir, D., Fernie, A.R., and Fait, A. (2016). Metabolite levels in winter wheat: Genetic architecture and the role of key metabolites in breeding. *PLOS ONE,* 11(5): e0155351. https://doi.org/10.1371/journal.pone.0155351

Twaij, B.M. and Hasan, M.N. (2022). Bioactive secondary metabolites from plant sources: types, synthesis, and their therapeutic uses. *International Journal of Plant Biology*, 13(1): 4–14. https://doi.org/10.3390/ijpb13010003

Vagndorf, N., Nielsen, N.H., and Andersen, J.R. (2018). QTL mapping and genomic prediction for resistance traits in barley and wheat. *Euphytica,* 214(8): 135. https://doi.org/10.1038/s41598-020-63862-3

Wang, M., Zhao, F., and Li, D. (2024). Integrating machine learning with QTL analysis to dissect complex traits in plants. *Briefings in Bioinformatics*, 25(2): bbad012. https://doi.org/10.1093/bib/bbad012

Wen, W., Li, D., Li, X., Gao, Y., Li, W., Li, H., Liu, J., et al.(2015). Genetic architecture of metabolic variation in maize. *Proceedings of the National Academy of Sciences of the United States of America*, 112(50): E806–E814.

Yali, W. (2022). Molecular markers: Their importance, types, and applications in modern agriculture. *Agriculture, Forestry and Fisheries,* 11(1): 8–14. https://doi.org/10.11648/j.aff.20221101.12

Xu, Z., Peters, R.J., Weirather, J., Luo, H., Liao, B., Zhang, X., Zhu, Y., Ji, A., Zhang, B., Hu, S., Au, K.F., Song, J., and Chen, S. (2016). Full-length transcriptome sequences and splice variants obtained by a combination of sequencing platforms applied to *Salvia miltiorrhiza* roots. *Plant Journal*, 85(6): 717–730. https://doi.org/10.1111/tpj.13150

Yu, D., Huang, R., Yu, S., Wei, J., Chen, L., Liu, Y., and Zhang, X. (2023). Construction of the first high-density genetic linkage map and QTL mapping of flavonoid and leaf-size related traits in *Epimedium*. *BMC Plant Biology,* 23: 278. https://doi.org/10.1186/s12870-023-04257-0

Zhang, L., Chen, Y., Wang, Y., and Huang, J. (2024). Multi-omics dissection of secondary metabolic pathways in crops. *The Plant Cell*, 36(1): 23–40. https://doi.org/10.1093/plcell/koae001

Zhou, Y., Massonnet, M., San Clemente, H., Hasing, T., Jiao, C., Kumar, M.S., Fennell, A., and Giovannoni, J.J. (2019). The genetic architecture of secondary metabolite diversity in *Arabidopsis thaliana*. *Nature Communications*, 10: 1–13. https://doi.org/10.1038/s41467-019-13093-7

Zhu, F., Wen, W., Cheng, Y., Alseekh, S., and Fernie, A.R. (2023). Integrating multi-omics data accelerates elucidation of plant primary and secondary metabolic pathways. *aBIOTECH,* 4(1): 47–56. https://doi.org/10.1007/s42994-022-00091-4

7 Molecular Markers Associated with Secondary Metabolism and Value Addition Using Biotechnological Interventions

Anshul Jaiswal, R. Sagar, Mohammad Sirazuddin Ansari, Aakansha Pandey, Deepankar Yadav, and Richa Rawat

7.1 INTRODUCTION

Plants have a complex system of metabolic pathways that regulate their growth, development, and adaptations to environmental factors (Liao et al. 2023). Plant metabolism is broadly categorized into primary and secondary metabolic pathways (Sangwan et al. 2018). Primary metabolism includes vital life-sustaining processes like respiration, photosynthesis, and the production of essential organic molecules that ensure plant growth and development. Conversely, secondary metabolism produces specialized compounds called plant secondary metabolites (PSMs). They are not directly involved in growth; however, they play crucial roles in plant survival, defence, and adaptations to environmental stresses. These specialized compounds include phenolics, terpenoids, alkaloids, etc. They enhance plant resilience by mitigating the oxidative stress, deterring herbivores, protecting against pathogens, and facilitating beneficial plant–microbe interactions (Isah 2019; Saddhe et al. 2021).

Beyond their ecological role, PSMs hold significant economic value in pharmaceuticals, nutraceuticals, cosmetics, and agriculture (Twaij & Hasan 2022). Their medicinal properties make them essential for the prevention and treatment of disease (Lee et al. 2019; Abbas et al. 2022). For instance, phenolic antioxidants aid in diabetes management and cancer risk reduction (Yi et al. 2005), alkaloids like berberine enhance insulin sensitivity (Debnath et al. 2018), and terpenoids and carotenoids support cardiovascular health and combat oxidative stress (Styrczewska et al. 2013). Growing demand in nutraceuticals further highlights the need for sustainable secondary metabolite (SM) production (Grimble 2003; Ranzato et al. 2014).

As the demand for high-value bioactive compounds derived from PSMs continues to grow, molecular markers have become crucial tools for improving the genetics of medicinal plants (Roberts & Mattoo 2019). These markers allow for the exact identification of genetic variables linked to specific SM biosynthesis, enhancing the selection of high-yielding plant varieties (Kowsalya et al. 2025). Derived from natural genetic variations, molecular markers assist in gene mapping associated with biosynthetic pathways, thereby accelerating breeding efforts and biotechnological advancements in medicinal plant development (Thottappilly et al. 2000).

The discovery of molecular markers associated with specific SMs allows for the targeted regulation of their biosynthetic pathways using genome-editing tools. Advances in genome sequencing technology have revolutionized crop genetic engineering, particularly the breeding strategies. The development of these markers has enabled the generation of high-resolution genetic and linkage

DOI: 10.1201/9781003542681-7

maps, aiding in the precise identification of quantitative trait loci (QTLs) associated with significant metabolic traits. Once target genes are identified, sophisticated genome-editing techniques can be applied to introduce precise modifications, optimizing trait expression with enhanced accuracy and efficiency (Kumar et al. 2024). Progress in metabolic engineering and genome-editing technologies has increased the possibility for optimizing the production of SMs (Sinha et al. 2019). Unlike primary metabolism, which is governed at the genetic level, secondary metabolism is primarily regulated through enzymatic activity and subcellular compartmentalization, making conventional breeding approaches challenging while opening avenues for biotechnological interventions. Precise gene-editing tools, including CRISPR/Cas9, enable targeted modifications in biosynthetic pathways to increase metabolite yields, while metabolic engineering strategies involving the introduction of foreign genes or overexpression of key biosynthetic enzymes optimize these pathways (Mipeshwaree et al. 2023). Moreover, tissue culture has long been a vital method for plant propagation and, in recent decades, has played a key role in biotechnology applications, including transgenesis, to advance both fundamental and applied plant research (Altpeter et al. 2016). It is also essential for successful plant gene editing, as reagents like CRISPR/Cas9 and single-guide RNAs (sgRNAs) are introduced into cultured cells to induce precise DNA sequence modifications (Zhang et al. 2014). While some plant species and genotypes have been successfully regenerated from edited or transformed cells, the process can be time-consuming and may lead to unintended genomic and epigenomic alterations. Consequently, tissue culture remains a major bottleneck in producing gene-edited plants and developing novel traits to enhance crop varieties. Despite these advancements, challenges such as low transformation efficiency, unintended mutations, and metabolic imbalances leading to the accumulation of toxic intermediates persist. Overcoming these limitations, integrated multi-omics approaches, computational modelling, and synthetic biology tools for precision metabolic engineering are needed (Helmy et al. 2020). By incorporating molecular markers and biotechnological innovations, researchers and plant breeders can significantly enhance the yield and potential of SMs, offering a promising role in the drugs and nutraceutical industries to ensure a sustainable supply of bioactive compounds with improved therapeutic properties (Bomzan et al. 2022).

This study emphasized the significance of PSMs, focusing on the pivotal role of molecular markers, genome-editing tools, and tissue culture techniques in enhancing their production. With the increasing demand for bioactive compounds in pharmaceuticals and biotechnology, it explores sustainable strategies to improve medicinal plant quality. Specifically, the objective of this synthesis is to explore the role of molecular markers in identifying genetic factors linked to SM biosynthesis and their application in marker-assisted breeding (MAB) for precise trait selection. Furthermore, it examines the potential of genome-editing tools to modify metabolite-associated genes while integrating tissue culture techniques for stable trait propagation with the help of diverse case studies. Additionally, the present synthesis underscores the significance of omics and bioinformatics tools in enhancing genome-editing precision and addressing challenges in optimizing SM production. Through these advanced approaches, the study aims to accelerate plant improvement programmes, ensuring a sustainable and efficient supply of high-value bioactive compounds for therapeutic and commercial applications.

7.2 SECONDARY METABOLITES

7.2.1 Classification and Their Significance

According to the British Nutrition Foundation (Goldberg, 2003), PSMs are classified into four main groups: terpenoids (carotenoids, sterols, cardiac glycosides, volatiles), phenolics (flavonoids, tannins, lignans, etc.), nitrogen-containing compounds (alkaloids, cyanogenic glucosides), and sulphur-containing compounds (glucosinolates, glutathione, phytoalexins, lectins) (Mazid et al. 2011), as shown in Figure 7.1.

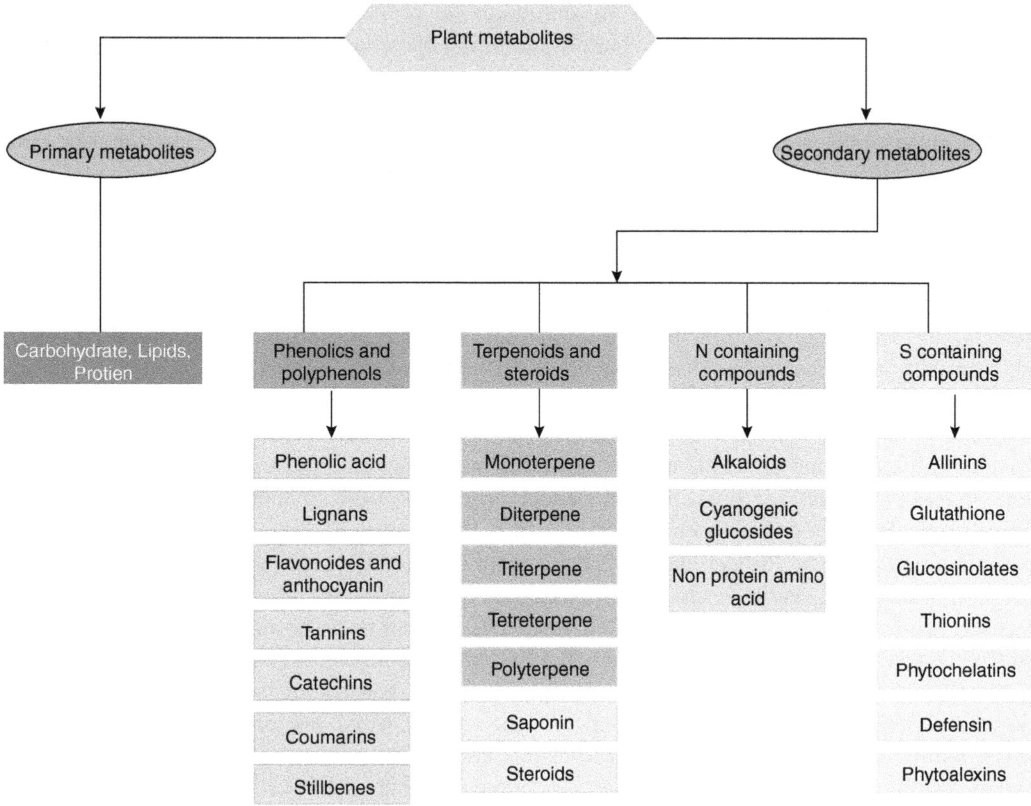

FIGURE 7.1 The four major types of secondary metabolites.

7.2.2 Biological Significance

Numerous SMs are produced by plants and are vital to their survival, especially for communication and defence. These bioactive constituents play multiple ecological roles, such as protection against biotic and abiotic stressors (Divekar et al. 2022). They either act as deterrents or directly suppress the growth of harmful organisms. For example, alkaloids and terpenes possess toxic properties that repel insects and suppress microbial growth (Zaynab et al. 2018). Phenolics, produced via the shikimic acid pathway, strengthen plant resistance by forming physical barriers or disrupting pathogen metabolism (Tak & Kumar 2020). Additionally, nitrogen- and sulphur-containing compounds derived from amino acids enhance plant immunity against microbial infections (Harborne 1988; Wink 2008). Some specialized metabolites, such as phytoalexins and phytoanticipins, act as antimicrobial agents. Phytoalexins are produced as a defence mechanism following pathogen invasion, while phytoanticipins are pre-formed compounds stored in plant tissues that become active when damage occurs (Tiku 2020). SMs are important for plant signalling and environmental interactions in addition to defence.

Volatile organic compounds (VOCs), including terpenoids, help attract pollinators and beneficial microbes while repelling herbivores (Harborne 1993). Flavonoids aid in root–microbe interactions by modulating the rhizosphere environment and facilitating nutrient uptake (Bais et al. 2002). Such interactions contribute to crop resilience by promoting symbiotic relationships with beneficial microbes. Furthermore, certain SMs regulate stress responses, such as phenolics protecting against UV radiation, while alkaloids and flavonoids mitigate oxidative stress (Schäfer & Wink 2009).

Many of these compounds exhibit strong antimicrobial properties, such as saponins disrupting fungal membranes (Yu et al. 2022) and tannins inhibiting bacterial enzymes (Pierpoint 2000). Some flavonoids and polyphenols also interfere with viral replication, demonstrating their potential in plant disease management (Wang et al. 2007; Chen et al. 2009; Yan et al. 2010). SMs are essential for plant survival because of their functions in defence, signalling, and stress adaptation. They may also help sustainable agriculture by reducing the need for pesticides and enhancing crop resilience.

7.2.3 Commercial Significance

PSMs are vital across multiple industries due to their diverse biological activities and wide-ranging applications. They play a significant role in pharmaceuticals, farming, cosmetics, food industries, etc. contributing to advancements in these fields (Elshafie et al. 2023).

Pharmaceuticals: Many PSMs serve as the foundation for drug development. Alkaloids like morphine (analgesic) and quinine (antimalarial) are widely used, while flavonoids and phenolics provide antioxidant and anti-inflammatory benefits in treating cardiovascular diseases and cancer. Terpenoids exhibit antimicrobial and anticancer properties, making them crucial in therapeutic formulations (Balunas & Kinghorn 2005; Hussein & El-Anssary 2019).

Agriculture and Crop Protection: PSMs act as natural pesticides, herbicides, and insect repellents, reducing reliance on synthetic chemicals. Alkaloids and terpenoids protect crops from pests and diseases, while some PSMs enhance plant resilience against drought and temperature extremes, improving agricultural sustainability (Cannell 1998; Savithramma et al. 2011).

Food and Beverages: PSMs contribute to flavour, aroma, and colour in food products. Phenolic compounds like tannins and flavonoids enhance the taste of fruits, tea, and wine (Tomás & Espin 2001), while terpenoids are used as essential oils and flavour enhancers. Additionally, alkaloids such as caffeine and theobromine are bioactive compounds that significantly influence the stimulant properties of tea, coffee, and chocolate (Pichersky et al. 2006; Elshafie et al. 2023).

Cosmetics and Personal Care: The cosmetic industry extensively incorporates PSMs due to their antioxidant and anti-ageing properties (Bjørklund et al. 2022). By shielding the skin from UV rays and oxidative stress, flavonoids and polyphenols enhance the general health of the skin. Additionally, terpenoid-based essential oils are widely utilized in perfumes, lotions, and shampoos for their therapeutic effects and aromatic qualities (Elshafie et al. 2023).

7.3 MOLECULAR MARKERS

Molecular markers, which are specific DNA sequences, have transformed plant research by offering a precise, efficient, and cost-effective alternative to conventional marker systems. In genetic engineering, these markers are linked to particular traits, such as SM production, allowing researchers to track and modify related genes. They act as essential tools for detecting and selecting individuals in a population that possesses genes linked to the synthesis of targeted SMs (Mipeshwaree et al. 2023). Unlike phenotypic and biochemical markers, these molecular markers remain unaffected by environmental factors, age, or physiological conditions, making them highly reliable for genetic analysis and breeding (Mirzaei 2021). Their ability to detect genetic variations at any stage of plant development has significantly enhanced breeding efficiency, conservation efforts, and medicinal plant research.

The development of polymerase chain reaction (PCR) technology in the 1980s, followed by advancements in genomics, bioinformatics, and proteomics in the 1990s, has greatly expanded the applications of DNA-based markers, particularly in major genome sequencing projects such as those of humans, Arabidopsis, and rice (Li et al. 2006). Earlier electrophoresis-based protein markers in the 1950s laid the foundation for understanding genetic variation at the protein level, which later evolved into modern molecular marker technologies (Mobasheri et al. 2017). These advancements have been crucial in ensuring the quality and safety evaluation of medicinal products and nutraceuticals, thereby increasing the commercial value and medical potential of herbal products.

Additionally, by enabling targeted trait selection, identifying important genes involved in the production of SMs, and improving genetic classification, molecular MAB has transformed the breeding of medicinal plants (Holland 2004). This strategy has solved issues associated with conventional breeding approaches like extended breeding cycles and genetic variability, resulting in the production of high-yielding plant varieties with stable metabolite profiles (Ahmar et al. 2020). Furthermore, MAB contributes significantly to the protection of endangered medicinal species, ensuring long-term production to fulfil the expanding global need for plant-derived therapeutic substances.

7.3.1 Types of Molecular Markers

Molecular markers are broadly classified into morphological, biochemical, and DNA-based categories. Among these, DNA-based markers, such as simple sequence repeats (SSRs), single nucleotide polymorphism (SNPs), and amplified fragment length polymorphisms (AFLPs), are preferred for their high reproducibility and efficacy in genetic mapping (Joshi et al. 1999). Breeders can precisely recognize and select plants displaying desired qualities well before they reach maturity due to these markers, which help in marker-assisted selection (MAS). Furthermore, advancements in high-efficiency genotyping and omics technologies, such as genomics, transcriptomics, and

TABLE 7.1
Classification and Brief Description of Different Molecular Markers

Marker Type	Description	Advantages	Limitations	Reference
Restriction Fragment Length Polymorphism (RFLP)	Uses restriction enzymes to cut DNA at specific sites, producing unique fragment patterns.	High accuracy, co-dominant inheritance, useful for genome mapping.	Labour-intensive, requires large amounts of high-quality DNA, not widely used in modern research.	Landry et al. 1987
Random Amplified Polymorphic DNA (RAPD)	Uses short primers to amplify random DNA sequences, revealing genetic polymorphisms.	Simple, cost-effective, does not require prior genome information.	Dominant marker, low reproducibility, limited use in precise genetic mapping.	Agarwal et al. 2008
Simple Sequence Repeats (SSR)	Also known as microsatellites, consists of repeating nucleotides sequences dispersed in the genome.	Highly polymorphic, co-dominant, reproducible, useful for genetic mapping.	Requires prior sequence data, expensive development process.	Nadeem et al. 2018
Amplified Fragment Length Polymorphism (AFLP)	Combines restriction enzyme digestion and PCR amplification to generate high-resolution genetic fingerprints.	High reproducibility, multiplexing capability, ideal for genetic diversity studies.	Technically demanding, requires multiple steps.	Mirzaei 2021
Inter-Simple Sequence Repeats (ISSR)	Amplifies genomic regions between microsatellites, revealing polymorphisms.	High reproducibility, informative, useful for population genetics.	Dominant marker, requires optimization for different species.	Amom & Nongdam, 2017
Single Nucleotide Polymorphism (SNP)	Identifies single-base changes in DNA sequences, providing high precision.	Most abundant marker, high-throughput, widely used in genomics and breeding.	High initial cost, requires advanced sequencing technologies.	Kuzay et al. 2019

metabolomics, have significantly advanced secondary metabolism research by permitting precise gene identification and pathway elucidation (Pourmohammad 2013). Consequently, DNA-based molecular markers have emerged as indispensable tools in genomics, providing a comprehensive array of methodologies optimized to diverse research applications (Table 7.1).

7.3.2 Role of Molecular Markers in Secondary Metabolism

Molecular markers have become essential in medicinal plant research by enabling precise genomic analysis, accelerating targeted breeding programmes, and supporting biodiversity conservation (Bhargava & Ohri 2016). They play an important role in the genetic improvement as well as sustainable production of medicinal plants while ensuring the authenticity and effectiveness of SMs.

The analysis of genetic diversity and phylogenetic relationships through these markers facilitates accurate species classification, reveals evolutionary linkages, and aids in the formulation of successful conservation plans. Techniques such as QTL mapping give information on the genetic regulation of complex traits, including SM biosynthesis, stress resilience, and overall plant productivity. Such insights are crucial for developing plant varieties enriched with therapeutically important compounds like alkaloids and flavonoids (Chukwu et al. 2019; Coates et al. 2018; Kuzay et al. 2019).

Advances in DNA fingerprinting methods, including AFLP and SSRs, have greatly enhanced the authentication of medicinal plant resources by preventing adulteration and preserving pharmaceutical integrity. Concurrently, breakthroughs in whole-genome and next-generation sequencing (NGS) technologies have transformed precision breeding, enabling the rapid development of high-yield, disease-resistant, and bioactive-rich varieties (Goodwin et al. 2016; Mirzaei 2021).

Molecular markers have also transformed the study of secondary metabolism. The creation of detailed genetic and linkage maps has facilitated the identification of QTLs associated with agronomic traits. Once key genes are identified, modern genome-editing tools, including homologous recombination (HR), zinc finger nucleases (ZFNs), transcription activator-like effector nucleases (TALENs), and the CRISPR/Cas system, can be applied for precise genetic modifications. Techniques like RNA interference and site-directed mutagenesis enhance SM biosynthesis and mitigate yield constraints (Amom & Nongdam 2017; Mipeshwaree et al. 2023; Datta et al. 2024).

MAS has emerged as a prominent basis of modern plant breeding by integrating DNA marker–trait associations into the selection process. This approach begins with crossing parent plants that exhibit contrasting characteristics, followed by self-pollination to generate a diverse progeny. As shown in Figure 7.2, DNA is then extracted and subjected to PCR and agarose gel electrophoresis analysis to identify particular genetic markers associated with desired characteristics. Plants carrying these markers are selected and cultivated, assuring the inheritance of superior characteristics in subsequent generations. MAS not only streamlines breeding but also integrates genetic, morphological, and chemical analyses (Collard & Mackill 2008; Hasan et al. 2024).

Various molecular markers have been extensively utilized to distinguish medicinal plant species and evaluate genetic diversity. For instance, random amplified polymorphic DNA (RAPD) markers have been instrumental in identifying species such as *Allium schoenoprasum* (Khapilina et al. 2021), *Codonopsis pilosula* (Zhang et al. 2024), and *Andrographis paniculata* (Padmesh et al. 1999). Likewise, inter-simple sequence repeat (ISSR) markers have demonstrated their effectiveness in differentiating *Arabidopsis thaliana* (Barth et al. 2002). Additionally, restriction fragment length polymorphism (RFLP), along with RAPD techniques, has been employed to investigate both intra- and inter-species diversity in medicinal plants such as *Glycyrrhiza* and *Echinacea* (Ganie et al. 2015; Sorkheh et al. 2019). Moreover, AFLP and SNP markers have provided critical insights into genetic relationships and gene expression regulation in various species, further demonstrating the versatility of molecular markers in plant research (Cheng et al. 2002; Joshi et al. 2004; Bhargava & Ohri 2016; Sumathi et al. 2018). Integrating molecular markers with advanced genetic analysis techniques enhances the precision and efficiency of breeding programmes. This comprehensive

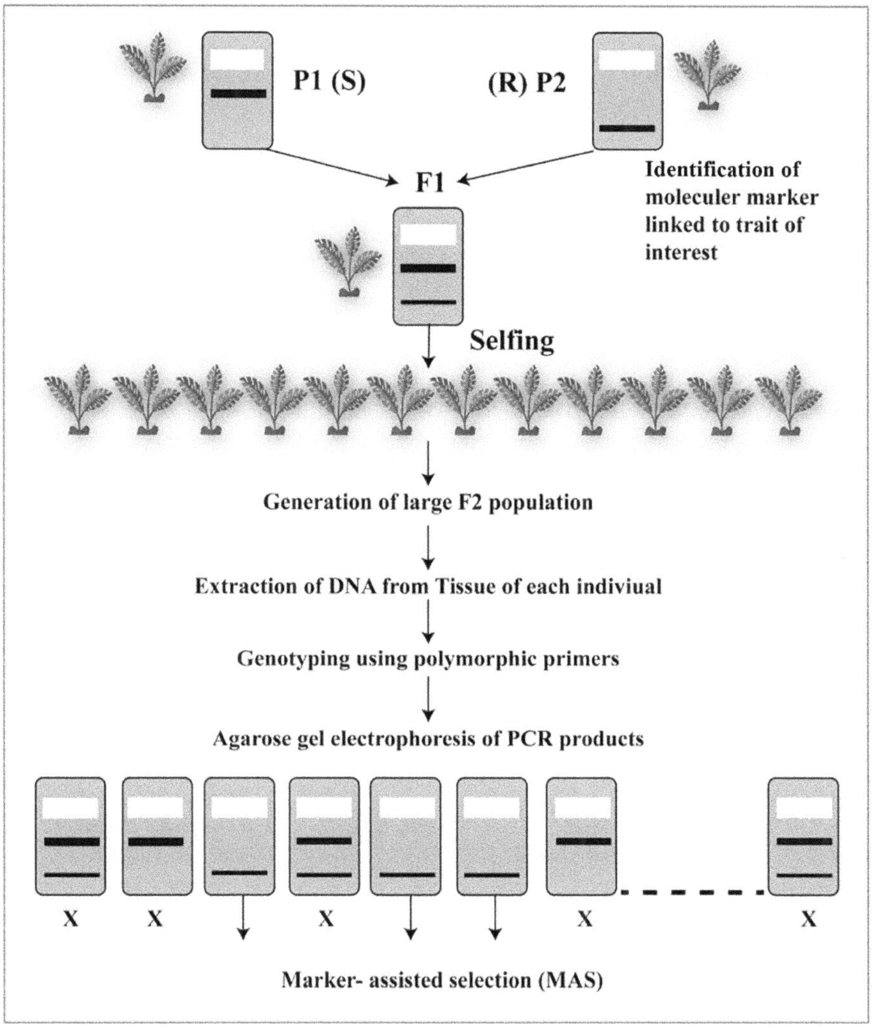

FIGURE 7.2 Schematic representation of marker-assisted selection (MAS) of particular trait in plant breeding programme.

approach facilitates the development of superior medicinal plant varieties with desired traits, paving the way for more targeted breeding strategies and improved SM production.

Five key factors must be considered when using DNA markers in MAS, as outlined below (Mackill & Ni 2001; Mohler & Singrün 2004).

- *Marker Reliability:* Markers must be closely linked (preferably <5 cm) to target genes for accurate phenotypic prediction. Using flanking or intragenic markers improves reliability.
- *DNA Requirements:* Sufficient high-quality DNA is needed, which can be challenging and costly to obtain.
- *Technical Procedure:* The marker assay should be simple, rapid, and high-throughput for efficient analysis.
- *Marker Polymorphism:* High polymorphism within the breeding population is crucial for effective genotype discrimination, particularly in core breeding materials.
- *Cost-Effectiveness:* The entire MAS process, including marker assay, must be economically viable.

TABLE 7.2
Molecular Marker Associated with Bioactive Compounds of Some Medicinal Plants and Their Metabolic Pathways

Medicinal Plants	Molecular Markers	Metabolic Pathway	Bioactive Compounds	Significance	Reference
Andrographis paniculata	RAPD, ISSR	Pathway involved in diterpenoid lactone synthesis	**Andrographolide**	Anti-inflammatory, hepatoprotective	Padmesh et al. 1999
Glycyrrhiza species (G. glabra, G. uralensis)	RFLP, RAPD, SSR	Triterpenoid saponin biosynthesis	**Glycyrrhizin**	Hepatoprotective, antiviral	Sorkheh et al. 2019
Withania somnifera	AFLP, ISSR, SNP	Steroidal lactone biosynthesis	**Withanolides**	Immunomodulatory, anticancer	Joshi et al. 2004
Hypericum perforatum	SSR, ISSR	Flavonoid and naphthodianthrone biosynthesis	**Hypericin**	Antidepressant, antiviral	Morshedloo et al. 2015
Salvia miltiorrhiza	RAPD, ISSR	Phenolic acid and diterpene biosynthesis	**Salvianolic acids, Tanshinones**	Cardiovascular health, antioxidant properties	Song et al. 2010
Rauvolfia serpentina (Indian Snakeroot)	RAPD, ISSR	Indole alkaloid biosynthesis	Reserpine	Antihypertensive, sedative	Faisal et al. 2012
Curcuma longa	SSR, ISSR, SNP	Curcuminoid biosynthesis	Curcumin	Anti-inflammatory, antioxidant, antimicrobial	Ayer 2017
Epimedium sagittatum	SSR, EST	flavonoid biosynthesis	Epimedin C	Cardiovascular protection, anticancerous, anti-inflammatory and antioxidant properties	Zeng et al. 2010

Table 7.2 represents detailed case studies that link molecular markers to specific metabolite pathways in various medicinal plants. By incorporating these cutting-edge approaches, researchers can develop superior medicinal plant varieties that ensure a sustainable supply of bioactive compounds, ultimately contributing to advancements in pharmaceuticals, healthcare, and biotechnology (Jiang 2013; Kowsalya et al. 2025).

7.4 GENOME EDITING

Genome editing is an advanced biotechnological method that enables precise modifications of specific genomic sequences (Chen & Gao 2014). This approach employs engineered endonucleases to recognize and cleave targeted DNA regions, creating single- or double-stranded breaks (DSBs; Zhang et al. 2018). These breaks are then repaired by the cell through non-homologous end joining (NHEJ), which can result in gene knockouts, insertions or deletions, or modifications, or by homology-directed repair (HDR), which allows for exact alterations (Tomar et al. 2019). Plants have been subjected to a variety of genetic manipulation approaches, including gene silencing, transgene overexpression, multiple transgene expression, and increased synthesis of bioactive natural compounds (Kaushik 2019). Of these, CRISPR-Cas9 has emerged as one of the most popular technologies.

7.4.1 CRISPR/Cas9

The CRISPR/Cas9 system is adapted from an indigenous defence mechanism in certain bacteria and archaea (Loureiro & da Silva 2019). In these organisms, fragments of invading pathogen DNA are incorporated into the CRISPR loci and transcribed into CRISPR RNA (crRNA), guiding the degradation of pathogen transcripts (Rodríguez et al. 2019). In this system, a single-guide RNA directs the Cas9 nuclease to the target gene loci, where it induces double-strand breaks. The subsequent repair process often leads to insertion or deletion mutations (Hille & Charpentier 2016), as shown in Figure 7.3 (Zhang et al. 2018; Tomar et al. 2019). Although inappropriate cleavage by Cas9 is a concern, this can be minimized by selecting highly specific target sites using tools such as CRISPR-Plant, GT-scan, and CRISPR direct (Minkenberg et al. 2018). Overall, CRISPR/Cas9 technology enables targeted modifications of genes involved in SM biosynthesis, allowing scientists to enhance or reduce the production of these compounds in plants. Table 7.3 shows case studies of gene editing in medicinal plants: target genes, affected SMs, and editing strategies.

7.4.2 CRISPR/Cas9 Editing in Important SMs

7.4.2.1 Phenolic Compounds

Polyphenols, including flavonoids and tannins, serve as antioxidants in plants. In *Salvia miltiorrhiza*, CRISPR/Cas9-mediated SmRAS mutation reduced phenolic acid content, increasing precursor

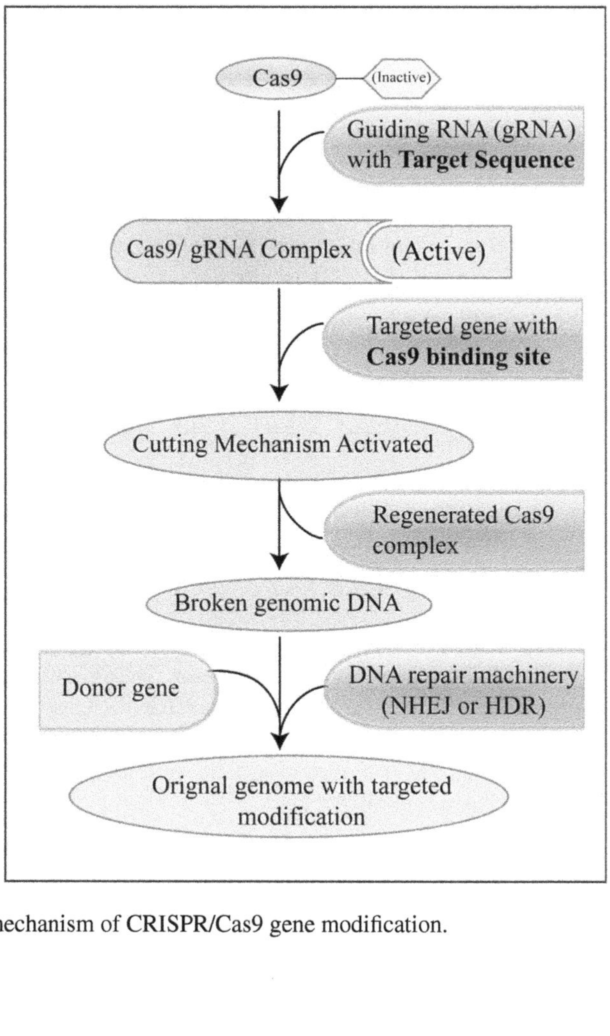

FIGURE 7.3 Basic mechanism of CRISPR/Cas9 gene modification.

TABLE 7.3
Recent Progress in CRISPR/Cas9 Genome-editing Tool, Targeted Genes, and Specific Editing Type for the Production of Secondary Metabolites in Some Medicinal Plants

Medicinal Plant	Secondary Metabolite(s) Affected	Target Gene(s)	Editing Type	Reference
Salvia miltiorrhiza	Rosmarinic acid, phenolic acids, diterpenoids	*SmRAS*	Knockout	Zhou et al. 2018
	Salvianolic acid B	*SmLAC*	Knockout	Zhou et al. 2021
	Tanshinones (diterpenes)	*SmCPS1*	Knockout	Luo et al. 2014
	Phenolic acids	*bZIP2*	Knockout	Shi et al. 2021
Papaver somniferum	Benzyl isoquinoline alkaloids (BIAs)	*4'OMT, 4'OMT2*	Knockout	Alagoz et al. 2016
Nicotiana tabacum	Nicotine	*BBL*	Knockout	Schachtsiek & Stehle 2019
Camellia sinensis	Caffeine	*CsHB1*	Directed mutagenesis	Ma et al. 2021
Symphytum officinale	Pyrrolizidine alkaloids	*HSS*	Directed mutagenesis	Zakaria et al. 2021
Atropa belladonna	Hyoscyamine	*AbH6H*	Targeted mutagenesis	Zeng et al. 2021
Dioscorea zingiberensis	Diosgenin	*Dzfps*	Targeted mutagenesis	Feng et al. 2018

levels (Zhou et al. 2018). Targeting bZIP2 in *Salvia miltiorrhiza* increased phenolic acid content, improving medicinal herb quality (Shi et al. 2021). Knocking out SmLAC genes reduced lignin and phenolic acid biosynthesis (Zhou et al. 2021).

7.4.2.2 Flavonoids

Flavonoids influence colour, aroma, and antioxidant properties in plants (Wang et al. 2018). CRISPR/Cas9-edited genes such as OsF3'H and OsTTG1 in rice enhanced anthocyanin biosynthesis (Jung et al. 2019; Yang et al. 2021). Editing PSY1, MYB12, and SRG1 in tomatoes produced diverse fruit colours (Yang et al. 2023). Phloridzin, crucial in apples, was reduced using CRISPR/Cas9 to investigate its role in growth and metabolism (Miranda et al. 2023). Editing FtMYB45 in Tartary buckwheat enhanced flavonoid content (Wen et al. 2022). Targeted mutagenesis of GmUGT in soybeans increased insect resistance (Zhang et al. 2022). Editing VvBZIP36 in grapevines resulted in anthocyanin accumulation (Tu et al. 2022). Simultaneous CRISPR/Cas9-mediated modification of multiple genes involved in isoflavonoid biosynthesis in soybeans significantly improved their resistance against soybean mosaic virus (Zhang et al. 2020).

7.4.2.3 Alkaloids

Benzyl isoquinoline alkaloids (BIAs) in *Papaver somniferum* were reduced by knocking out 4'OMT and 4'OMT2 genes (Alagoz et al. 2016). In *Nicotiana tabacum*, CRISPR/Cas9 editing of BBL genes generated nicotine-free plants (Schachtsiek & Stehle 2019). CRISPR/Cas9 knockout of SlS5aR2 in tomatoes decreased α-tomatine levels (Akiyama et al. 2019), while targeting StSSR2 in potatoes lowered SGA content (Zheng et al. 2021). Editing the HSS gene in *Symphytum officinale* reduced pyrrolizidine alkaloid toxicity (Zakaria et al. 2021).

7.4.2.4 Oils

CRISPR/Cas9 has improved oil composition in various crops. In *Camelina sativa*, targeting FAD2 genes increased oleic acid content by up to 60% while reducing linolenic acid levels (Jiang et al.

2017). Editing FAE1 genes altered fatty acid composition, reducing very long-chain fatty acids (Ozseyhan et al. 2018). In peanuts, modifying AhFAD2 increased oleic acid content, aiding breeding programmes (Yuan et al. 2019). CRISPR/Cas9 also enhanced oil quality in soybean by modifying fatty acid biosynthesis genes.

7.4.2.5 Terpenes

Saponins in *Medicago* species vary in bioactivity hemolytic sapogenins have pharmacological effects, while soyasapogenols do not (D'Addabbo et al. 2020). Genes such as *CYP93E2* and *CYP72A61* regulate soyasapogenol B biosynthesis. In *Medicago truncatula*, CYP93E2 was knocked out using CRISPR/Cas, which shifted production to useful haemolytic sapogenins (Confalonieri et al. 2021). In *Cichorium intybus* (chicory), inulin extraction requires removing bitter sesquiterpene lactones (STLs). CRISPR/Cas9 editing of *CiGAS*, the key STL biosynthesis enzyme, eliminated STL production (Cankar et al. 2021). In *Salvia miltiorrhiza*, tanshinones (bioactive diterpenes) were completely removed in homozygous *SmCPS1* CRISPR/Cas9 mutants without affecting phenolic acids. This demonstrates the usefulness of CRISPR/Cas9 in the study of SMs (Luo et al. 2014). These results demonstrate that CRISPR/Cas9 is an effective metabolic engineering tool that can be used to improve plant characteristics for use in medicine, nutrition, and agriculture.

7.5 PLANT TISSUE CULTURE

Plant tissue culture (PTC), initially conceptualized by Haberlandt in 1902 and significantly advanced by Murashige and Skoog in 1962, has become a pivotal tool in plant biotechnology. It provides controlled conditions for the development and modifications of plant cells, facilitating various applications in agriculture, pharmaceuticals, and conservation. This technique enables large-scale propagation of economically important crops and medicinal herbs, supports genetic screening for desirable traits, and aids in the preservation of rare and endangered plant species (Sehgal & Khan 2020; Erdem & Uysal 2021).

The fundamental idea behind PTC is totipotency, which is the capacity of a single cell to develop into an entire plant. This principle is exploited through various in vitro propagation techniques, including organogenesis and somatic embryogenesis, both of which depend on the regulation of plant growth regulators (Finer & Dhillon 2016). In vitro propagation typically involves several steps: selection of explants, establishment of callus or cell suspension cultures, regeneration of plant structures, shoot and root proliferation, and acclimatization (Aasim et al. 2019; Hussain et al. 2012; Pradhan et al. 2022).

One of the prominent uses of PTC lies in promoting the synthesis of SMs. These bioactive molecules, vital for plant survival, defence, and environmental adaptation, are typically produced under stress stimuli (Chandana et al. 2018; Erb & Kliebenstein 2020). To enhance their accumulation, both abiotic elicitors such as $AgNO_3$, ethyl alcohol, and methyl jasmonate as well as biotic elicitors including chitin, yeast extract, etc. have been effectively utilized (Kaur & Pati 2018; Piątczak et al. 2019). Additionally, bioreactors provide scalable and cost-effective platforms for the mass production of valuable compounds. Moreover, PTC serves as an essential platform for genetic engineering by providing an efficient means of introducing and propagating genetically modified traits. The integration of genetic engineering techniques like CRISPR-Cas9 with tissue culture enables precise genome modifications, enhancing SM production and improving plant stress resilience, establishing a strong foundation in modern biotechnology (Shahzad et al. 2017; Abahmane 2020). Furthermore, Agrobacterium-mediated hairy root culture using elicitors, bioreactors, artificial polyploidy, and CRISPR-Cas9 genome editing is an effective method for producing SMs on a large scale (Li et al. 2017; Supriya et al. 2020). Using tissue culture techniques, a large body of research shows that different SMs may be successfully produced from a variety of plant sources. The synthesis of these chemicals from different medicinal plants using these techniques is shown in Table 7.4.

TABLE 7.4
Production of Secondary Metabolites through Plant Tissue Culture in Different Medicinal Plants

Plant Species	Explant	Secondary Metabolites	Reference
Salvia bulleyana	Leaves and shoots	Phenolics including caffeic acid, rosmarinic acid and its hexoside, salvianolic acid, salvianolic acid K, salvianolic acid F isomer I & II, caffeic acid derivatives I and II, and methyl rosmarinate	Wojciechowska et al. 2020
Rehmannia elata	Seeds, leaves, and shoots	Iridoid glycosides (aucubin, harpagide, harpagoside) and phenylethanoid glycosides (verbascoside and isoverbascoside)	Piatczak et al. 2019
Linum flavum	Hypocotyls	Podophyllotoxin-related compounds (6-methoxy podophyllotoxin, podophyllotoxin, deoxypodophyllotoxin)	Renouard et al. 2018
Cullen corylifolium	Cotyledon callus cultures	Psoralen, daidzein, and genistein	Singh et al. 2020
Centella asiatica	Petioles and leaves	Triterpenoids such as madecassoside, asiaticoside, madecassic acid, and asiatic acid	Baek et al. 2020
Sphagneticola calendulacea	Nodes, internodes, and leaves	Phenolic acids (caffeic, syringic, p-coumaric, ferulic, salicylic) and flavonoids (rutin, myricetin, kaempferol)	Kundu et al. 2018
Psoralea drupacea	Cotyledon-derived callus, seeds, leaves, internodes, roots	Meroterpene bakuchiol	Lystvan et al. 2010
Pyrenacantha volubilis	Radicle-derived roots	Monoterpene-derived indole alkaloid camptothecin	Hima et al. 2020
Mentha spicata	Leaves and shoots	Phenolic acids including rosmarinic acid, caffeic acid, lithospermic acid, chlorogenic acid, and cinnamic acid	Yousefian et al. 2020
Carthamus tinctorius	Callus and suspension cultures	Flavonoids, aromatic glycosides	Liu et al. 2021
Artemisia arborescens	Nodal segments	Phenolics, tannins, and essential oils	Riahi et al. 2022

7.6 OMICS APPROACHES AND MARKER DEVELOPMENT

An integrated framework for comprehending gene function within intricate metabolic and physiological networks is offered by high-throughput omics technologies, such as transcriptomics, proteomics, metabolomics, and genomics (Trethewey 2001). Statistical tools like Principal Component Analysis (PCA) and correlation network analysis help predict novel gene functions based on observed metabolic variations (Rischer et al. 2006).

7.6.1 Unveiling Gene Functions through Metabolomics

Metabolomics serves as a crucial tool in functional genomics due to the smaller number of metabolites compared to genes or proteins and their direct physiological relevance (Trethewey 2001). However, the challenge lies in the absence of a direct gene-to-metabolite link. Raamsdonk et al. (2001) addressed this through metabolic compensation studies in silent mutants, showing that genes with similar functions exhibit similar metabolic alterations. This approach, combined with targeted metabolomic analysis in transgenic plants, has been crucial in validating enzyme roles (Guo et al. 2001). Comprehensive transcriptomic and metabolic analyses in *Arabidopsis* mutants have revealed the regulatory roles of transcription factors like Myb28 and Myb29 in controlling aliphatic glucosinolate formation. Moreover, BL-SOM clustering has been instrumental in uncovering previously unannotated sulfotransferase genes associated with glucosinolate production (Hirai et al. 2005). In addition, research on PAP1 overexpression in Arabidopsis has demonstrated its pivotal involvement in anthocyanin biosynthetic pathways (Tohge et al. 2005).

7.6.2 Discovering Novel Metabolic Pathways

Although tracer techniques have mapped major SM pathways, many crucial downstream steps remain unidentified (Hartmann 2007). This knowledge gap is particularly evident in terpenoid biosynthesis, where only upstream pathways are well-documented (Lichtenthaler 2000). Functional genomic tools are essential for accelerating the discovery of novel pathways. These tools not only identify genes involved in biosynthesis but also elucidate the origin and fate of metabolites. The integration of transcriptomic and metabolomic data in functional genomics enables the prediction of previously unknown metabolic steps (Tohge et al. 2005). Weckwerth and Fiehn (2002) highlighted the potential of metabolomics in identifying novel pathways. Gene-to-metabolite correlation networks derived from transcriptomic and metabolomic data can provide insights into the pathways of both known and previously uncharacterized compounds (Rischer et al. 2006). However, research in this area remains limited. Farag et al. (2008) identified a methylated isoflavone pathway in *Medicago truncatula* using metabolomics, while Keurentjes et al. (2006) demonstrated the efficacy of untargeted metabolomics via high-performance liquid chromatography quadrupole time-of-flight mass spectrometry (HPLC-QTOF MS) in Arabidopsis, detecting new metabolites and previously unknown biosynthetic pathways.

7.6.3 Deciphering Complex Metabolic Networks

Plant secondary metabolism involves intricate interactions among genes, transcription factors, enzymes, metabolites, and transporters (Yang et al. 2012). Changes at any level can induce widespread, nonlinear effects across the system. While transcriptomic studies have advanced gene-to-gene network analysis (Chang et al. 2019), integrating metabolomics enhances the accuracy of metabolic network inference (Savoi et al. 2022). Large-scale metabolite profiling allows systematic analysis of pathway interactions and condition-specific metabolic changes (Cakır et al. 2009; Volkova et al. 2020). Techniques such as BL-SOM (Batch Learning Self-Organizing Maps) efficiently cluster genes and metabolites to uncover functional relationships. As an advanced multivariate analysis approach, BL-SOM arranges metabolites and transcripts into cells on a two-dimensional lattice, where those exhibiting similar patterns are grouped into the same or neighbouring cells. This facilitates effective clustering and enhances pattern recognition (Mounet et al. 2009). Despite these advancements, further computational improvements are needed to fully unravel these complex networks (Cakır et al. 2009).

7.6.4 Bioinformatics in Omics Data Analysis

Integrating systems biology knowledge with conventional genetic engineering techniques streamlines efforts at the wet lab stage by eliminating redundant tasks. Utilizing bioinformatics tools allows for the efficient refinement of target genes or enzymes in metabolic engineering processes (Abdi et al. 2024). These analytical approaches are also useful for examining proteins that participate in secondary metabolic pathways. Secondary structure prediction aids in recognizing random coil regions, which play a crucial role in defining protein architecture. Furthermore, the developed model provides valuable theoretical insights, offering physicochemical and structural data for the functional assessment of target proteins. Additionally, biotechnological research can leverage these findings to manipulate biosynthetic pathways, enhancing SM production (Mehboob-ur-Rahman et al. 2016; Ma et al. 2018).

Tools like MetaboAnalyst, KaPPA-View, PRIMe, Babelomics, and Cytoscape support large-scale data visualization and network construction (Tokimatsu et al. 2005; Akiyama et al. 2008; Medina et al. 2010). Specialized platforms like AraCyc for *Arabidopsis* and CathaCyc for *Catharanthus roseus* provide curated metabolic pathway databases (Zhang et al. 2005; Van Moerkercke et al. 2013). These tools, combined with metabolic engineering strategies, enable precise pathway modifications to enhance SM biosynthesis for pharmaceutical and industrial applications.

7.7 CHALLENGES IN MARKER DEVELOPMENT AND GENE EDITING

The development of molecular markers along with the adoption of precise gene-editing tools holds immense promise for enhancing the synthesis of valuable SMs in plants. Nonetheless, multiple obstacles still need to be addressed to achieve this potential completely.

Current molecular marker technologies, while powerful, face limitations that hinder their widespread application in metabolic engineering. One major challenge is species specificity. The development of effective markers is often unique to each plant species, requiring significant investment and expertise. This specificity limits the transferability of MAS strategies across different crops. Another issue is reproducibility. Some marker systems, especially those based on PCR amplification (such as RAPDs and AFLPs), can show poor reproducibility (Jones et al. 1997) due to their reliance on arbitrary primer binding. This inconsistency hampers the reliability of results across different laboratories or experiments.

MAS may incur higher costs than conventional methods, especially regarding initial setup and labour expenses.

7.7.1 Challenges in Gene Editing

Off-target interaction is a common technological challenge in editing genomes. These unintended effects occur due to variations in the repeat variable dinucleotide (RVD) system's ability to accurately recognize and bind to particular target sequences. When off-target cleavage is significant, it may exceed the DNA's repair mechanisms, potentially resulting in cell or organism death (Carroll 2011). The challenges associated with genome editing across various systems are briefly outlined below.

- *Unintended Mutations and Gene Knockouts*
  CRISPR/Cas9-induced DSBs can be repaired through HR or NHEJ. While HR allows for precise gene replacement, NHEJ often introduces random nucleotide insertions or deletions, leading to frameshift mutations that may cause unintended gene knockouts. This unpredictability remains a major limitation in gene-editing applications (Weeks et al. 2016; Yan et al. 2020).
- *Promoter Selection and Gene Regulation*
  The efficacy of sgRNA expression is largely influenced by promoter selection, which differs among plant taxa. While endogenous U6 promoters have been successfully used in crops like grapes and cotton, their effectiveness is inconsistent across different species. Additionally, the deployment of tissue-specific promoters necessitates careful optimization to achieve precise and regulated expression of genes within designated plant tissues (Jiang et al. 2013; Feder et al. 2020; Ren et al. 2021).
- *Challenges in Editing Polyploid Genomes*
  Polyploid plants have complex genomes with multiple gene copies, making it difficult to introduce precise mutations. Achieving the desired genetic modification requires editing all alleles of a target gene, which is a labour-intensive and technically demanding process. The

presence of paralogous genes also complicates gene dosage effects, further limiting successful genome editing in polyploid species (Li et al. 2017; Schaart et al. 2021).

- *Low Transformation Success Rate and Delivery Barriers*
  Conventional gene delivery approaches like Agrobacterium-mediated transformation and particle bombardment often face challenges such as low regeneration rates of transgenic plants and inconsistent stable gene incorporation. Emerging strategies, including ribonucleoprotein (RNP) complex delivery and protoplast transfection, offer potential advantages by minimizing off-target modifications. However, these techniques are still costly, technically demanding, and limited in their applicability to specific plant species (Woo et al. 2015; Liang et al. 2018; Mao et al. 2019).
- *Off-Target Effects and Genome Instability*
  CRISPR/Cas9 can inadvertently bind to non-target genomic sites, leading to unintended chromosomal rearrangements such as deletions, inversions, and translocations. These off-target events may cause genotoxic effects that negatively impact plant growth and development. Strategies such as modified guide RNA (gRNA) designs and Cas9 variants are being explored to enhance target specificity and minimize unintended mutations (Brunet & Jasin 2018; Blattner et al. 2020; Yin et al. 2022).
- *Cas9 Variants and Targeting Limitations*
  The widely used SpCas9 enzyme is efficient at targeting GC-rich sequences but has difficulty editing AT-rich regions, leading to reduced editing efficiency. High-fidelity Cas9 variants such as eSpCas9(1.1), SpCas9-HF2, and HypaCas9 have been engineered to improve target specificity and reduce off-target effects. However, these variants often show lower editing efficiency, necessitating further optimization to balance precision and effectiveness (Kleinstiver et al. 2016; Chen et al. 2017; Xu et al. 2019).

7.8 FUTURE IMPLICATIONS

The future of medicinal plants holds great potential, with half a million plant species across the globe yet to be fully studied for their medicinal properties. These unexplored plants could contribute significantly to both present and future medical research. Throughout history, medicinal plants have been integral to human civilization, including their role in religious and ceremonial traditions. In the coming years, genetic engineering combined with mutation techniques will become a vital tool for plant breeders. Many plant species have traditionally been developed by combining selective breeding and mutation. New opportunities for using mutant breeding in crop improvement have come from recent advancements in in vitro cultivation and molecular technologies.

Among recent technological innovations, heavy ionizing radiation has emerged as a potent mutagenic tool due to its ability to induce a broad range of genetic variations and achieve high mutation frequencies (Abe et al. 2021). The Food and Agriculture Organization (FAO) has officially documented 3,248 mutant cultivars, underscoring the profound contribution of induced mutagenesis to crop improvement programmes (Kharkwal 2023). Furthermore, in vitro culture techniques allow the precise treatment of small plant tissues and callus with mutagens, thereby facilitating the genetic enhancement of crop species (Xu et al. 2022). Continued progress in in vitro cell selection methods aimed at improving disease resistance holds substantial promise. Additionally, molecular marker technologies are anticipated to become increasingly pivotal in advancing research on medicinal plants. They enable precise genetic identification, facilitate the assessment of genetic diversity, and accelerate the selection of desirable traits, such as enhanced medicinal properties, disease resistance, and improved yield. The integration of molecular markers with genome-editing techniques will further refine breeding strategies, leading to more efficient and targeted improvements in medicinal plants (Ma et al. 2018; Rahman et al. 2016).

PTC represents a cornerstone of modern biotechnology-driven breeding strategies, serving a crucial function in the preservation and large-scale propagation of medicinal and aromatic plant species. Strategies such as artificial polyploidy induction (Supriya et al. 2020) and the use of bioreactors (Li et al. 2017) are critical in improving these plants. An equally significant strategy involves Agrobacterium-mediated genetic transformation, which facilitates the improvement of medicinal plants by upregulating genes involved in SM biosynthesis and suppressing those linked to unwanted compound formation. Moreover, approaches such as hairy root culture establishment and the introduction of genome-editing systems like CRISPR-Cas9, TALENs, and ZFNs provide additional avenues for genetic enhancement of medicinal and aromatic species. Among these, CRISPR-Cas9 has emerged as a highly efficient and precise technology for generating targeted mutations and fine-tuning the biochemical profiles of plants (Feng et al. 2018; Zhou et al. 2018). To ensure the quality and consistency of herbal products, Good Agricultural Practices (GAP) (Chan et al. 2012) now govern the global cultivation of medicinal plants.

7.9 CONCLUSION

The emergence of advanced molecular marker systems, precise genome-editing techniques, and PTC methodologies has transformed research and biotechnological applications related to SM production in medicinal plants. By identifying key genetic factors associated with SM biosynthesis, molecular markers enable precise trait selection and accelerate breeding programmes, ensuring higher yields and quality of bioactive compounds. Additionally, CRISPR-Cas9 and other genome-editing technologies offer precise modifications to metabolic pathways, optimizing the production of SMs. PTC further enhances SM production by providing a controlled environment for elicitor treatments, genetic modifications, and large-scale propagation. Integrating multi-omics approaches, bioinformatics, and computational modelling facilitates deeper insights into metabolic pathways, allowing for targeted genetic interventions to improve medicinal plant varieties sustainably. Despite these advancements, challenges such as low transformation efficiency, off-target gene-editing effects, and species-specific marker limitations persist. Addressing these challenges through improved bioinformatics tools, high-throughput screening, and synthetic biology will be crucial for future research. With the increasing global need for bioactive molecules sourced from plants, the integration of molecular marker analysis, genome-editing technologies, and tissue culture approaches offers a powerful route to strengthen the therapeutic and commercial value of medicinal species. Advancements in metabolic engineering and synthetic biology are anticipated to open new avenues for the sustainable, large-scale synthesis of valuable phytochemicals, maintaining their vital role across pharmaceutical, nutraceutical, and biotechnological industries.

ACKNOWLEDGEMENTS

AJ is thankful to the University Grants Commission, New Delhi, for providing the Joint CSIR-UGC Fellowship. RS acknowledges the Science and Engineering Research Board (SERB), New Delhi, for the financial support with file number: EEQ/2021/000356. We also acknowledge IOE, Institute of Science, Banaras Hindu University, for financial assistance.

REFERENCES

Aasim, M. Sameeullah, M. Karataş, M. Bakirci, S. Bakhsh, A. & Akhtar, M. S. (2019). An insight into biotechnological approaches used for the improvement of SMs from the medicinal aquatic plant, water hyssop (*Bacopa monnieri* L.). In Akhtar, M. S. & Swamy, M. K. eds. *Natural bio-active compounds: volume 3: biotechnology, bioengineering, and molecular approaches* (pp. 123–152). Singapore: Springer Singapore. https://doi.org/10.1007/978-981-13-7438-8_5

Abahmane, L. (2020). A comparative study between temporary immersion system and semi-solid cultures on shoot multiplication and plantlets production of two Moroccan date palm (*Phoenix dactylifera* l.) varieties in vitro. *Notulae Scientia Biologicae*, *12*(2), 277–288.

Abbas, F. O'Neill Rothenberg, D. Zhou, Y. Ke, Y. & Wang, H. C. (2022). Volatile organic compounds as mediators of plant communication and adaptation to climate change. *Physiologia Plantarum, 174*(6), e13840.

Abdi, G. Jain, M. Barwant, M. Tendulkar, R. Tendulkar, M. Tariq, M. & Amir, A. (2024). Unveiling the dynamic role of bioinformatics in automation for efficient and accurate data processing and interpretation. In Singh, V. & Kumar, A. eds. *Advances in bioinformatics* (pp. 279–319). Singapore: Springer Nature Singapore.□□

Abe, T. Ichida, H. Hayashi, Y. Morita, R. Shirakawa, Y. Ishii, K. ... & Okumoto, Y. (2021). ("Distinctive development of embryo and endosperm caused by ... – Springer") Ion beam mutagenesis-an innovative and effective method for plant breeding and gene discovery. In Sivasankar, S. Ellis, N. Jankuloski, L. & Ingelbrecht, I. eds. *Mutation breeding, genetic diversity and crop adaptation to climate change* (pp. 411–423). Wallingford: CAB International.

Agarwal, M., Shrivastava, N., & Padh, H. (2008). Advances in molecular marker techniques and their applications in plant sciences. *Plant Cell Reports, 27*(4), 617–631.

Ahmar, S. Gill, R. A. Jung, K. H. Faheem, A. Qasim, M. U. Mubeen, M. & Zhou, W. (2020). Conventional and molecular techniques from simple breeding to speed breeding in crop plants: recent advances and future outlook. *International Journal of Molecular Sciences, 21*(7), 2590.

Akiyama, K. Chikayama, E. Yuasa, H. Shimada, Y. Tohge, T. Shinozaki, K. ... & Saito, K. (2008). PRIMe: a Web site that assembles tools for metabolomics and transcriptomics. *In Silico Biology*, 8(3–4), 339–345.

Akiyama, R. Lee, H. J. Nakayasu, M. Osakabe, K. Osakabe, Y. Umemoto, N. ... & Mizutani, M. (2019). Characterization of steroid 5α-reductase involved in α-tomatine biosynthesis in tomatoes. *Plant Biotechnology, 36*(4), 253–263.

Alagoz, Y. Gurkok, T. Zhang, B. & Unver, T. (2016). Manipulating the biosynthesis of bioactive compound alkaloids for next-generation metabolic engineering in opium poppy using CRISPR-Cas 9 genome editing technology. *Scientific Reports, 6*(1), 30910.

Alam, M. A. Subhan, N. Hossain, H. Hossain, M. Reza, H. M. Rahman, M. M. & Ullah, M. O. (2016). Hydroxycinnamic acid derivatives: a potential class of natural compounds for the management of lipid metabolism and obesity. *Nutrition & Metabolism, 13*, 1–13.

Altpeter, F. Springer, N. M. Bartley, L. E. Blechl, A. E. Brutnell, T. P. Citovsky, V. ... & Stewart Jr, C. N. (2016). Advancing crop transformation in the era of genome editing. *The Plant Cell, 28*(7), 1510–1520.

Amom, T. & Nongdam, P. (2017). The use of molecular marker methods in plants: a review. *International Journal of Current Research and Review, 9*(17), 1–7.

Ayer, D. K. (2017). Breeding for quality improvement in turmeric (Curcuma longa L.): A review. *Advances in Plants & Agriculture Research, 6*(6), 201–204.

Baek, S. Ho, T.T. Lee, H. Jung, G. Kim, Y. E. Jeong, C. S. & Park, S. Y. (2020). Enhanced biosynthesis of triterpenoids in *Centella asiatica* hairy root culture by precursor feeding and elicitation. *Plant Biotechnology Reports, 14*, 45–53.

Bais, H. P. Walker, T. S. Stermitz, F. R. Hufbauer, R. A. & Vivanco, J. M. (2002). Enantiomeric-dependent phytotoxic and antimicrobial activity of (±)-catechin. A rhizo secreted racemic mixture from spotted knapweed. *Plant Physiology, 128*(4), 1173–1179.

Balunas, M. J. & Kinghorn, A. D. (2005). Drug discovery from medicinal plants. *Life Sciences, 78*(5), 431–441.

Barth, S. Melchinger, A. E. & Lübberstedt, T. H. (2002). Genetic diversity in *Arabidopsis thaliana* L. Heynh. investigated by cleaved amplified polymorphic sequence (CAPS) and inter-simple sequence repeat (ISSR) markers. *Molecular Ecology, 11*(3), 495–505.

Bhargava, A. & Ohri, D. (2016). Origin of genetic variability and improvement of quinoa (*Chenopodium quinoa* Willd.). *Gene Pool Diversity and Crop Improvement, 1*, 241–270.

Bjørklund, G. Shanaida, M. Lysiuk, R. Butnariu, M. Peana, M. Sarac, I. ... & Chirumbolo, S. (2022). Natural compounds and products from an anti-aging perspective. *Molecules, 27*(20), 7084.

Blattner, G. Cavazza, A. Thrasher, A. J. & Turchiano, G. (2020). Gene editing and genotoxicity: targeting the off-targets. *Frontiers in Genome Editing, 2*, 613252.

Bomzan, D. P. Shilpashree, H. B. & Nagegowda, D. A. (2022). Agrobacterium-mediated in planta transformation in periwinkle. In Courdavault, V. & Besseau, S. eds. *Catharanthus roseus: methods and protocols* (pp. 301–315). New York: Springer US.

Brunet, E. & Jasin, M. (2018). Induction of chromosomal translocations with CRISPR-Cas9 and other nucleases: understanding the repair mechanisms that give rise to translocations. *Chromosome Translocation, 1044*, 15–25.

Cakır, T. Hendriks, M. M. Westerhuis, J. A. & Smilde, A. K. (2009). Metabolic network discovery through reverse engineering of metabolome data. *Metabolomics*, *5*, 318–329.

Cankar, K. Bundock, P. Sevenier, R. Häkkinen, S. T. Hakkert, J. C. Beekwilder, J. ... & Bosch, D. (2021). Inactivation of the germacrene A synthase genes by CRISPR/Cas9 eliminates the biosynthesis of sesquiterpene lactones in *Cichorium intybus* L. *Plant Biotechnology Journal*, *19*(12), 2442–2453.

Cannell, R. J. (1998). How to approach the isolation of a natural product. In Cannell, R. J. P. eds. *Natural products isolation* (pp. 1–51). Totowa, NJ: Humana Press.

Carroll, D. (2011). Genome engineering with zinc-finger nucleases. *Genetics*, *188*(4), 773–782.

Chan, K. Shaw, D. Simmonds, M. S. Leon, C. J. Xu, Q. Lu, A. ... & Duez, P. (2012). Good practice in reviewing and publishing studies on herbal medicine, with special emphasis on traditional Chinese medicine and Chinese materia medica. *Journal of Ethnopharmacology*, *140*(3), 469–475.

Chandana, B. C. Kumari Nagaveni, H. C. Lakshmana, D. Shashikala, S. K. & Heena, M. S. (2018). Role of PTC in micropropagation, SMs production and conservation of some endangered medicinal crops. *Journal of Pharmacognosy and Phytochemistry*, *3*, 246–251.

Chang, Y. M. Lin, H. H. Liu, W. Y. Yu, C. P. Chen, H. J. Wartini, P. P. ... & Li, W. H. (2019). "Comparative transcriptomics method to infer gene coexpression networks and its applications to maize and rice leaf transcriptomes." ("Comparative transcriptomics method to infer gene coexpression networks ...") *Proceedings of the National Academy of Sciences of the United States America*, *116*(8), 3091–3099.

Chen, J. S. Dagdas, Y. S. Kleinstiver, B. P. Welch, M. M. Sousa, A. A. Harrington, L. B. ... & Doudna, J. A. (2017). Enhanced proofreading governs CRISPR–Cas9 targeting accuracy. *Nature*, *550*(7676), 407–410.

Chen, K. & Gao, C. (2014). Targeted genome modification technologies and their applications in crop improvements. *Plant Cell Reports*, *33*, 575–583.

Chen, X. Cromer, B. A. & Lynch, J. W. (2009). "Molecular determinants of β-carboline inhibition of the glycine receptor." ("Role of SMs in plant defense against pathogens") *Journal of Neurochemistry*, *110*(5), 1685–1694.

Cheng, Z. Sameshima, K. & Chen, J. K. (2002). Genetic variability and relationship of kenaf germplasm based on RAPD analysis. *China's Fiber Crops*, *24*(1), 1–11.

Chukwu, S. C. Rafii, M. Y. Ramlee, S. I. Ismail, S. I. Oladosu, Y. Okporie, E. ... & Jalloh, M. (2019). Marker-assisted selection and gene pyramiding for resistance to bacterial leaf blight disease of rice (*Oryza sativa* L.). *Biotechnology & Biotechnological Equipment*, *33*(1), 440–455.

Coates, D. J. Byrne, M. & Moritz, C. (2018). "Genetic diversity and conservation units: dealing with the species-population continuum in the age of genomics. *Frontiers in Ecology and Evolution*, *6*, 165.

Collard, B. C. & Mackill, D. J. (2008). Marker-assisted selection: an approach for precision plant breeding in the twenty-first century. *Philosophical Transactions of the Royal Society B: Biological Sciences*, *363*(1491), 557–572.

Confalonieri, M. Carelli, M. Gianoglio, S. Moglia, A. Biazzi, E. & Tava, A. (2021). CRISPR/Cas9-mediated targeted mutagenesis of CYP93E2 modulates the triterpene saponin biosynthesis in *Medicago truncatula*. *Frontiers in Plant Science*, *12*, 690231.

D'Addabbo, T. Argentieri, M. P. Żuchowski, J. Biazzi, E. Tava, A. Oleszek, W. & Avato, P. (2020). Activity of saponins from Medicago species against phytoparasitic nematodes. *Plants*, *9*(4), 443.

Datta, R., Kapadia, C., Danish, S. & Singh, S. eds. 2024. *Non-coding RNA in plants: modulation and stress responses*. Amsterdam: Elsevier.

Debnath, B. Singh, W. S. Das, M. Goswami, S. Singh, M. K. Maiti, D. & Manna, K. (2018). Role of plant alkaloids on human health: a review of biological activities. *Materials Today Chemistry*, *9*, 56–72.

Divekar, P. A. Narayana, S. Divekar, B. A. Kumar, R. Gadratagi, B. G. Ray, A. ... & Behera, T. K. (2022). Plant SMs as defense tools against herbivores for sustainable crop protection. *International journal of molecular sciences*, *23*(5), 2690.

Elshafie, H. S. Camele, I. & Mohamed, A. A. (2023). A comprehensive review on the biological, agricultural and pharmaceutical properties of SMs based-plant origin. *International Journal of Molecular Sciences*, *24*(4), 3266.

Erb, M. & Kliebenstein, D. J. (2020). Plant SMs as defenses, regulators, and primary metabolites: the blurred functional trichotomy. *Plant Physiology*, *184*(1), 39–52.

Erdem, M. & Uysal, H. (2021). Sentetik tohum. *Frontiers in Life Sciences and Related Technologies*, *2*(2), 68–74.

Farag, M. A. Huhman, D. V. Dixon, R. A. & Sumner, L. W. (2008). Metabolomics reveals novel pathways and differential mechanistic and elicitor-specific responses in phenylpropanoid and isoflavonoid biosynthesis in *Medicago truncatula* cell cultures. *Plant Physiology*, *146*(2), 387.

Faisal, M., Alatar, A. A., Ahmad, N., Anis, M., & Hegazy, A. K. (2012). An efficient and reproducible method for in vitro clonal multiplication of *Rauvolfia tetraphylla* L. and evaluation of genetic stability using DNA-based markers. *Applied Biochemistry and Biotechnology, 168*(7), 1739–1752.

Feder, A. Jensen, S. Wang, A. Courtney, L. Middleton, L. Van Eck, J. ... & Giovannoni, J. J. (2020). Tomato fruit as a model for tissue-specific gene silencing in crop plants. *Horticulture Research, 7*, 142.

Feng, S. Song, W. Fu, R. Zhang, H. Xu, A. & Li, J. (2018). Application of the CRISPR/Cas9 system in *Dioscorea zingiberensis. Plant Cell, Tissue and Organ Culture (PCTOC), 135*, 133–141.

Finer, J. & Dhillon, T. (2016). Transgenic plant production. *Plant biotechnology genetics: principles techniques application* (pp. 245–274).

Ganie, S. H. Upadhyay, P. Das, S. & Sharma, M. P. (2015). Authentication of medicinal plants by DNA markers. *Plant Gene, 4*, 83–99.

Goldberg G. 2003. *Plants: diet and health: the report of a British Nutrition Foundation task force.* Oxford: Blackwell Publishing Ltd.

Goodwin, S. McPherson, J. D. & McCombie, W. R. (2016). Coming of age: ten years of next-generation sequencing technologies. *Nature Reviews Genetics, 17*(6), 333–351.

Grimble, R. F. (2003). Nutritional therapy for cancer cachexia. *Gut, 52*(10), 1391–1392.

Guo, D. Chen, F. Inoue, K. Blount, J. W. & Dixon, R. A. (2001).Downregulation of caffeic acid 3-O-methyltransferase and caffeoyl CoA 3-O-methyltransferase in transgenic alfalfa: impacts on lignin structure and implications for the biosynthesis of G and S lignin. *The Plant Cell, 13*(1), 73–88.

Harborne, B. (1988). *Introduction to ecological biochemistry*, third ed. New York: Academic Press.

Harborne, J. B. (1993 [1986]). *The flavonoids: advances in research since*. London: Chapman & Hall, London.

Hartmann, T. (2007). From waste products to ecochemicals: fifty years research of plant secondary metabolism. *Phytochemistry, 68*(22-24), 2831–2846.

Hasan, N. Laskar, R. A. Farooqui, S. A. Naaz, N. Sharma, N. Budakoti, M. ... & Bhinda, M. S. (2024). Genetic improvement of medicinal and aromatic plant species: breeding techniques, conservative practices and future prospects. *Crop Design, 3*, 100080.

Helmy, M. Smith, D. & Selvarajoo, K. (2020). Systems biology approaches integrated with artificial intelligence for optimized metabolic engineering. *Metabolic Engineering Communications, 11*, e00149.

Hille, F. & Charpentier, E. (2016). CRISPR-Cas: biology, mechanisms and relevance. *Philosophical Transactions of the Royal Society B: Biological Sciences, 371*(1707), 20150496.

Hima, S., Midhu, C. K., Krishnakumar, G., & Satheeshkumar, K. (2020). In vitro seedlings as dynamic explants for establishment of root cultures of *Pyrenacantha volubilis* hook. for camptothecin production. *Proceedings of the National Academy of Sciences, India Section B: Biological Sciences, 90*(2), 405–413.

Hirai, M. Y. Klein, M. Fujikawa, Y. Yano, M. Goodenowe, D. B. Yamazaki, Y. ... & Saito, K. (2005). Elucidation of gene-to-gene and metabolite-to-gene networks in Arabidopsis by integration of metabolomics and transcriptomics. *Journal of Biological Chemistry, 280*(27), 25590–25595.

Holland, J. B. (2004, September). Implementation of molecular markers for quantitative traits in breeding programs—challenges and opportunities. In *Proceedings of the 4th international crop science congress* (Vol. 26, pp. 1–13). Gosford, Australia: Regional Institute.

Hussain, M. S. Fareed, S. Ansari, S. Rahman, M. A. Ahmad, I. Z. & Saeed, M. (2012). Current approaches toward production of secondary plant metabolites. *Journal of Pharmacy and Bioallied Sciences, 4*(1), 10–20.

Hussein, R. A. & El-Anssary, A. A. (2019). Plants SMs: the key drivers of the pharmacological actions of medicinal plants. *Herbal Medicine, 1*(3), 11–30.

Isah, T. (2019). Stress and defense responses in plant SMs production. *Biological Research, 52*, 39.

Jiang, G. L. (2013). Molecular markers and marker-assisted breeding in plants. *Plant Breeding from Laboratories to Fields, 3*, 45–83.

Jiang, W. Bikard, D. Cox, D. Zhang, F. & Marraffini, L. A. (2013). CRISPR-assisted editing of bacterial genomes. *Nature Biotechnology, 31*(3), 233.

Jiang, W. Z. Henry, I. M. Lynagh, P. G. Comai, L. Cahoon, E. B. & Weeks, D. P. (2017). Significant enhancement of fatty acid composition in seeds of the allohexaploid, *Camelina sativa*, using CRISPR/Cas9 gene editing. *Plant Biotechnology Journal, 15*(5), 648–657.

Jones, C. J. Edwards, K. J. Castaglione, S. Winfield, M. O. Sala, F. Van de Wiel, C. ... & Karp, A. (1997). Reproducibility testing of RAPD, AFLP and SSR markers in plants by a network of European laboratories. *Molecular Breeding, 3*, 381–390.

Joshi, K. Chavan, P. Warude, D. & Patwardhan, B. (2004). Molecular markers in herbal drug technology. *Current Science*, *87*, 159–165.

Joshi, S. P. Ranjekar, P. K. & Gupta, V. S. (1999). Molecular markers in plant genome analysis. *Current Science*, *77*, 230–240.

Jung, Y. J. Lee, H. J. Kim, J. H. Kim, D. H. Kim, H. K. Cho, Y. G. ... & Kang, K. K. (2019). CRISPR/Cas9-targeted mutagenesis of F3′ H, DFR and LDOX, genes related to anthocyanin biosynthesis in black rice (*Oryza sativa* L.). *Plant Biotechnology Reports*, *13*, 521–531.

Kaur, K. & Pati, P. K. (2018). Stress-induced metabolite production utilizing plant hairy roots. In: Srivastava, V. Mehrotra, S. & Mishra, S. eds. *Hairy roots: an effective tool of plant biotechnology* (pp. 123–145). Singapore: Springer.

Kaushik, P. (2019). *Application of conventional, biotechnological and genomics approaches for eggplant (solanum melongena. L). Breeding with a focus on bioactive phenolics* (Doctoral dissertation, Universitat Politècnica de València).

Keurentjes, J. J. Fu, J. De Vos, C. R. Lommen, A. Hall, R. D. Bino, R. J. ... & Koornneef, M. (2006). The genetics of plant metabolism. *Nature Genetics*, *38*(7), 842–849.

Khapilina, O. Turzhanova, A. Danilova, A. Tumenbayeva, A. Shevtsov, V. Kotukhov, Y. & Kalendar, R. (2021). Primer binding site (PBS) profiling of genetic diversity of natural populations of endemic species *Allium ledebourianum* schult. *BioTech*, *10*(4), 23.

Kharkwal, M. C. (2023). Role of mutation breeding in crop improvement with special reference to Indian subcontinent. In Penna, S. & Mohan Jain, S. eds. *Mutation breeding for sustainable food production and climate resilience* (pp. 355–428). Singapore: Springer Nature Singapore.

Kleinstiver, B. P. Pattanayak, V. Prew, M. S. Tsai, S. Q. Nguyen, N. T. Zheng, Z. & Joung, J. K. (2016). High-fidelity CRISPR–Cas9 nucleases with no detectable genome-wide off-target effects. *Nature*, *529*(7587), 490–495.

Kowsalya, K. Halka, J. Ramalashmi, K. Anand, A. V. Razia, M. Anand, M. ... & Arun, M. (2025). Molecular marker-assisted breeding in medicinal plants for enhanced SMs production. In Chen, J-S. ed. *Biotechnology, multiple omics, and precision breeding in medicinal plants* (pp. 160–176). Boca Raton: CRC Press.

Kumar, R. Das, S. P. Choudhury, B. U. Kumar, A. Prakash, N. R. Verma, R. ... & Mishra, V. K. (2024). Advances in genomic tools for plant breeding: harnessing DNA molecular markers, genomic selection, and genome editing. *Biological Research*, *57*(1), 80.

Kundu, S. Salma, U. Ali, M. N. Hazra, A. K. & Mandal, N. (2018). Development of transgenic hairy roots and augmentation of SMs by precursor feeding in *Sphagneticola calendulacea* (L.) Pruski. *Industrial Crops and Products*, *121*, 206–215.

Kuzay, S. Xu, Y. Zhang, J. Katz, A. Pearce, S. Su, Z. ... & Dubcovsky, J. (2019). Identification of a candidate gene for a QTL for spikelet number per spike on wheat chromosome arm 7AL by high-resolution genetic mapping. *Theoretical and Applied Genetics*, *132*, 2689–2705.

Landry, B. S. Kesseli, R. V. Farrara, B. & Michelmore, R. W. (1987). A genetic map of lettuce (*Lactuca sativa* L.) with restriction fragment length polymorphism, isozyme, disease resistance and morphological markers. *Genetics*, *116*(2), 331–337.

Lee, M. H. Jeon, H. S. Kim, S. H. Chung, J. H. Roppolo, D. Lee, H. J. ... & Park, O. K. (2019). Lignin-based barrier restricts pathogens to the infection site and confers resistance in plants. *The EMBO Journal*, *38*(23), e101948.

Li, B. Cui, G. Shen, G. Zhan, Z. Huang, L. Chen, J. & Qi, X. (2017). Targeted mutagenesis in the medicinal plant *Salvia miltiorrhiza*. *Scientific Reports*, *7*(1), 43320.

Li, X. Duan, X. Jiang, H. Sun, Y. Tang, Y. Yuan, Z. ... & Zhang, D. (2006). Genome-wide analysis of basic/helix-loop-helix transcription factor family in rice and Arabidopsis. *Plant Physiology*, *141*(4), 1167–1184.

Liang, Z. Chen, K. Zhang, Y. Liu, J. Yin, K. Qiu, J. L. & Gao, C. (2018). Genome editing of bread wheat using biolistic delivery of CRISPR/Cas9 in vitro transcripts or ribonucleoproteins. *Nature Protocols*, *13*(3), 413–430.

Liao, Z. Zhou, Z. Li, Y. & Zhang, Y. (2023). Plant metabolism and synthetic biology. *Synthetic and Systems Biotechnology*, *8*(3), 563–576.

Lichtenthaler, H. K. (2000). Non-mevalonate isoprenoid biosynthesis: enzymes, genes and inhibitors. *Biochemical Society Transactions*, *28*(6), 785–789.

Liu, N. Wu, X. Y. SONG, Y. D. GAO, W. & HUANG, L. Q. (2021). Tissue culture of safflower and analysis of SMs in suspension cells. *China Journal of Chinese Materia Medica*, *46*, 4380–4388.

Loureiro, A. & da Silva, G. J. (2019). CRISPR-Cas: converting a bacterial defence mechanism into a state-of-the-art genetic manipulation tool. *Antibiotics*, *8*(1), 18.

Luo, H. Zhu, Y. Song, J. Xu, L. Sun, C. Zhang, X. ... & Chen, S. (2014). Transcriptional data mining of *Salvia miltiorrhiza* in response to methyl jasmonate to examine the mechanism of bioactive compound biosynthesis and regulation. *Physiologia Plantarum*, *152*(2), 241–255.

Lystvan, K. Belokurova, V. Sheludko, Y. Ingham, J. L. Prykhodko, V. Kishchenko, O. ... & Kuchuk, M. (2010). Production of bakuchiol by in vitro systems of Psoralea drupacea Bge. *Plant Cell, Tissue and Organ Culture (PCTOC)*, *101*, 99–103.

Ma, W. Kang, X. Liu, P. Zhang, Y. Lin, X. Li, B. & Chen, Z. (2021). The analysis of transcription factor CsHB1 effects on caffeine accumulation in tea callus through CRISPR/Cas9 mediated gene editing. *Process Biochemistry*, *101*, 304–311.

Ma, Y. Liu, Y. & Cheng, J. (2018). Protein secondary structure prediction based on data partition and semi-random subspace method. *Scientific Reports*, *8*(1), 9856.

Mackill, D. J. & Ni, J. (2001). Molecular mapping and marker-assisted selection for major-gene traits in rice. In Khush, G.S., Brar, D.S. & Hardy, B. eds. *Rice genetics IV* (pp. 137–151). Enfield, NH, USA: Science Publishers, Inc. Printed in India.

Mao, Y. Botella, J. R. Liu, Y. & Zhu, J. K. (2019). Gene editing in plants: progress and challenges. *National Science Review*, *6*(3), 421–437.

Mazid, M., Khan, T. A., & Mohammad, F. (2011). Role of secondary metabolites in defense mechanisms of plants. *Biology and Medicine Journal*, *3*(2), 232–249. Aligarh, India: Biology and Medicine Journal.

Medina, I. Carbonell, J. ½. Pulido, L. Madeira, S. C. Goetz, S. Conesa, A. ... & Dopazo, J. ½. (2010). Babelomics: an integrative platform for the analysis of transcriptomics, proteomics and genomic data with advanced functional profiling. *Nucleic Acids Research*, *38*(suppl_2), W210–W213.

Mehboob-ur-Rahman, T. S. Mahmood-ur-Rahman, M. A. I. & Zafar, Y. (2016). Bioinformatics: a way forward to explore "plant omics". In Abdurakhmonov, I. Y. ed. *Bioinformatics-updated features and applications* (p. 203). Croatia: Intech.

Minkenberg, B. Zhang, J. Xie, K. & Yang, Y. (2018). CRISPR-PLANT v2: an online resource for highly specific guide RNA spacers based on improved off-target analysis. *Plant Biotechnology Journal*, *17*(1), 5.

Mipeshwaree Devi, A. Khedashwori Devi, K. Premi Devi, P. Lakshmipriyari Devi, M. & Das, S. (2023). Metabolic engineering of plant SMs: prospects and its technological challenges. *Frontiers in Plant Science*, *14*, 1171154.

Miranda, S. Piazza, S. Nuzzo, F. Li, M. Lagrèze, J. Mithöfer, A. ... & Martens, S. (2023). CRISPR/Cas9 genome-editing applied to MdPGT1 in apple results in reduced foliar phloridzin without impacting plant growth. *The Plant Journal*, *113*(1), 92–105.

Mirzaei, S. (2021). Application of molecular markers in plant sciences; An overview. *Central Asian Journal of Plant Science Innovation*, *1*(4), 192–200.

Mobasheri, A. Bay-Jensen, A. C. Van Spil, W. E. Larkin, J. & Levesque, M. C. (2017). Osteoarthritis Year in Review 2016: biomarkers (biochemical markers). *Osteoarthritis and Cartilage*, *25*(2), 199–208.

Mohler, V. & Singrün, C. (2004). General considerations: marker-assisted selection. In Lörz, H. & Wenzel, G. eds. *Molecular marker systems in plant breeding and crop improvement* (pp. 305–317). Berlin; Heidelberg: Springer.

Morshedloo, M. R., Ebadi, A., Maggi, F., Fattahi, R., Yazdani, D., & Jafari, M. (2015). Chemical characterization of the essential oil compositions from Iranian populations of *Hypericum perforatum* L. *Industrial Crops and Products, 76*, 565–573.

Mounet, F. Moing, A. Garcia, V. Petit, J. Maucourt, M. Deborde, C. ... & Lemaire-Chamley, M. (2009). Gene and metabolite regulatory network analysis of early developing fruit tissues highlights new candidate genes for the control of tomato fruit composition and development. *Plant Physiology*, *149*(3), 1505–1528.

Nadeem, M. A. Nawaz, M. A. Shahid, M. Q. Doğan, Y. Comertpay, G. Yıldız, M. ... & Baloch, F. S. (2018). DNA molecular markers in plant breeding: current status and recent advancements in genomic selection and genome editing. *Biotechnology & Biotechnological Equipment*, *32*(2), 261–285.

Ozseyhan, M. E. Kang, J. Mu, X. & Lu, C. (2018). Mutagenesis of the FAE1 genes significantly changes fatty acid composition in seeds of *Camelina sativa*. *Plant Physiology and Biochemistry*, *123*, 1–7.

Padmesh, P. Sabu, K. K. Seeni, S. & Pushpangadan, P. (1999). The use of RAPD in assessing genetic variability in *Andrographis paniculata* Nees, a hepatoprotective drug. *Current Science*, *3*, 833–835.

Piątczak, E. Jeleń, A. Makowczyńska, J. Zielińska, S. Kuźma, Ł. & Balcerczak, E. (2019). Establishment of hairy root cultures of *Rehmannia elata* NE Brown ex Prain and production of iridoid and phenylethanoid glycosides. *Industrial Crops and Products*, *137*, 308–314.

Pichersky, E. Noel, J. P. & Dudareva, N. (2006). Biosynthesis of plant volatiles: nature's diversity and ingenuity. *Science*, *311*(5762), 808–811.

Pierpoint, W. S. (2000). Why should plants make medicines?. *Biochemist*, *22*(6), 38–40.

Pourmohammad, A. (2013). Application of molecular markers in medicinal plant studies. *Acta Universitatis Sapientiae, Agricultura et Environment*, *5*(1), 80–90.

Pradhan, S. Tarai, R. K. & Panigrahi, I. (2022). Plant growth regulators in passion fruit. In Singh, R. & Kumar, S. eds. *Plant growth regulators in tropical and Sub-tropical fruit crops* (pp. 505–519). Boca Raton, FL: CRC Press.

Raamsdonk, L. M. Teusink, B. Broadhurst, D. Zhang, N. Hayes, A. Walsh, M. C. ... & Oliver, S. G. (2001). A functional genomics strategy that uses metabolome data to reveal the phenotype of silent mutations. *Nature Biotechnology*, *19*(1), 45–50.

Ranzato, E. Martinotti, S. Calabrese, C. M. & Calabrese, G. (2014). Role of nutraceuticals in cancer therapy. *Journal of Food Research*, *3*(4), 18–27.

Ren, C. Liu, Y. Guo, Y. Duan, W. Fan, P. Li, S. & Liang, Z. (2021). Optimizing the CRISPR/Cas9 system for genome editing in grape by using grape promoters. *Horticulture research*, *8*, 52.

Renouard, S. Corbin, C. Drouet, S. Medvedec, B. Doussot, J. Colas, C. ... & Hano, C. (2018). Investigation of *Linum flavum* (L.) hairy root cultures for the production of anticancer aryltetralin lignans. *International Journal of Molecular Sciences*, *19*(4), 990.

Riahi, L. Chograni, H. Ben Rejeb, F. Ben Romdhane, M. Masmoudi, A. S. & Cherif, A. (2022). Efficient in vitro regeneration of the endangered species *Artemisia arborescens* L. through direct organogenesis and impact on SMs production. *Horticulture, Environment, and Biotechnology*, *63*(3), 439–450.

Rischer, H. Orešič, M. Seppänen-Laakso, T. Katajamaa, M. Lammertyn, F. Ardiles-Diaz, W. ... & Goossens, A. (2006). Gene-to-metabolite networks for terpenoid indole alkaloid biosynthesis in *Catharanthus roseus* cells. *Proceedings of the National Academy of Sciences of the United States of America*, *103*(14), 5614–5619.

Roberts, D. P. & Mattoo, A. K. (2019). Sustainable crop production systems and human nutrition. *Frontiers in Sustainable Food Systems*, *3*, 72.

Rodríguez, D. R. Ramírez-Solís, R. Garza-Elizondo, M. A. de Lourdes Garza-Rodriguez, M. & Barrera-Saldana, H. A. (2019). Genome editing: a perspective on the application of CRISPR/Cas9 to study human diseases. *International Journal of Molecular Medicine*, *43*(4), 1559–1574.

Saddhe, A. A. Manuka, R. & Penna, S. (2021). Plant sugars: homeostasis and transport under abiotic stress in plants. *Physiologia Plantarum*, *171*(4), 739–755.

Sangwan, N. S. Jadaun, J. S. Tripathi, S. Mishra, B. Narnoliya, L. K. & Sangwan, R. S. (2018). Plant metabolic engineering. In Barh, D. Azevedo, V. & Kumar, M. eds. *Omics technologies and bio-engineering* (pp. 143–175). London: Academic Press.

Savithramma, N. Rao, M. L. & Suhrulatha, D. (2011). Screening of medicinal plants for SMs. *Middle-East Journal of Scientific Research*, *8*(3), 579–584.

Savoi, S. Santiago, A. Orduña, L. & Matus, J. T. (2022). Transcriptomic and metabolomic integration as a resource in grapevine to study fruit metabolite quality traits. *Frontiers in Plant Science*, *13*, 937927.

Schaart, J. G. van de Wiel, C. C. & Smulders, M. J. (2021). Genome editing of polyploid crops: prospects, achievements and bottlenecks. *Transgenic Research*, *30*(4), 337–351.

Schachtsiek, J. & Stehle, F. (2019). Nicotine-free, nontransgenic tobacco (*Nicotiana tabacum* l.) edited by CRISPR-Cas9. *Plant Biotechnology Journal*, *17*(12), 2228–2230.

Schäfer, H. & Wink, M. (2009). Medicinally important SMs in recombinant microorganisms or plants: progress in alkaloid biosynthesis. *Biotechnology Journal: Healthcare Nutrition Technology*, *4*(12), 1684–1703.

Sehgal, D. & Khan, T. (2020). PTC: beyond being a tool for genetic engineering. In Varjani, S. Pandey, A. & Gnansounou, E. eds. *Environmental microbiology and biotechnology: volume 1: biovalorization of solid wastes and wastewater treatment* (pp. 175–200). Singapore: Springer.

Shahzad, A. Parveen, S. Sharma, S. Shaheen, A. Saeed, T. Yadav, V. ... & Upadhyay, A. (2017). PTC: applications in plant improvement and conservation. In Shahzad, A. Sharma, S. and Siddiqui, A. eds. *Plant biotechnology: principles and applications* (pp. 37–72). Singapore: Springer.

Shi, M. Du, Z. Hua, Q. & Kai, G. (2021). CRISPR/Cas9-mediated targeted mutagenesis of bZIP2 in *Salvia miltiorrhiza* leads to promoted phenolic acid biosynthesis. *Industrial Crops and Products, 167*, 113560.

Singh, T. Yadav, R. & Agrawal, V. (2020). Effective protocol for isolation and marked enhancement of psoralen, daidzein and genistein in the cotyledon callus cultures of *Cullen corylifolium* (L.) Medik. *Industrial Crops and Products, 143*, 111905.

Sinha, S. Sandhu, K. Bisht, N. Naliwal, T. Saini, I. & Kaushik, P. (2019). Ascertaining the paradigm of secondary metabolism enhancement through gene level modification in therapeutic plants. *Journal of Young Pharmacists, 11*(4), 337–343.

Song, Z., Li, X., Wang, H., & Wang, J. (2010). Genetic diversity and population structure of Salvia miltiorrhiza Bge in China revealed by ISSR and SRAP. *Genetica, 138*(2), 241–249.

Sorkheh, K. Zolfaghari, M. & Ercisli, S. (2019). Reliability authentication of *Glycyrrhiza glabra* L. populations from south Iran using SSR and SNP-based markers. *Proceedings of the National Academy of Sciences, India Section B: Biological Sciences, 89*, 1283–1294.

Styrczewska, M. Kulma, A. Kostyn, K. Hasiewicz-Derkacz, K. & Szopa, J. (2013). Flax terpenoid pathway as a source of health promoting compounds. *Mini Reviews in Medicinal Chemistry, 13*(3), 353–364.

Sumathi, M. Bachpai, V. K. W. Mayavel, A. Dasgupta, M. G. Nagarajan, B. Rajasugunasekar, D. ... & Yasodha, R. (2018). Genetic linkage map and QTL identification for adventitious rooting traits in red gum eucalypts. *3 Biotech, 8*, 1–10.

Supriya, R. Kala, R. G. & Thulaseedharan, A. (2020). Hairy root culture: secondary metabolite production in a biotechnological perspective. In Murthy, K. ed. *Plant metabolites: methods, applications and prospects* (pp. 89–110). Singapore: Springer.

Tak, Y. & Kumar, M. (2020). Phenolics: a key defence secondary metabolite to counter biotic stress. In Lone, A. M. Shuab, R. & Kamili, M. eds. *Plant phenolics in sustainable agriculture: volume 1* (pp. 309–329). Singapore: Springer.

Thottappilly, G. Magonouna, H. D. & Omitogun, O. G. (2000). The use of DNA markers for rapid improvement of crops in Africa. *African Crop Science Journal, 8*(1), 99–108.

Tiku, A. R. (2020). Antimicrobial compounds (phytoanticipins and phytoalexins) and their role in plant defense. In Mérillon, J. M. & Ramawat, K. G. eds. *Co-evolution of SMs* (pp. 845–868). Cham: Springer.

Tohge, T. Nishiyama, Y. Hirai, M. Y. Yano, M. Nakajima, J. I. Awazuhara, M. ... & Saito, K. (2005). Functional genomics by integrated analysis of metabolome and transcriptome of Arabidopsis plants over-expressing an MYB transcription factor. *The Plant Journal, 42*(2), 218–235.

Tokimatsu, T. Sakurai, N. Suzuki, H. Ohta, H. Nishitani, K. Koyama, T. ... & Shibata, D. (2005). KaPPA-view: a web-based analysis tool for integration of transcript and metabolite data on plant metabolic pathway maps. *Plant physiology, 138*(3), 1289–1300.

Tomar, V. Saini, S. S. Juneja, K. Agrawal, P. K. & Sircar, D. (2019). Transgenic technologies and their potential applications in horticultural crop improvement. In Sharma, Y. K. ed. *Advances in plant transgenics: methods and applications* (pp. 189–212). Singapore: Springer.

Tomás-Barberán, F. A. & Espín, J. C. (2001). Phenolic compounds and related enzymes as determinants of quality in fruits and vegetables. *Journal of the Science of Food and Agriculture, 81*(9), 853–876.

Trethewey, R. N. (2001). Gene discovery via metabolic profiling. *Current Opinion in Biotechnology, 12*(2), 135–138.

Tu, M. Fang, J. Zhao, R. Liu, X. Yin, W. Wang, Y. ... & Fang, Y. (2022). CRISPR/Cas9-mediated mutagenesis of VvbZIP36 promotes anthocyanin accumulation in grapevine (*Vitis vinifera*). *Horticulture Research, 9*, uhac022.

Twaij, B. M. & Hasan, M. N. (2022). Bioactive SMs from plant sources: types, synthesis, and their therapeutic uses. *International Journal of Plant Biology, 13*(1), 4–14.

Van Moerkercke, A. Fabris, M. Pollier, J. Baart, G. J. Rombauts, S. Hasnain, G. ... & Goossens, A. (2013). CathaCyc, a metabolic pathway database built from *Catharanthus roseus* RNA-Seq data. *Plant and Cell Physiology, 54*(5), 673–685.

Volkova, S. Matos, M. R. Mattanovich, M. & Marín de Mas, I. (2020). Metabolic modelling as a framework for metabolomics data integration and analysis. *Metabolites, 10*(8), 303.

Wang, W. Li, Y. Dang, P. Zhao, S. Lai, D. & Zhou, L. (2018). Rice SMs: structures, roles, biosynthesis, and metabolic regulation. *Molecules*, *23*(12), 3098.

Wang, Y. H. Zhang, Z. K. Yang, F. M. Sun, Q. Y. He, H. P. Di, Y. T. ... & Hao, X. J. (2007). Benzylphenethylamine alkaloids from Hosta plantaginea with inhibitory activity against tobacco mosaic virus and acetylcholinesterase. *The Journal of Natural Products*, *70*(9), 1458–1461.

Weckwerth, W. & Fiehn, O. (2002). Can we discover novel pathways using metabolomic analysis?. *Current Opinion in Biotechnology*, *13*(2), 156–160.

Weeks, D. P. Spalding, M. H. & Yang, B. (2016). Use of designer nucleases for targeted gene and genome editing in plants. *Plant biotechnology journal*, *14*(2), 483–495.

Wen, D. Wu, L. Wang, M. Yang, W. Wang, X. Ma, W. ... & Shi, Y. (2022). CRISPR/Cas9-mediated targeted mutagenesis of FtMYB45 promotes flavonoid biosynthesis in tartary buckwheat (*Fagopyrum tataricum*). *Frontiers in Plant Science*, *13*, 879390.

Wink, M. (2008). Plant secondary metabolism: diversity, function and its evolution. *Natural Product Communications*, *3*(8), 300801.

Wojciechowska, M. Owczarek, A. Kiss, A. K. Grąbkowska, R. Olszewska, M. A. & Grzegorczyk-Karolak, I. (2020). Establishment of hairy root cultures of *Salvia bulleyana* Diels for production of polyphenolic compounds. *Journal of Biotechnology*, *318*, 10–19.

Woo, J. W. Kim, J. Kwon, S. I. Corvalán, C. Cho, S. W. Kim, H. ... & Kim, J. S. (2015). DNA-free genome editing in plants with preassembled CRISPR-Cas9 ribonucleoproteins. *Nature Biotechnology*, *33*(11), 1162–1164.

Xu, J. Beleski, D. G. & Vendrame, W. A. (2022). Effects of culture methods and plant growth regulators on in vitro propagation of *Brassavola nodosa* (L.) Lindl. hybrid. *In Vitro Cellular & Developmental Biology-Plant*, *58*(6), 931–941.

Xu, W. Song, W. Yang, Y. Wu, Y. Lv, X. Yuan, S. ... & Yang, J. (2019). Multiplex nucleotide editing by high-fidelity Cas9 variants with improved efficiency in rice. *BMC Plant Biology*, *19*, 1–10.

Yan, M. Y. Li, S. S. Ding, X. Y. Guo, X. P. Jin, Q. & Sun, Y. C. (2020). A CRISPR-assisted nonhomologous end-joining strategy for efficient genome editing in *Mycobacterium tuberculosis*. *MBio*, *11*(1), 10–1128.

Yan, X. H. Chen, J. Di, Y. T. Fang, X. Dong, J. H. Sang, P. ... & Hao, X. J. (2010). Anti-tobacco mosaic virus (TMV) Quassinoids from *Brucea javanica* (L.) Merr. *Journal of Agricultural and Food Chemistry*, *58*(3), 1572–1577.

Yang, C. Q. Fang, X. Wu, X. M. Mao, Y. B. Wang, L. J. & Chen, X. Y. (2012). Transcriptional regulation of plant secondary metabolism F. *Journal of Integrative Plant Biology*, *54*(10), 703–712.

Yang, T. Ali, M. Lin, L. Li, P. He, H. Zhu, Q. ... & Deng, L. (2023). Recoloring tomato fruit by CRISPR/Cas9-mediated multiplex gene editing. *Horticulture Research*, *10*(1), uhac214.

Yang, X. Wang, J. Xia, X. Zhang, Z. He, J. Nong, B. ... & Deng, G. (2021). OsTTG1, a WD40 repeat gene, regulates anthocyanin biosynthesis in rice. *The Plant Journal*, *107*(1), 198–214.

Yi, W. Fischer, J. Krewer, G. & Akoh, C. C. (2005). Phenolic compounds from blueberries can inhibit colon cancer cell proliferation and induce apoptosis. *Journal of Agricultural and Food Chemistry*, *53*(18), 7320–7329.

Yin, J. Lu, R. Xin, C. Wang, Y. Ling, X. Li, D. ... & Hu, J. (2022). Cas9 exo-endonuclease eliminates chromosomal translocations during genome editing. *Nature Communications*, *13*(1), 1204.

Yousefian, S. Lohrasebi, T. Farhadpour, M. & Haghbeen, K. (2020). Production of phenolic acids in hairy root cultures of medicinal plant *Mentha spicata* L. in response to elicitors. *Molecular Biology Research Communications*, *9*(1), 23–32.

Yu, Z. Wu, X. & He, J. (2022). Study on the antifungal activity and mechanism of tea saponin from *Camellia oleifera* cake. *European Food Research and Technology*, *248*, 2321–2333.

Yuan, M. Zhu, J. Gong, L. He, L. Lee, C. Han, S. ... & He, G. (2019). Mutagenesis of FAD2 genes in peanut with CRISPR/Cas9 based gene editing. *BMC Biotechnology*, *19*, 1–7.

Zakaria, M. M. Schemmerling, B. & Ober, D. (2021). CRISPR/Cas9-mediated genome editing in comfrey (*Symphytum officinale*) hairy roots results in the complete eradication of pyrrolizidine alkaloids. *Molecules*, *26*(6), 1498.

Zaynab, M. Fatima, M. Abbas, S. Sharif, Y. Umair, M. Zafar, M. H. & Bahadar, K. (2018). Role of SMs in plant defense against pathogens. *Microbial Pathogenesis*, *124*, 198–202.

Zeng, L., Zhang, Q., Jiang, C., Zheng, Y., Zuo, Y., Qin, J., ... & Deng, H. (2021). Development of *Atropa bellado*nna L. plants with high-yield hyoscyamine and without its derivatives using the CRISPR/Cas9 system. *International Journal of Molecular Sciences*, *22*(4), 1731.
Zeng, S. Xiao, G. Guo, J. Fei, Z. Xu, Y. Roe, B. A. & Wang, Y. (2010). Development of a EST dataset and characterization of EST-SSRs in a traditional Chinese medicinal plant, *Epimedium sagittatum* (Sieb. Et Zucc.) Maxim. *BMC Genomics*, *11*, 1–11.
Zhang, D. Tian, H. Liu, X. Zhang, Y. Li, H. & Zhan, H. (2024). Development and application of simple sequence repeat markers based on whole-genome sequencing in *Codonopsis lanceolata*. *Genetic Resources and Crop Evolution*, *71*(2), 651–664.
Zhang, D. Wang, Z. Wang, N. Gao, Y. Liu, Y. Wu, Y. ... & Liu, B. (2014). Tissue culture-induced heritable genomic variation in rice, and their phenotypic implications. *PLoS ONE*, *9*(5), e96879.
Zhang, P. Du, H. Wang, J. Pu, Y. Yang, C. Yan, R. ... & Yu, D. (2020). Multiplex CRISPR/Cas9-mediated metabolic engineering increases soya bean isoflavone content and resistance to soya bean mosaic virus. *Plant Biotechnology Journal*, *18*(6), 1384–1395.
Zhang, P. Foerster, H. Tissier, C. P. Mueller, L. Paley, S. Karp, P. D. & Rhee, S. Y. (2005). MetaCyc and AraCyc: metabolic pathway databases for plant research. *Plant Physiology*, *138*(1), 27–37.
Zhang, Y. Guo, W. Chen, L. Shen, X. Yang, H. Fang, Y. ... & Jiao, Y. (2022). CRISPR/Cas9-mediated targeted mutagenesis of GmUGT enhanced soybean resistance against leaf-chewing insects through flavonoids biosynthesis. *Frontiers in Plant Science*, *13*, 802716.
Zhang, Y. Massel, K. Godwin, I. D. & Gao, C. (2018). Applications and potential of genome editing in crop improvement. *Genome Biology*, *19*(1), 210.
Zheng, Z. Ye, G. Zhou, Y. Pu, X. Su, W. & Wang, J. (2021). Editing sterol side chain reductase 2 gene (StSSR2) via CRISPR/Cas9 reduces the total steroidal glycoalkaloids in potato. *All Life*, *14*(1), 401–413.
Zhou, Z. Li, Q. Xiao, L. Wang, Y. Feng, J. Bu, Q. ... & Zhang, L. (2021). Multiplexed CRISPR/Cas9-mediated knockout of laccase genes in *Salvia miltiorrhiza* revealed their roles in growth, development, and metabolism. *Frontiers in Plant Science*, *12*, 647768.
Zhou, Z. Tan, H. Li, Q. Chen, J. Gao, S. Wang, Y. ... & Zhang, L. (2018). CRISPR/Cas9-mediated efficient targeted mutagenesis of RAS in *Salvia miltiorrhiza*. *Phytochemistry*, *148*, 63–70.

8 From Genotyping to Chemotyping of Medicinal Plants

Current Status and Perspective

Shreya Arora and Kalpana Katiyar

8.1 INTRODUCTION

Medicinal plants have been fundamental to both traditional and contemporary medicine, supplying a wide range of bioactive compounds with potential therapeutic effects. It is essential to comprehend the genetic and biochemical foundations of these plants to enhance their quality, consistency, and effectiveness. Medicinal plants are vital sources of bioactive compounds with significant pharmacological value. Understanding their genetic characteristics and chemical properties is essential for proper identification, authentication, conservation, and utilization. This integrated approach allows for the identification of genetic markers associated with metabolic characteristics, aiding in marker-assisted breeding, authentication, and quality assurance. Genotyping sheds light on the genetic diversity within and between species through molecular markers such as simple sequence repeats (SSRs), single nucleotide polymorphisms (SNPs), amplified fragment length polymorphisms (AFLPs), and, more recently, high-throughput genome sequencing technologies. These methods enable scientists to trace the lineage of plants, pinpoint elite genotypes, and link specific genes to the production of valued phytochemicals (Kane and Cronk, 2008). Conversely, chemotyping entails the profiling of metabolites via techniques such as gas chromatography-mass spectrometry (GC-MS), liquid chromatography-mass spectrometry (LC-MS), and nuclear magnetic resonance (NMR) spectroscopy. This process is crucial for establishing the phytochemical profile and correlating metabolic patterns with genotypic differences (Fiehn, 2002). Through metabolomics, particular chemotypes associated with high medicinal value or bioactivity can be recognized, resulting in the creation of improved cultivars. The combination of genotypic and chemotypic information has led to the identification of quantitative trait locis (QTLs) and gene–metabolite networks, promoting advancements in metabolic engineering and synthetic biology within medicinal plant research (Verpoorte et al., 2000). Furthermore, the utilization of bioinformatics and machine learning methodologies has enhanced the prediction of metabolic pathways and regulatory components linked to secondary metabolite production (Patra et al., 2013).

This chapter presents an in-depth overview of the latest methodologies, case studies, and future directions in the genotyping and chemotyping of medicinal plants, highlighting their intersection for practical applications in plant-based drug development, preservation, and commercial growth.

8.2 GENOTYPING IN MEDICINAL PLANTS

Genotyping is the process of determining genetic variations among individual organisms by analysing their DNA sequences. Advanced high-throughput genotyping technologies have revolutionized the way plant genomes are characterized. Commonly used molecular marker systems are mentioned in

 DOI: 10.1201/9781003542681-8

TABLE 8.1
Basis of Molecular Markers with Their Application in Plant Genetic Improvement

Marker Type	Principle	Applications
SSR (Simple Sequence Repeat)	PCR amplification of repeat motifs	Genetic diversity, cultivar identification
SNP (Single Nucleotide Polymorphism)	Single-base-pair variations	GWAS, trait-marker association
AFLP (Amplified Fragment Length Polymorphism)	Restriction digestion followed by PCR	Genotyping in non-model species
NGS (Next-Generation Sequencing)	Massive parallel sequencing	Whole-genome and transcriptome analysis

Table 8.1. These markers play a crucial role in detecting allelic variations that affect the expression of biosynthetic genes (Patra et al., 2013).

The genotyping of medicinal plants serves as a crucial method for comprehending the genetic variations that affect the generation of bioactive secondary metabolites, which are key to their medicinal effects. This method is vital for pinpointing particular genetic markers linked to favourable traits, including an increased production of medicinal compounds. Techniques such as SSR markers, SNP analysis, and Genome-Wide Association Studies (GWAS) have been utilized to evaluate genetic diversity, identify cultivars, and facilitate marker-assisted selection for traits associated with the biosynthesis of secondary metabolites (Sharma et al., 2023; Mipeshwaree Devi et al., 2023). By offering a deeper understanding of the genetic basis underlying medicinal properties, genotyping is essential for enhancing breeding, conservation efforts, and quality control of medicinal plants (Klein-Junior et al., 2021). In addition, analysing genetic structure with tools such as GENEALEX, and Arlequin offers valuable information about gene flow, bottleneck occurrences, and genetic drift, which are essential for prioritizing the conservation of endangered medicinal plants according to frameworks like the IUCN Red List (Smouse and Peakall, 2012).

8.2.1 Role in Authentication and Quality Control of Medicinal Plants

Ensuring accurate identification of medicinal plants is crucial for guaranteeing the effectiveness, safety, and traceability of herbal medicinal products. Traditional methods of morphological classification often lack reliability due to phenotypic variation, hidden species, and practices of adulteration. Consequently, molecular genotyping methods offer a reliable and reproducible approach for species identification (Techen et al., 2014).

DNA barcoding, which employs universal gene regions such as rbcL, matK, and the Internal Transcribed Spacer (ITS), facilitates quick and accurate taxonomic identification at the species level. The DNA barcoding process includes DNA extraction, polymerase chain reaction (PCR) amplification, sequencing, and comparative analysis with established databases like Barcode of Life Data System (BOLD) and GenBank.

An illustrative case shows the use of DNA barcoding in distinguishing *Rauvolfia serpentina* from its morphologically similar adulterants (*Catharanthus roseus, Alstonia scholaris*), where traditional morphology-based techniques were ineffective (Yu et al., 2021). By integrating genotyping technologies, the herbal industry can adopt validated traceability systems that ensure uniformity across batches, pharmacological effectiveness, and adherence to regulations, protecting public health.

8.2.2 Techniques for Genotyping

Genotyping methods play a crucial role in the molecular profiling of medicinal plants, offering essential insights into their genetic variation, evolutionary connections, and proper identification. Several sophisticated techniques, such as DNA barcoding, molecular markers (including SNPs,

SSRs, random amplified polymorphic DNA (RAPD), and AFLP), whole-genome sequencing (WGS), and transcriptomics, are commonly employed for genotyping medicinal flora.

8.2.2.1 DNA Barcoding and Molecular Markers

DNA barcoding is a prominent method used to recognize and authenticate plant species through standardized genetic markers. Frequently utilized loci in barcoding encompass rbcL, matK, and the ITS regions, which assist in differentiating species even when their morphological characteristics are alike (Kress et al., 2009). This technique has been vital in validating medicinal plants, particularly in instances of adulteration or misclassification (Kumar et al., 2018). RAPD and inter simple sequence repeat (ISSR) are another frequently used techniques that generate distinct DNA profiles. They have been used in wide range of medicinal plant species due to their simple and convenient delivery of the results (Kumar et al., 2014; Boomibalagan et al., 2021; Singh et al., 2022). RAPD markers are valuable for studying genetic relationships and identifying variations within populations. Similarly, AFLP markers provide a high-throughput methodology by integrating restriction enzyme digestion with selective PCR amplification, making them particularly advantageous for large-scale genotyping of medicinal plants (Paun and Schonswetter, 2012; Adhikari et al., 2022). The assessment of genetic diversity in *Withania somnifera* (Ashwagandha) utilized SSR markers, revealing considerable population variation and offering insights to guide conservation efforts (Chawla et al., 2025).

8.2.2.2 Whole-Genome Sequencing and Transcriptomics

WGS facilitates a thorough examination of the entire genetic structure of medicinal plants, revealing both coding and non-coding regions, structural variations, and regulatory elements. Advances in high-throughput sequencing technology permit the sequencing of plant genomes with exceptional detail, providing significant knowledge about genes linked to secondary metabolite synthesis, disease resistance, and stress resilience. For instance, WGS of *Panax ginseng* has provided essential information regarding the genetic foundation of ginsenoside synthesis, vital for the medicinal characteristics of ginseng products (Xu et al., 2017). Furthermore, WGS supports the identification of genes responsible for secondary metabolite production, a central focus in improving medicinal plants for pharmaceutical uses. Transcriptomics works alongside WGS by examining gene expression under various environmental conditions, developmental phases, and biotic or abiotic stressors. RNA-Seq technology enables an in-depth profiling of gene expression, aiding in the discovery of genes involved in the production of secondary metabolites, such as alkaloids, flavonoids, and terpenoids (Tyagi et al., 2022). For example, RNA-Seq conducted on *C. roseus* led to the discovery of genes linked to vinblastine biosynthesis, an important anti-cancer agent (Verma et al., 2014).

8.2.2.3 Implementation of Bioinformatics Tools for Genetic Profiling

The vast amount of data generated from genomic and transcriptomic investigations requires bioinformatics tools for proficient analysis, interpretation, and visualization. Bioinformatics platforms such as Basic Local Alignment Search Tool (BLAST), GENEIOUS, and Galaxy carry out sequence alignment, functional annotation, and comparative genomics, assisting researchers in comprehending the functional components of plant genomes (Zahoor et al., 2019). Specialized genetic profiling tools like STRUCTURE and Trait Analysis by aSSociation, Evolution, and Linkage (TASSEL) are employed to evaluate population structure, genetic diversity, and marker–trait associations, facilitating the identification of loci associated with significant traits such as medicinal value, disease resistance, and yield (Khan, 2013; Porras-Hurtado et al., 2013).

8.3 CHEMOTYPING IN MEDICINAL PLANTS

Chemotyping encompasses an extensive analysis of phytochemical components, especially secondary metabolites such as alkaloids, flavonoids, terpenoids, and phenolics (Fiehn, 2002). These

TABLE 8.2
Comparative Analytical Approaches for Secondary Metabolite Characterization

Technique	Principle	Applications
GC-MS (Gas Chromatography-Mass Spectrometry)	Volatile compound analysis	Essential oil profiling
LC-MS (Liquid Chromatography-Mass Spectrometry)	Separation of non-volatile compounds	Alkaloid and flavonoid profiling
NMR (Nuclear Magnetic Resonance)	Structural elucidation of molecules	Metabolite identification
FTIR (Fourier Transform Infrared Spectroscopy)	Infrared absorption spectra	Functional group analysis

Source: Assouguem et al. (2025).

metabolites contribute to the therapeutic properties of the plant and act as chemical indicators for taxonomic categorization and quality assessment. Various analytical platforms used for the identification, quantification, and assessment of secondary metabolites are mentioned in Table 8.2.

Chemotyping enables the categorization of species into specific chemovars, each possessing a distinctive chemical makeup. The chemical profiles can be affected by both genetic and environmental elements (Verpoorte et al., 2000). The integration of genotyping and chemotyping allows for a better understanding of genotype–phenotype connections (Dixon and Strack, 2003). This combination aids in uncovering biosynthetic pathways, identifying transcriptional regulators, and applying precision breeding techniques. Such combined methods improve metabolic engineering, conservation efforts, and the creation of standardized medicinal products. Multi-omics datasets can be further examined using bioinformatics tools and statistical approaches to pinpoint candidate genes and regulatory networks involved in secondary metabolism (Jamil et al., 2020).

8.3.1 Significance of Secondary Metabolites in Medicinal Effectiveness

Secondary metabolites are unique bioactive substances produced by plants through intricate biosynthetic processes, often in response to environmental factors, pathogens, or abiotic pressures. Unlike primary metabolites such as carbohydrates, amino acids, and nucleotides, secondary metabolites are not crucial for basic plant survival; however, they play vital roles in ecological interactions and defence mechanisms. From a medicinal standpoint, secondary metabolites, including alkaloids, terpenoids, flavonoids, saponins, tannins, and phenolic acids, are the pharmacologically active components that contribute to the healing properties of herbal products. For instance, vinca alkaloids like vincristine and vinblastine, which are derived from *C. roseus*, display strong anti-mitotic effects and are widely used in cancer treatment regimens (Mendonce et al., 2025). Likewise, flavonoids obtained from *Camellia sinensis* exhibit significant antioxidant and anti-inflammatory properties, highlighting the essential medicinal importance of polyphenolic secondary metabolites (Panche et al., 2016). Conducting quantitative and qualitative assessments of these metabolites through chemotyping ensures that the medicinal plant material retains its intended pharmacological properties across various batches, cultivation scenarios, and geographical locations. Thus, the precise characterisation of secondary metabolites is critical for standardization, dosage calculations, and quality control in phytotherapeutic practices.

8.3.2 Chemotaxonomy and Its Role in Plant Classification

Chemotaxonomy is a scientific field that combines the profiling of chemical constituents with plant systematics, utilizing the composition of secondary metabolites as molecular indicators to clarify phylogenetic relationships, distinguish closely related species, and confirm botanical identities. This methodology offers enhanced objectivity and molecular clarity compared to traditional morphological taxonomy, which can be complicated by phenotypic variability and environmental factors. Profiling of terpenoids and alkaloids is frequently used to establish chemotaxonomic classifications (Amin and Park, 2025). An exemplary case of chemotaxonomy was observed in *Centella asiatica*, where variations in triterpenoid content facilitated the identification of distinct chemotypes that correspond to different therapeutic attributes (Sun et al., 2020). Furthermore, chemical profiling based on flavonoids has aided in refining the taxonomy of species within the Lamiaceae family, which is essential for the standardization of medicinal applications (Moshari-Nasirkandi et al., 2023). Consequently, chemotaxonomy functions as a potent integrative approach for validating medicinal plants, assessing biodiversity, and conducting evolutionary biology research, thus connecting phytochemistry and taxonomy into a cohesive scientific platform.

8.3.3 Techniques for Chemotyping

Chemotyping utilizes advanced analytical and computational methods to methodically examine the chemical components of medicinal plants. Precise chemotyping is essential for maintaining therapeutic uniformity, verifying plant identity, and investigating plant biodiversity (Amin and Park, 2025).

8.3.3.1 Chromatographic and Spectrometric Techniques

Chromatography and mass spectrometry are fundamental techniques in contemporary chemotyping methods. They facilitate the qualitative and quantitative analysis of intricate mixtures of secondary metabolites.

1. *High-Performance Liquid Chromatography (HPLC):* HPLC isolates compounds based on their interactions with both stationary and mobile phases, allowing for the profiling of phenolics, alkaloids, flavonoids, and other significant phytochemicals.
2. *Gas Chromatography-Mass Spectrometry:* GC-MS is particularly effective for analysing volatile and semi-volatile secondary metabolites like essential oils and terpenes. This method offers both separation (GC) and structural characterization (MS) of phytochemicals.
3. *Liquid Chromatography-Mass Spectrometry:* LC-MS merges the separating capabilities of liquid chromatography with the detection sensitivity of mass spectrometry. It is particularly efficient in profiling non-volatile, thermally sensitive secondary metabolites such as glycosides, saponins, and alkaloids.
4. *NMR Spectroscopy:* NMR generates non-targeted, reproducible chemical fingerprints by examining the magnetic properties of nuclei. It provides detailed structural information about complex mixtures of metabolites without the need for prior separation.

8.3.3.2 Metabolomics Approaches and Phytochemical Profiling

Metabolomics provides a comprehensive perspective on the various small molecules present in a sample. It includes both targeted (specific metabolites) and untargeted (comprehensive profiling) approaches.

1. *Untargeted Metabolomics*
   Untargeted methods seek to capture the broadest spectrum of metabolites without any bias. These approaches are employed during the discovery phase to detect unknown bioactive compounds.

Application Example: Untargeted metabolomics using LC-MS on *C. asiatica* differentiated its chemotypes through their triterpenoid saponin profiles, uncovering unique pharmacological potentials (Rafi et al., 2024).

2. *Targeted Metabolomics*
   In targeted metabolomics, specific known metabolites are quantitatively measured to confirm and standardize certain therapeutic compounds found in medicinal plants.
   Application Example: The targeted analysis of curcuminoids in *Curcuma longa* ensures the standardization of turmeric products for applications related to anti-inflammatory effects (Osorio-Tobón et al., 2016).
3. *Multivariate Statistical Analysis*
   Statistical techniques such as principal component analysis (PCA) and partial least squares-discriminant analysis (PLS-DA) are employed to analyse complex metabolomic datasets, highlighting group distinctions and identifying relevant biomarkers (Worley and Powers, 2013).

8.3.4 Environmental and Genetic Influences on Chemotypes

The generation of secondary metabolites in medicinal herbs is affected by both genetic and environmental aspects. Gaining insight into these influences is crucial for enhancing the quality and uniformity of herbal products. Differences in metabolite compositions can arise among individuals of the same species, influenced by environmental factors and genetic characteristics. In this discussion, we will explore the impacts of soil, climate, and geographic factors, as well as the genetic and epigenetic control of secondary metabolites. Environmental factors like soil composition, climate, and geographical location play a crucial role in determining the chemical makeup of medicinal plants. These elements lead to differences in secondary metabolites such as alkaloids, flavonoids, terpenoids, and phenolic compounds, which are often associated with the plant's defensive strategies against abiotic stresses, including drought, temperature changes, and nutrient deficiencies.

1. *Soil Composition:* The pH level, nutrient content, and texture of the soil affect how minerals and nutrients become available, which can influence plant metabolic processes (Maddala et al., 2024). For example, soil with lower iron content had a positive influence on colour content of saffron flower in Moroccan saffron (Chaouqi et al., 2023).
2. *Climate:* Factors such as temperature, humidity, and patterns of precipitation directly influence plant growth and the production of metabolites (Jangpangi et al., 2025).
3. *Geographical Conditions:* Plants found in various geographic areas show different chemotypes due to the differences in regional climates. For instance, *Lavandula angustifolia* (lavender) plants from elevated altitudes produce more phenolic compounds because of the more severe conditions present at these heights (Demasi et al., 2018).

8.4 ADVANCEMENTS IN PLANT BIOTECHNOLOGY AND OMICS TECHNOLOGIES ABOUT SECONDARY METABOLITES

8.4.1 Genomic Methods for Investigating of Pathways

The completion of plant genome sequencing has advanced our capability to examine the genetic foundations of secondary metabolism. Genomic strategies offer valuable insights into the biosynthetic gene clusters (BGCs) that govern the synthesis of secondary metabolites such as alkaloids, terpenoids, and flavonoids. The rise of high-throughput sequencing has facilitated the identification and detailed analysis of genes involved in the creation of secondary metabolites. GWAS has become an essential tool for discovering QTLs linked to metabolite production (He and Gai, 2023).

8.4.2 Transcriptomics and Gene Expression Analysis

Transcriptomics, especially utilizing RNA sequencing (RNA-Seq), enables a thorough examination of gene expression across various conditions, which is vital for understanding how genes regulate secondary metabolism. Transcription factors are crucial in governing the biosynthesis of secondary metabolites. By analyzing transcriptomic data, significant transcription factors involved in the production of secondary metabolites can be identified. These findings have enriched our understanding of how environmental factors like light, temperature, and stress affect metabolite accumulation (Sao et al., 2022).

8.4.3 Metabolomics and Chemical Profiling

Metabolomics serves as a vital tool for identifying and quantifying secondary metabolites found in plants. It utilizes advanced analytical techniques such as LC-MS, GC-MS, and NMR spectroscopy to profile these metabolites.

Metabolic Fingerprinting: Metabolomic profiling enables the identification of plants based on their distinct chemical signatures. This is especially useful for chemotaxonomy, where metabolic profiles help classify plants into chemovars or varieties that possess unique medicinal attributes (Fiehn, 2002; Amin and Park, 2025). The quantification of metabolites allows researchers to evaluate the levels of bioactive compounds under varying cultivation conditions, which can be leveraged to optimize growth and environmental factors to boost the production of specific secondary metabolites (Waris et al., 2022). In the instance of *Ginkgo biloba*, a plant recognized for its medicinal benefits, such as enhancing cognitive function, metabolomic methods like LC-MS have been employed to analyse its specific chemical composition, particularly the flavonoid glycosides and terpenoids. This profiling has provided significant insights into the environmental and genetic factors that influence the production of bioactive compounds, thus aiding in quality control for commercial herbal products (Yao et al., 2012).

8.4.4 Multi-Omics Integration

Combining multi-omics data—merging genomics, transcriptomics, proteomics, and metabolomics—offers a thorough approach to investigating the biosynthesis of secondary metabolites. This methodology allows researchers to explore how genetic, transcriptional, and biochemical networks interact to regulate metabolite production. Systems biology, enhanced by multi-omics data, allows researchers to model and simulate plant metabolic pathways in silico. This facilitates the identification of critical regulatory nodes and metabolic bottlenecks that govern metabolite production. Through metabolic engineering, plants can be tailored to overproduce specific metabolites or redirect metabolic flux toward desired products (Palazzotto and Webar, 2018). Artificial Intelligence (AI) and Machine Learning: The combination of AI and machine learning with multi-omics data has resulted in the creation of predictive models for secondary metabolite production. These models optimize metabolic pathways by forecasting how alterations in one pathway will influence the overall metabolic network (Chi et al., 2024). A study focused on boosting alkaloid biosynthesis in *Nicotiana benthamiana* utilized a multi-omics strategy to assess genomic, transcriptomic, and metabolomic information. This comprehensive approach facilitated the discovery of crucial metabolic genes associated with alkaloid production, and researchers leveraged this information to genetically modify the plant to increase alkaloid synthesis for pharmaceutical purposes (Kalalagh et al., 2025).

8.5 CASE STUDIES: MULTI-OMICS IN MAJOR MEDICINAL PLANTS

Integrating genotypic and chemotypic information has effectively associated genetic differences with the synthesis of secondary metabolites in medicinal plants. Various case studies display the effectiveness of this combined approach in enhancing plant breeding, ensuring quality control, and discovering new bioactive compounds.

- *Ginseng (P. ginseng):* A comprehensive study on molecular insights of fruit development has been carried out. The result provided the integrative network of transcriptome and metabolome during fruit maturation. Using the generated database, the metabolic and transcriptomic atlas of associated with primary, secondary, and hormone signalling pathways could be generated (Bian et al., 2025).
- *C. roseus*: Monoterpenes-indole alkaloid (MIA) is an anticancer compound present abundantly in the plant. The single-cell transcriptome and single-cell metabolome analyses revealed that the biosynthesis and transport of MIA are cell-type specific and the genes of the MIA pathway are expressed in the distinct localization (Li et al., 2023).
- *Cannabis sativa*: A comprehensive genetic and physical map of *C. sativa* reveals key molecular determinants of cannabinoid diversity, highlighting the complex genomic architecture and recombination patterns that distinguish drug-type and hemp-type chemotypes (Laverty et al., 2019).
- In *Salvia miltiorrhiza* (Danshen), Liu et al. (2025) created a platform (SmilODB) using various databases such as genome assemblies, transcriptomics databases, proteomics databases and metabolomic data to provide the integrative network for the translational research.

8.6 ELITE GENOTYPES FOR COMMERCIAL PRODUCTION OF SECONDARY METABOLITES

Identifying elite genotypes is vital for enhancing the production of secondary metabolites in medicinal plants. Through the processes of genotyping and chemotyping, it is possible to select plants with superior chemical profiles for commercial farming.

Genomic Selection: The use of genomics and metabolomics techniques aids in pinpointing specific genetic markers associated with high-yielding genotypes, such as those that produce elevated levels of bioactive compounds. This facilitates the creation of improved cultivars that possess enhanced pharmacological properties (Alami et al., 2022).

Metabolite Profiling and Breeding: Techniques for metabolite profiling, like GC-MS and LC-MS, are employed to evaluate the chemical composition of various genotypes. When combined with QTL mapping, breeders can identify plants that have optimal levels of bioactive metabolites. In the case of *P. ginseng*, genetic and chemical assessments enabled the identification of elite genotypes featuring high ginsenoside levels. Marker-assisted selection was utilised to breed ginseng plants that produce increased amounts of these bioactive compounds, thereby improving the plant's medicinal effectiveness (Selwal et al., 2023).

8.7 APPLICATIONS, CHALLENGES, AND FUTURE AVENUES

The combination of biotechnology and omics technologies has important implications in the realm of secondary metabolites.

Metabolic Engineering: A key area of focus is the application of CRISPR-Cas9 and synthetic biology techniques to alter plant genomes and boost metabolite generation. By integrating heterologous pathways from microorganisms or other plant species, it becomes feasible to produce valuable compounds in plants that are otherwise challenging to source (Das et al., 2024).

Pharmacogenomics: Discovering genetic markers associated with particular metabolites will support the formulation of standardized medicinal products with reliable therapeutic effectiveness. Pharmacogenomics enables the customisation of plant-based treatments according to individual genetic profiles, maximising drug effectiveness and safety (Chaachouay and Zidane, 2024; Qiu et al., 2023).

Sustainability: Biotechnological methods can lessen the reliance on the wild harvesting of medicinal plants, aiding conservation initiatives. Moreover, systems based on bioreactors are being developed for the in vitro production of secondary metabolites, providing a sustainable alternative to conventional cultivation practices (Patra et al., 2013).

Combining genomic and metabolomic information in the study of secondary metabolites offers a valuable opportunity to enhance our comprehension of medicinal plants.

Nevertheless, there are considerable obstacles in merging intricate omics data collections, developing comprehensive databases, and utilizing innovative technologies such as CRISPR and AI. This segment focuses on major challenges and future directions for exploration in the area of secondary metabolites in medicinal plants. Although progress has been made in genomics and metabolomics, merging these two fields remains a challenging endeavour due to various barriers. Multi-omics investigations generate extensive and complex datasets that necessitate advanced computational tools for their analysis and integration. While genomic data reveals the genetic structure, metabolomic data sheds light on the plant's bioactive compounds. However, linking these levels of information has been difficult owing to the complexity surrounding plant secondary metabolism and its regulation. Many biosynthetic genes related to secondary metabolites have been identified, yet their complete roles and interactions within metabolic pathways often remain ambiguous. This deficiency in functional annotation in genomic databases leads to gaps in understanding how genetic variations affect metabolite profiles.

The combination of genomic and metabolomic data demands the creation of advanced bioinformatics tools and models driven by data to forecast secondary metabolite production based on genetic diversity. Creating international databases that merge genotypic and chemotypic data is crucial for standardising and assuring the quality of medicinal plant products. The Medicinal Plant Genomic Resource Database (MPGRD) represents an effort to compile genetic data for medicinal plants. Nevertheless, it is still in an underdeveloped state, with numerous medicinal plants lacking comprehensive genetic and chemical profiles essential for extensive research (Hou et al., 2020). Establishing thorough and standardised global databases would facilitate improved cross-referencing of plant data and amplify research into the medicinal attributes of plants by providing a stronger foundation for analysis and comparison. Innovative technologies like CRISPR/Cas9 gene editing and AI-based metabolomics present promising approaches to address the challenges in secondary metabolite research. These advanced tools allow for more accurate manipulation of plant genetic characteristics and facilitate the efficient analysis of large metabolomic datasets.

CRISPR Technology in Plant Modification: Utilising CRISPR technology provides the capability for targeted gene alteration, enabling scientists to modify genes that play a role in the biosynthesis of secondary metabolites. By knocking out or enhancing particular genes, researchers can optimize the yield of valuable secondary metabolites in medicinal plants. AI algorithms are progressively being utilized to analyse metabolomic data, allowing for the prediction of chemical profiles in plants and the identification of bioactive compounds. Machine learning techniques can reveal trends in intricate metabolomic data that might otherwise remain hidden, accelerating the discovery of secondary metabolites. AI can enhance the understanding of how genetic and environmental elements affect the synthesis of secondary metabolites (Zhang et al., 2025).

8.8 CONCLUSION

Research into secondary metabolites in medicinal plants has made remarkable progress because of advancements in genomics, chemotyping, and multi-omics techniques. These improvements have led to a deeper understanding of the genetic regulation of bioactive compounds and have facilitated the identification, standardisation, and enhancement of medicinal plants for pharmaceutical use. The combination of genetic information with chemical profiles is transforming the comprehension of plant biology and its relevance to therapeutic applications. Numerous breakthroughs have fuelled the expansion of research on secondary metabolites.

Integration of Genomics and Metabolomics: Merging genomic and metabolomic information has allowed for a more thorough understanding of the biosynthetic pathways responsible for secondary metabolites. It also helps identify genetic loci linked to metabolite production, enabling more effective breeding and selection of plants with improved medicinal traits. Advanced analytical methods such as HPLC, GC-MS, and LC-MS provide detailed characterizations of complex mixtures of secondary metabolites. When these techniques are combined with MS and NMR spectroscopy, they allow for accurate identification and quantification of bioactive compounds, ensuring the reliability and authenticity of herbal products.

Artificial Intelligence and Machine Learning: The application of machine learning algorithms and AI models to metabolomics data analysis is on the rise. These technologies can estimate the pharmacological properties of uncharacterized compounds by examining extensive datasets of genomic and metabolomic information, thereby accelerating the process of drug discovery.

CRISPR and Metabolic Engineering: The CRISPR-Cas9 system enables precise genome editing of medicinal plants to boost the production of secondary metabolites. This advancement offers potential for sustainable and efficient generation of bioactive compounds with therapeutic relevance. These advancements carry significant implications for quality assurance, authentication, and the enhancement of medicinal plants. They also present improved solutions to challenges such as species adulteration and variability in therapeutic efficacy.

REFERENCES

Adhikari, S., Biswas, A., Saha, S., Bandyopadhyay, T.K. and Ghosh, P., 2022. AFLP-based assessment of genetic variation in certain Indian elite cultivars of Cymbopogon species. *Journal of Applied Research on Medicinal and Aromatic Plants*, *29*, 100372.

Alami, M.M., Ouyang, Z., Zhang, Y., Shu, S., Yang, G., Mei, Z. and Wang, X., 2022. The current developments in medicinal plant genomics enabled the diversification of secondary metabolites' biosynthesis. *International Journal of Molecular Sciences*, *23*(24), pp.15932–15958.

Amin, A. and Park, S., 2025. Chemotaxonomy, an efficient tool for medicinal plant identification: current trends and limitations. *Plants*, *14*(14), pp.2234–2256.

Assouguem, A., Annemer, S., Mohammed, K. and Lazraq, A., 2025. Innovative approaches in the extraction, identification, and application of secondary metabolites from plants. *Phyton*, *94*(6), pp.1631–1654.

Bian, X., Qiu, Y., Zhao, X., Wei, H., Zhang, Y., Zhang, W., Wang, J., Sun, X. and Sun, X., 2025. Multi-omics dissection of metabolic and transcriptional regulation underlying fruit maturation in Panax ginseng. *BMC Plant Biology*, *25*(1), 1133.

Boomibalagan, P., Subramanian, S.R., Rajasekharan, P.E., Karpakal, S., Veeranan, U., Saminathan, E., Narayanan, V. and Kathiresan, D., 2021. Genetic relationship and polymorphism of selected medicinal plants of Asclepiadaceae using RAPD molecular analysis method. *Ecological Genetics and Genomics*, *21*, 100101.

Chaachouay, N. and Zidane, L., 2024. Plant-derived natural products: a source for drug discovery and development. *Drugs and Drug Candidates*, *3*(1), pp.184–207.

Chaouqi, S., Moratalla-López, N., Alonso, G.L., Lorenzo, C., Zouahri, A., Asserar, N., Haidar, E.M. and Guedira, T., 2023. Effect of soil composition on secondary metabolites of Moroccan saffron (*Crocus sativus* L.). *Plants*, *12*(4), p.711–728.

Chawla, R., Sharma, H., Dadheech, A., Jattan, M., Devi, S., Kumar, P. and Singh, T., 2025. Genetic insights into diversity and population structure of ashwagandha (*Withania somnifera* (L.) Dunal) using EST-SSR, ISSR and SSR markers: implications for enhancing agricultural and industrial value. *Industrial Crops and Products*, *224*, 120242.

Chi, J., Shu, J., Li, M., Mudappathi, R., Jin, Y., Lewis, F., Boon, A., Qin, X., Liu, L. and Gu, H., 2024. Artificial intelligence in metabolomics: a current review. *TrAC Trends in Analytical Chemistry*, *178*, 117852.

Das, S., Kwon, M. and Kim, J.Y., 2024. Enhancement of specialized metabolites using CRISPR/Cas gene editing technology in medicinal plants. *Frontiers in Plant Science*, *15*, 1279738.

Demasi, S., Caser, M., Lonati, M., Cioni, P.L., Pistelli, L., Najar, B. and Scariot, V., 2018. Latitude and altitude influence secondary metabolite production in peripheral alpine populations of the Mediterranean species *Lavandula angustifolia* Mill. *Frontiers in Plant Science*, *9*, p.983.

Dixon, R. A. and Strack, D., 2003 Phytochemistry meets genome analysis, and beyond. *Phytochemistry*, *62*(6), pp.815–816.

Fiehn, O., 2002. Metabolomics–the link between genotypes and phenotypes. *Plant Molecular Biology*, *48*(1), pp.155–171.

He, J. and Gai, J., 2023. Genome-wide association studies (GWAS). In *Plant Genotyping: Methods and Protocols* (pp. 123–146). New York: Springer US.

Hou, J.I., Zhang, J., Zhang, B., Jin, X., Zhang, H. and Jin, Z., 2020. Transcriptional analysis of metabolic pathways and regulatory mechanisms of essential oil biosynthesis in the leaves of *Cinnamomum camphora* (L.) Presl. *Frontiers in Genetics*, *11*, 598714.

Jamil, I.N., Remali, J., Azizan, K.A., Nor Muhammad, N.A., Arita, M., Goh, H.H. and Aizat, W.M., 2020. Systematic multi-omics integration (MOI) approach in plant systems biology. *Frontiers in Plant Science*, *11*, p.944.

Jangpangi, D., Patni, B., Chandola, V. and Chandra, S., 2025. Medicinal plants in a changing climate: understanding the links between environmental stress and secondary metabolite synthesis. *Frontiers in Plant Science*, *16*, 1587337.

Kalalagh, K.F., Papon, N., Courdavault, V., van der Krol, S., Kappers, I.F. and Kashkooli, A.B., 2025. Metabolic engineering in *Nicotiana benthamiana*. *aBIOTECH*, *4*, pp.1–25.

Kane, N.C. and Cronk, Q., 2008. Botany without borders: barcoding in focus. *Molecular Ecology*, *17*(24), pp.5175–5176.

Khan, M., 2013. Association mapping using TASSEL software. *Plant Breed Genomics*. Available at: http://articles.extension.org/pages/62755/association-mapping-using-tasselsoftware (verified 7 January 2017).

Klein-Junior, L.C., de Souza, M.R., Viaene, J., Bresolin, T.M., de Gasper, A.L., Henriques, A.T. and Vander Heyden, Y., 2021. Quality control of herbal medicines: from traditional techniques to state-of-the-art approaches. *Planta Medica*, *87*(12/13), pp.964–988.

Kress, W.J., Erickson, D.L., Jones, F.A., Swenson, N.G. Perez, R., Sanjur, O. and Bermingham, E., 2009. Plant DNA barcodes and a community phylogeny of a tropical forest dynamics plot in Panama. *Proceedings of the National Academy of Sciences*, *106*(44), pp.18621–18626.

Kumar, A., Mishra, P., Singh, S.C. and Sundaresan, V., 2014. Efficiency of ISSR and RAPD markers in genetic divergence analysis and conservation management of Justicia adhatoda L., a medicinal plant. *Plant Systematics and Evolution*, *300*(6), pp.1409–1420.

Kumar, J.S., Seethapathy, V.K.G.S., Ravikanth, R.G.G. and Shaanker, R.U., 2018. Assessment of adulteration in raw herbal trade of important medicinal plants of India using DNA barcoding. *3 Biotech*, *8*(3), pp.1–8.

Laverty, K.U., Stout, J.M., Sullivan, M.J., Shah, H., Gill, N., Holbrook, L., Deikus, G., Sebra, R., Hughes, T.R., Page, J.E. and Van Bakel, H., 2019. A physical and genetic map of *Cannabis sativa* identifies extensive rearrangements at the THC/CBD acid synthase loci. *Genome Research*, *29*(1), pp.146–156.

Li, C., Wood, J.C., Vu, A.H., Hamilton, J.P., Rodriguez Lopez, C.E., Payne, R.M., Serna Guerrero, D.A., Gase, K., Yamamoto, K., Vaillancourt, B. and Caputi, L., 2023. Single-cell multi-omics in the medicinal plant *Catharanthus roseus*. *Nature Chemical Biology*, *19*(8), pp.1031–1041.

Liu, Y.H., Zeng, W.Q., Tao, S.F., Du, Y.N., Yu, F.H., Qi, Z.C. and Yang, D.F., 2025. SmilODB: a multi-omics database for the medicinal plant danshen (*Salvia miltiorrhiza*, Lamiaceae). *Frontiers in Plant Science*, *16*, 1586268.

Maddala, V.K.S., Singh, S. and Bandi, J., 2024. Soil properties influencing nutrient signaling in plant soil systems. In Sharma, A. and Kumar, R. eds. *Essential minerals in plant-soil systems* (pp. 373–393). Amsterdam: Elsevier.

Mendonce, K.C., Palani, N., Rajadesingu, S., Radhakrishnan, K., Ayyar, M. and Priya, L.S., 2025. Pharmacological potential of bioactive compounds in *Catharanthus roseus* extract: a comprehensive review. *Toxicology Reports*, *12*, 101998.

Mipeshwaree Devi, A., Khedashwori Devi, K., Premi Devi, P., Lakshmipriyari Devi, M. and Das, S., 2023. Metabolic engineering of plant secondary metabolites: prospects and its technological challenges. *Frontiers in Plant Science*, *14*, 1171154.

Moshari-Nasirkandi, A., Alirezalu, A., Alipour, H. and Amato, J., 2023. Screening of 20 species from Lamiaceae family based on phytochemical analysis, antioxidant activity and HPLC profiling. *Scientific Reports*, *13*(1), 16987.

Osorio-Tobón, J.F., Carvalho, P.I., Barbero, G.F., Nogueira, G.C., Rostagno, M.A. and de Almeida Meireles, M.A., 2016. Fast analysis of curcuminoids from turmeric (*Curcuma longa* L.) by high-performance liquid chromatography using a fused-core column. *Food Chemistry*, *200*, pp.167–174.

Palazzotto, E. and Weber, T., 2018. Omics and multi-omics approaches to study the biosynthesis of secondary metabolites in microorganisms. *Current Opinion in Microbiology*, *45*, pp.109–116.

Panche, A.N., Diwan, A.D. and Chandra, S. R., 2016. Flavonoids: An overview. *Journal of Nutritional Science*, *5*, e47.

Patra, B., Schluttenhofer, C., Wu, Y., Pattanaik, S. and Yuan, L., 2013. Transcriptional regulation of secondary metabolite biosynthesis in plants. *Biochimica et Biophysica Acta (BBA): Gene Regulatory Mechanisms*, *1829*(11), pp.1236–1247.

Paun, O. and Schönswetter, P., 2012. Amplified fragment length polymorphism: an invaluable fingerprinting technique for genomic, transcriptomic, and epigenetic studies. In Rajora, S. ed. *Plant DNA Fingerprinting and Barcoding: Methods and Protocols* (pp. 75–87). Totowa, NJ: Humana Press.

Porras-Hurtado, L., Ruiz, Y., Santos, C., Phillips, C., Carracedo, Á. and Lareu, M.V., 2013. An overview of STRUCTURE: applications, parameter settings, and supporting software. *Frontiers in Genetics*, *4*, p.98.

Qiu, S., Cai, Y., Yao, H., Lin, C., Xie, Y., Tang, S. and Zhang, A., 2023. Small molecule metabolites: discovery of biomarkers and therapeutic targets. *Signal Transduction and Targeted Therapy*, *8*(1), p.132.

Rafi, M., Madya, M.M., Karomah, A.H., Septaningsih, D.A., Ridwan, T., Rohaeti, E., Aisyah, S. and Idroes, R., 2024. LC-HRMS-based metabolomics for profiling the metabolites in different plant parts of *Centella asiatica*. *HAYATI Journal of Biosciences*, *31*(6), pp.1106–1115.

Sao, P., Parveen, R., Khattri, A., Sharma, S., Tiwari, N. and Singh, S., 2022. Understanding the metabolomics of medicinal plants under environmental pollution: challenges and opportunities. In Sharma, R. and Pandey, A. eds. *Environmental Pollution and Medicinal Plants* (pp.193–205). Singapore: Springer.

Selwal, N., Rahayu, F., Herwati, A., Latifah, E., Suhara, C., Suastika, I.B.K., Mahayu, W.M. and Wani, A.K., 2023. Enhancing secondary metabolite production in plants: exploring traditional and modern strategies. *Journal of Agriculture and Food Research*, *14*, 100702.

Sharma, R.S., Vaidya, N., Maloo, S.R., Kumar, A., Sharma, S., Ramkrishnan, R.S. and Kumari, V., 2023. Application of molecular markers in assessment of genetic diversity of medicinal plants. In Singh, A. and Verma, P. eds. *Molecular Marker Techniques: A Potential Approach of Crop Improvement* (pp. 103–116). Singapore: Springer Nature Singapore.

Singh, R., Mahendra, P.S. and Sadhana, C., 2022. Genetic diversity analysis in five medicinal plants of Solanaceae family using RAPD markers. *International Journal of Pharmacy and Biological Sciences*, *12*(1), pp.120–125.

Smouse, R.P.P. and Peakall, R., 2012. GenAlEx 6.5: genetic analysis in Excel. Population genetic software for teaching and research—an update. *Bioinformatics*, *28*(19), pp.2537–2539.

Sun, B., Wu, L., Wu, Y., Zhang, C., Qin, L., Hayashi, M., Kudo, M., Gao, M. and Liu, T., 2020. Therapeutic potential of *Centella asiatica* and its triterpenes: a review. *Frontiers in Pharmacology*, *11*, 568032.

Techen, N., Parveen, I., Pan, Z. and Khan, I.A., 2014. DNA barcoding of medicinal plant material for identification. *Current Opinion in Biotechnology*, *25*, pp.103–110.

Tyagi, P., Singh, D., Mathur, S., Singh, A. and Ranjan, R., 2022. Upcoming progress of transcriptomics studies on plants: an overview. *Frontiers in Plant Science*, *13*, 1030890.

Verma, M., Ghangal, R., Sharma, R., Sinha, A.K. and Jain, M., 2014. Transcriptome analysis of *Catharanthus roseus* for gene discovery and expression profiling. *PLoS ONE*, *9*(7), e103583.

Verpoorte, R., Van der Heijden, R. and Memelink, J., 2000. Engineering the plant cell factory for secondary metabolite production. *Transgenic Research*, *9*(4), pp.323–343.

Waris, M., Kocak, E., Gonulalan, E.M., Demirezer, L.O., Kır, S. and Nemutlu, E., 2022. Metabolomics analysis insight into medicinal plant science. *TrAC Trends in Analytical Chemistry*, *157*, 116795.

Worley, B. and Powers, R., 2013. Multivariate analysis in metabolomics. *Current Metabolomics*, *1*(1), pp.92–107.

Xu, J., Chu, Y., Liao, B., Xiao, S., Yin, Q., Bai, R., Su, H., Dong, L., Li, X., Qian, J. and Zhang, J., 2017. Panax ginseng genome examination for ginsenoside biosynthesis. *Gigascience*, *6*(11), gix093.

Yao, X., Shang, E., Zhou, G., Tang, Y., Guo, S., Su, S., Jin, C., Qian, D., Qin, Y. and Duan, J.A., 2012. Comparative characterization of total flavonol glycosides and terpene lactones at different ages, from different cultivation sources and genders of *Ginkgo biloba* leaves. *International Journal of Molecular Sciences*, *13*(8), pp.10305–10315.

Yu, J., Wu, X.I., Liu, C., Newmaster, S., Ragupathy, S. and Kress, W.J., 2021. Progress in the use of DNA barcodes in the identification and classification of medicinal plants. *Ecotoxicology and Environmental Safety*, *208*, p.111691.

Zahoor, I., Shafi, A., Fazili, K.M. and Haq, E., 2019. Bioinformatics and medicinal plant research: current scenario. In Verma, S. and Singh, R. eds. *Essentials of Bioinformatics, Volume III: In Silico Life Sciences: Agriculture* (pp. 141–157). Cham: Springer International Publishing.

Zhang, D., Xu, F., Wang, F., Le, L. and Pu, L., 2025. Synthetic biology and artificial intelligence in crop improvement. *Plant Communications*, *6*(2), 100–118.

9 Clonal Fidelity in Micropropagated Medicinal Plant

Kajal Kumari, Pritom Biswas, Ankita Kumari, and Nitish Kumar

9.1 INTRODUCTION

When it comes to medicinal plants, clonal fidelity is particularly important in micropropagation, which is the process of asexually propagating plants using tissue culture techniques (Moraes et al., 2021; Sharma et al., 2014). In relation to the parent plant, clonal fidelity describes the genetic homogeneity of plants generated via tissue culture (micropropagation) (Alizadeh et al., 2015). The objective is to ensure that the cloned plant retains the parent plant's true-to-type characteristics. This is important in micropropagted medicinal plants as maintaining clonal fidelity ensures that the micropropagated plants will produce the same types and quantities of desired active compounds as the parent plant. Furthermore, some medicinal plants may possess unique genetic traits that contribute to their medicinal efficacy. By ensuring clonal fidelity, these traits can be preserved, minimizing the risk of genetic variation that could occur during propagation. In general, clonal integrity is essential for preserving medicinal plants' efficacy, safety, and consistency as well as their commercial viability (Awere et al., 2024).

Many medications contain natural compounds or secondary metabolites that are obtained from medicinal plants (Husen 2022; 2023; 2024a; 2024b; Husen & Pant, 2024a; 2024b; Pant et al., 2025; Pant & Husen, 2025). However, the bulk of therapeutic plants are still harvested from the wild, with only a small number being grown. Due to over-exploitation brought on by the increasing demand for these items, species are at risk, biodiversity is in jeopardy, and plant materials and products are being adulterated, all of which have an impact on ecosystems (Volenzo & Odiyo, 2020). For millennia, traditional medicine has made use of plant resources. Alternative biotechnological techniques, such as micropropagation, have been developed for a variety of medicinal plants to address the increasing demand for raw plant materials for the purpose of medical uses (Ramawat & Arora, 2021). A technique called micropropagation uses the *in vitro* culture process to multiply specific explants. Producing plants with the same genotype as a sole source plant—that is, the mother plant—is the aim of micropropagation. Cloning is a biological process that produces a population of plants known as "clones." The process of micropropagation for the *in vitro* generation of plantlets involves several steps. The following are this method's main advantages: (i) creating plants free of disease; (ii) preserving genetic homogeneity; (iii) conserving rare or endangered species; and (iv) developing excellent clones quickly. However, it is now well recognized that undifferentiated cells (i.e., calli), tissues, isolated protoplasts, and morphological characteristics of regenerated plants all exhibit genetic variability (Bidabadi & Jain, 2020).

In the micropropagation technique, both genetic and epigenetic variations may occur, potentially leading to phenotypic changes. These are collectively known as somaclonal variation (Sun et al., 2013). While in some cases, somaclonal variation may introduce beneficial traits such as enhanced

DOI: 10.1201/9781003542681-9

stress tolerance or improved secondary metabolite production, it is generally undesirable in medicinal plant propagation, where genetic fidelity is paramount. To detect such variations, various molecular marker-based techniques, including random amplified polymorphic DNA (RAPD), restriction fragment length polymorphism (RFLP), amplified fragment length polymorphism (AFLP), inter-simple sequence repeat (ISSR), and simple sequence Repeat (SSR), are commonly used for assessing genotypic differences (Grover & Sharma, 2016). Cytogenetic analysis, such as flow cytometry and chromosomal analysis, helps detect numerical and structural chromosomal variations (Sliwinska et al., 2022). Additionally, biochemical analyses, including high-performance liquid chromatography (HPLC) and gas chromatography-mass spectroscopy (GC-MS), are used to study and identify various biochemical compounds present in plant tissue cultures (Behera et al., 2024). Each tool has its advantages and limitations in assessing variations. To maintain clonal fidelity in micropropagation, several strategies can be employed:

- *Selection of Explant:* The choice of explant plays a crucial role in preserving clonal fidelity. Explants taken from meristematic tissues, such as shoot tips and apical meristems, are preferred due to their lower likelihood of genetic variation.
- *Choice of Plant Growth Regulators (PGRs):* Selecting an appropriate combination of PGRs is essential to ensure proper regeneration while minimizing the risk of somaclonal variation.
- *Minimising Subculturing:* Excessive subculturing, which involves transferring plant tissues to fresh culture media for continued growth and multiplication, should be minimized as it can increase the likelihood of genetic instability.

Implementing these strategies can help maintain the genetic integrity of micropropagated plants, ensuring consistency and stability in regenerated plantlets.

Beyond these fundamental approaches, cutting-edge procedures like cryopreservation and slow-growth storage have been investigated to maintain the genetic identity of therapeutic plants for extended periods of time. Clonal fidelity is protected via cryopreservation, which includes storing plant tissues for an extended period of time at extremely low temperatures. This process successfully reduces cellular and genetic alterations (Benelli, 2021). Epigenetic monitoring is another promising approach, as recent studies suggest that DNA methylation and histone modifications play crucial roles in regulating gene expression during micropropagation (Handy et al., 2011). By incorporating epigenetic assessments alongside molecular and biochemical analyses, researchers can gain deeper insights into the factors influencing clonal fidelity and develop improved protocols to mitigate variation.

9.2 MICROPROPAGATION AND ITS APPLICATION IN MEDICINAL PLANTS

In micropropagation, there are various stages or steps for the *in vitro* growth of plantlets.

Stage (i)—Explant Establishment: This is the first stage, where a suitable plant part (explant), such as shoot tips, organs, or axillary buds, is disinfected and cultured aseptically on a suitable medium.

Stage (ii)—Rapid Multiplication: In this stage, the primary explants that have successfully passed in stage (i) are transferred aseptically to a new medium to generate numerous clonal propagules.

Stage (iii)—Rooting: This stage focuses on inducing the establishment of fully developed plantlets, including shoot elongation and root formation.

Stage (iv)—Acclimatisation: In this stage, the *in vitro* regenerated plantlets undergo a gradual adaptation to acclimatize in the external environment.

Stage (v)—Hardening: In this final stage, the regenerated plantlets are transferred to soil under natural conditions. However, when plants grown *in vitro* are transplanted into soil, their survival rate often decreases.

Since the beginning of time, curiosity and purpose have driven the exploration of medicinal plants, which play a vital role in healthcare systems. Many modern medicines are derived from plants (Debnath et al., 2006). Reports indicate that approximately 25% of legal drugs contain plant extracts or are based on plant compounds. The pharmaceutical industry, along with the fragrance, taste, and colour ingredient industries, generates billions of dollars annually from plant-derived compounds. For example, aspirin, one of the most common and widely used painkillers, is derived from various species of *Salix* and *Spiraea*. Additionally, some of the most effective anticancer drugs, including vinblastine, taxol, vincristine, colchicine, saponins, and paclitaxel, are also plant-based products (Lemma et al., 2020).

The annual value of the medicinal plant trade in India is estimated to be around US$ 1 billion (Debnath et al., 2006). On a global scale, the production and export of medicinal plants are on the rise. According to a WHO estimate, the demand for raw materials derived from medicinal plants is expected to increase by more than $5 trillion by 2050, with an annual growth rate of 15% to 25%. Given this critical situation, it is essential to take initial action to conserve these important plant species using biotechnological methods, such as plant tissue culture techniques (Rasool Hassan., 2012). Tissue culture methods not only meet pharmaceutical and other commercial requirements but also provide a practical means for mass production and germplasm preservation of superior medicinal plants (Hussain et al., 2012). Another important approach is genetic improvement to enhance the yielding capacity of medicinal plants (Permadi et al., 2024). Additionally, micropropagation is a process of vegetative development and multiplication from viable, regenerating cells under aseptic and favourable conditions on an appropriate culture medium and enables the production of a large number of plants in a brief time (Zhou & Wu, 2006).

There are various applications of micropropagated technique for medicinal plants:

- Rapid multiplication of desired medicinal plants producing thousands of microprogramed plants from single parents.
- Regenerated plants which are produced from micropropagation technique have genetic uniformity.
- Micropropagation play key role in conserving rare and endangered medicinal plant species by preserving their genetic material in a controlled environment.
- Produce disease-free plants.
- Plant improvement through regeneration techniques combined with *in vitro* cell manipulation.
- Enhancement of plant propagation rate.
- Independent of seasonal influence.

9.3 FACTORS INFLUENCING CLONAL FIDELITY IN MICROPROPAGATED PLANTS

There are numerous factors which are involved in influencing the clonal fidelity in micropropagated plants, such as genetic factors like somaclonal variation, influence of culture conditions like media composition, lights temperature, PGRs. There are growing number of studies indicating that the variability observed in regenerated plants through *in vitro* technique is not only due to genetic but also epigenetic changes (Duta-Cornescu et al., 2023). Because they supply the vital nutrients—both macro and micronutrients—necessary for plant growth, culture media are crucial to the micropropagation process. Any imbalance in these nutrients can affect the plant health. Proper light and temperature also affect the micropropagation. For instance, quality and intensity of light affect the shoot regeneration and elongation (Cavallaro et al., 2022). Additionally, PGRs like auxin, cytokinin, and gibberellin can also affect the micropropagated plants. The concentration and combination of plant growth hormone influence the process of cell division, shoot elongation and root formation; for example, auxin promote root elongation, while cytokinin promote shoot elongation (Kurepa & Smalle, 2022; Sosnowski et al., 2023).

9.4 ASSESSMENT OF CLONAL FIDELITY

9.4.1 Molecular Markers

Molecular, cytogenic, and biochemical approaches have been used to assess clonal fidelity. These methods are limited in that they may be affected by developmental and environmental factors (Krishna & Singh, 2007). Molecular technology offers a separate method for characterizing compounds derived from medicinal plants. Cytogenetic techniques concentrate on looking at the chromosomes of plant cells to detect any differences that might have developed over the tissue culture procedure, while biochemical analysis examines the levels and activities of specific molecules within plant tissues (Kwasniewska & Bara, 2022).

Molecular markers are highly valued and have the capability to differentiate between genotypes in the field of genetic research. Anthocyanins, phenolic and, other secondary metabolites were the first molecules to be employed as markers to differentiate between various plant species (Kumari et al., 2024). Allozyme and isozymes, two types of enzyme markers, briefly gained popularity before more potent DNA markers that identify individual differences based on polymorphism in DNA sequence were available (Grover & Sharma, 2016).

Molecular markers are broadly classified into two categories: (i) non-polymerase chain reaction (PCR)-based approaches, which rely on nucleic acid hybridisation, and (ii) PCR-based approaches, which involve DNA amplification. PCR-based molecular markers include methods like RAPD and AFLP, whereas hybridization-based markers include RFLP (Grover & Sharma, 2016). As described in Figure 9.1, applications of molecular markers in the clonal fidelity assessment of various medicinal plants have been shown in Table 9.1.

9.4.1.1 Hybridization-Based Molecular Marker

RFLP: RFLP is a molecular marker that is based on hybridization. According to Botstein et al. (1980), it is the first DNA marker to produce a molecular map of the human genome. The favourable aspect of this marker is that it is co-dominant in nature. It has been extensively utilised in the creation of linkage maps in *Brassica* spp. (Sarwat et al., 2012) and *Olea* spp. (De la Rosa et al., 2004) because of its co-dominant nature. Herbal cures and aromatic herbs have been widely used for quick identification,

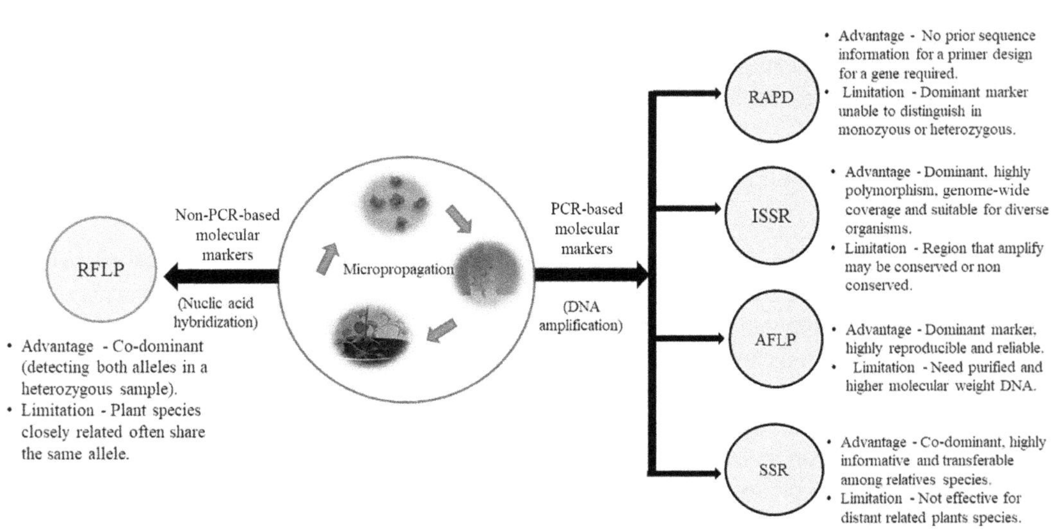

FIGURE 9.1 Molecular markers-assisted assessment of genetic integrity in micropropagated medicinal plants. RFLP: restriction fragment length polymorphism; RAPD: random amplified polymorphic DNA; ISSR: inter simple sequence repeat; AFLP: amplified length polymorphism; SSR simple sequence repeat.

TABLE 9.1
Molecular Markers and Their Applications in Clonal Fidelity Assessment of Medicinal Plants

Molecular Markers	Application	Plant	References
ISSR	Assessment of genetic fidelity in regenerated plantlets	*Morus* spp.	Rohela et al., 2018
ISSR	To find the genetic variability in regenerated plantlets	*Andrographis echioides*	Savitikadi et al., 2020
ISSR	To determine genetic diversity of plant which is derived from artificial selection	*Stevia rebaudiana*	Luz et al., 2020
RFLP	Assessment of clonal fidelity	*Solanum capsicoides*	Dharman and Anilkumar, 2020
SCAR	To develop in vitro protocol *Viola serpens*	*V. serpens*	Jha et al., 2020
SCAR	To identify the germplasm	*Moringa oleifera*	Ravi et al., 2021
TRAP	To find the genetic structure	*Saccharum* spp.	CADK Junior et al., 2020
RAPD	Evaluation of genetic stability of micropropagated plant	*Rhynchostylis retusa*	Oliya et al., 2021
AFLP and ISSR	Assessment of genetic clonal fidelity	*Acorus calamus*	Tikendra et al., 2022
RAPD and ISSR	Assessment of genetic homogeneity	*Thunbergia coccinea*	Sultana et al., 2022
ISSR and SCoT	Check genetic fidelity	*Solena amplexicaulis*	Koppula et al., 2023
RAPD	To check genetic fidelity of micropropagated shoot	*Curcuma longa*	Rani et al., 2024
AFLP	To check the genetic stability	*Solanum nigrum*	AL-Kiyyam et al., 2024
RAPD and ISSR	To assess the genetic fidelity of plants regenerated through in vitro techniques.	*S. rebaudiana*	Biswas et al., 2024
ISSR	Assessment of genetic fidelity	*Coleus forskohlii*	Badhepuri et al., 2025

AFLP = amplified fragment length polymorphism; ISSR = inter simple sequence repeats; RFLP = restriction fragment length polymorphism; RAPD = random amplified polymorphic DNA; SCAR = sequence characterized amplified region; SCoT = start codon targeted polymorphism; TRAP = target region amplification polymorphism.

and PCR processes need nanogram amounts of DNA to amplify and produce enough template DNA (Gnavi et al., 2010). Examining genetic differentiation in *Santalum* spp. (Byrne et al., 2003), studying the inheritance of chloroplast DNA in *Leucadendron* spp. (Pharmawati et al., 2004) and *Bupleurum* spp. (Xie et al., 2009), and enabling marker-assisted breeding and selection in a variety of medicinal plants result in improved traits like disease resistance and yield. It also plays a crucial role in elucidating evolutionary relationships and analysing population genetics among different species.

The initial challenge was the need for substantial amounts of DNA (5–20 mg). However, in the current era of high-throughput genomics, using 51% of this quantity as a template for PCR amplification is now more than sufficient. The limitation of this marker is that closely related species often share the same allele (Biswas & Kumar, 2023).

9.4.1.2 PCR-Based Molecular Markers

The DNA marker techniques facilitated by PCR application are widely used in plant genomics. The advancement in many areas of plant genetics has led to a better understanding of molecular markers. In modern research, every area of plant cultivation, development, and improvement is now significantly aided by creative approaches arising from recent advancements in molecular marker technology.

RAPD: A single 10-nucleotide sequence with 60% or higher GC content is used for DNA amplification and fingerprinting (DAF) in RAPD, a PCR-based marker (Williams et al., 1990). Only when the priming site appears twice in opposite orientation within about 2000 bases would PCR amplification be anticipated in this marker. Consequently, it is normal for amplified fragments from several species to exhibit correspondence when the same RAPD primers are used (Rieseberg and associates, 1996). Sequence variation at priming sites and/or length variation in the target sequence between primer binding sites are the causes of RAPD polymorphisms. By producing several bands with a single primer, this approach demonstrates that a comparatively small number of primers can be utilized to generate a high number of fragments. Typically, these pieces come from different genomic areas, which enable the quick analysis of numerous loci (Alizadeh et al., 2015).

Among the PCR-based molecular techniques, RAPD is easy to perform and does not require knowledge of the DNA sequence for the amplification. Because of its simplicity, it is used in taxonomic and systematic analyses of plants (Arif et al., 2010) and in detection of somaclonal variation in tissue culture. It has been used in various medicinal plants, that is, the RAPD fingerprinting technique assessed the genetic stability of *Bacopa monnieri* plant spp. (Ramesh et al., 2011). RAPD markers had confirmed the genetic fidelity of micropropagated clones in *Curcuma longa* (M Rani et al., 2024). These markers are highly polymorphic and serve as valuable tools for germplasm characterization, phylogenetic studies, clonal fidelity testing, genetic relationship analysis in *Gloriosa superba L* (Tilwari & Sharma, 2021). Many different loci have been simultaneously examined for polymorphism using this technique. In addition, RAPD requires less technical expertise.

ISSR: ISSR is a PCR-based method which is involved in amplifying the DNA segment located between two identical microsatellite repeat regions that are oriented in opposite directions. ISSR markers have provided valuable insights in genetic variability, molecular phylogeny, genome mapping, and evolutionary biology, particularly in studies of both closely related and distant plant species. The microsatellite repeats used as primers can consist of di-, tri-, tetra-, or penta-nucleotide sequences. The primer may be unanchored or, more commonly, anchored at the 3′ or 5′ end, incorporating one to four degenerate bases that extend into flanking sequences (Alizadeh et al., 2015). Use microsatellites, typically 16–25 base pairs long, as primers in a single-primer PCR reaction to target various genomic loci (Chuang et al., 2010). These markers are also used in detecting the genetic variation during the callus formation in *Zingiber officinale* (Rosli et al., 2025). These techniques used in *Pueraria tuberos* generated monomorphic bands and confirmed the genetic uniformity (Kanthaliya et al., 2024). These markers were used to assess the molecular diversity among 10 genotypes of *G. superba* representing different geographic and technical backgrounds (Tilwar & Sharma, 2021). The goal of this marker is to amplify inter-SSR sequence of varying lengths. The advantage of this marker is co-dominant in nature. This marker is highly polymorphic, which means it can detect a large amount of genetic variations useful for distinguishing between different genotypes in micropropagation.

AFLP: AFLP is a PCR-based technique that identifies genomic restricted fragments and can be applied to DNAs of any origin or complexity (Alizadeh et al., 2015). To amplify genomic segments from arbitrary places, these marker PCR amplifications are carried out under rigorous conditions using adapter-specific primers that contain random extensions of one to three nucleotides at the 3′ end. For every 4096 potential segments, a 3-nucleotide extension will yield 1. On a polyacrylamide sequencing gel, the resultant amplicons are examined and usually split into 40–200 different bands. It is anticipated that polymorphism will occur because of mutation in the restricted recognition sites (Grover et al., 2016).

Hence, with the help of AFLP marker checking the somaclonal variation in micropropagation regenerates in *Echinacea purpurea* (Chuang et al., 2010). AFLP plays an important role in stabilizing the genetic stability of cultivation as well as *in vitro* regenerated medicinal plant such as Aloe and Allium (Gantait et al., 2014). This marker has been widely used for DNA fingerprinting and the analysis of genetic diversity, as well as for constructing genetic maps, evaluating genetic relatedness, studying population genetic structure and mating systems, and predicting phytochemical markers for assessing genetic fidelity and species identification, including molecular tagging (Sarwat et al., 2012; Adhikari et al., 2022; Vir et al., 2024).

SSR: SSR markers, also known as sequence tagged microsatellite markers, are sequence specific marker. In this marker, primers are designed to be complementary to the flanking regions of the microsatellite locus which are conserved among genotypes of the same species. Despite the challenges of developing SSR markers being labour-intensive, time-consuming, and expensive, they offer several advantages over other markers, including codominance, locus specificity, and high reproducibility. SSR markers consist of repeating units with 1–6 bp in length. SSR markers have been extensively used in molecular evolution, systematic taxonomy, parentage analysis, genetic variation, linkage, comparative mapping, and functional diversity studies in plant species (Liu et al., 2017).

Medicinal plant species have also been studied using the SSR molecular marker technique. In *Psidium guajava*, 24 SSR markers were analysed in 36 varieties along with wild species (Kherwar et al., 2018). SSR is a locus-specific marker that is inherent in a co-dominant manner. Therefore, if the other allele of the gene is transferred from parent to progeny, the linked loci will also transfer which is traceable and better genetic map can be constructed during characterization of plant species.

9.4.2 Cytogenic Techniques

The genetic makeup of an organism can be modified by changes in the number and structure of chromosomes, as well as alterations in RNA and DNA content. Many researchers have used chromosomal analysis and variations in other nuclear components to identify differences in *in vitro* regenerants (Fiuk, et al., 2010). Cytological analysis, which involves observing conventionally stained, condensed somatic chromosomal aberrations using light microscopy, oil immersion, or other advanced microscopy techniques, has been employed by several researchers (Alizadeh et al., 2015). This technique was used in *Nardostachys jatamansi* for genetic analysis (Bose et al., 2016). Another technology of cytogenic technique is flow cytometry. This technique involves the preparation of an aqueous suspension of intact nuclei whose DNA is stained with a DNA fluorochrome and then suspended in a fluid stream to be passed through an electronic detection system. The most common application of flow cytometry is stabilizing genome size and ploidy. Flow cytometry and chromosomal analyses are used for the DNA content estimation in medicinal plants such as *Raphanus sativus* and *Crocus sativus* (Sliwinska et al., 2020).

9.4.3 Biochemical Analysis

Biochemical analysis for assessing the clonal fidelity of micropropagated medicinal plants involves the study and identification of various biochemical compounds present in plant tissue cultures propagated under controlled laboratory conditions (S Behera et al., 2024). Techniques such as HPLC and GC-MS are commonly used to evaluate the profiles of secondary metabolites. Many medicinal plants are valued for their secondary metabolites such as alkaloids, flavonoids, terpenoids, glycosides, and phenolics, which are often responsible for the plant's therapeutic properties. In *Evolvulus alsinoides*, secondary metabolite profiling identifies compounds such as propanoic acid, 2-furancarboxaldehyde, 2-phenylglycine, cyclopentasiloxane, benzophenone, furfural, betazole, n-decan, pentanoic acid, n-hexadecanoic acid, squalene, γ-sitosterol, and butanoic acid using the GC-MS technique (CO Awere et al., 2024). HPLC was used to identify phenolic compounds in *Zuccagnia punctata* (MA Álvarez et al., 2024). In *Rhodiola imbricata*, photochemical compounds were analysed with the help of HPLC (AK Bhardwaj et al., 2018).

9.5 STRATEGIES FOR MAINTAINING CLONAL FIDELITY

There are several strategies to maintain clonal fidelity in plant tissue culture, including selection of explants, which plays a crucial role in maintaining clonal fidelity. Explants taken from meristematic tissues, such as shoot tips and apical meristems, are ideal for preserving genetic stability.

These tissues are actively dividing and less likely to undergo genetic changes, which helps prevent somaclonal variation. The second choice of PGRs is another critical factor in maintaining clonal fidelity during plant tissue culture. PGRs influence various aspects of plant growth, such as shoot proliferation, root formation, and callus induction. Selecting the appropriate PGR combination is essential to ensure that the regenerated plants remain genetically identical to the mother plant and free from somaclonal variation. Common PGRs used in tissue culture include auxins and cytokinins, which play key roles in regulating plant growth (JC Cardoso et al., 2018).

Minimizing subculturing is also an important strategy for maintaining clonal fidelity. Subculturing involves transferring plant tissue from one culture medium to another, typically to promote continued growth or multiplication. While subculturing is necessary for propagation, frequent subculturing can increase the risk of somaclonal variation due to genetic mutations or changes that may occur during the process. Therefore, reducing the frequency of subculturing helps maintain the genetic integrity of the plants.

9.6 FUTURE PERSPECTIVES

The reliability of micropropagation in producing consistent, high-quality medicinal plants is crucial for the pharmaceutical and nutraceutical industries. Future research should focus on refining assessment methods, optimizing culture conditions, and exploring advanced biotechnological tools to enhance clonal fidelity (Moraes et al., 2021). One promising avenue is the integration of genomics and epigenomics to gain a deeper understanding of the molecular mechanisms underlying somaclonal variation. Techniques such as whole-genome sequencing, DNA methylation analysis, and transcriptomics can provide detailed insights into genetic and epigenetic changes occurring during micropropagation, allowing for the identification of specific markers associated with instability. Such markers can then be used for early screening of regenerants, ensuring that only true-to-type plants are selected for mass production (Alizadeh et al., 2015). Optimizing culture conditions is another critical area for future research. Careful control of PGRs, media composition, and environmental factors can minimize stress-induced variations. The use of bioreactors and automated micropropagation systems may offer more controlled and standardized culture environments, reducing the risk of somaclonal variation. Additionally, exploring the potential of cryopreservation for long-term storage of explants can help maintain genetic stability over extended periods (Moraes et al., 2021).

9.7 CONCLUSION

Ensuring clonal fidelity in micropropagated medicinal plants is essential for maintaining genetic stability and phytochemical consistency, which directly impact their therapeutic value. Somaclonal variations, induced by prolonged *in vitro* culture, plant growth regulators, and environmental factors, pose significant challenges in micropropagation. Employing molecular markers such as RAPD, ISSR, and SSR allows for effective genetic stability assessment, while biochemical techniques like HPLC and GC-MS help ensure phytochemical uniformity. Strategies, including optimized explant selection, controlled use of PGRs, and minimizing subculturing cycles, have been explored to mitigate variations. Additionally, cryopreservation and epigenetic monitoring present promising approaches for long-term genetic conservation. Advances in genome-editing technologies and high-throughput sequencing may further enhance the precision of clonal fidelity assessment, making micropropagation a more reliable tool for large-scale medicinal plant production. Future research should focus on refining these techniques and gaining deeper insights into the mechanisms of somaclonal variation to improve the stability and efficiency of *in vitro* propagation. By integrating molecular, biochemical, and advanced biotechnological approaches, the reliability of micropropagation can be significantly enhanced, ensuring the consistent production of high-quality medicinal plants with desired bioactive compounds.

REFERENCES

Adhikari, S., Biswas, A., Saha, S., Bandyopadhyay, T. K., & Ghosh, P. (2022). AFLP-based assessment of genetic variation in certain Indian elite cultivars of Cymbopogon species. *Journal of Applied Research on Medicinal and Aromatic Plants*, *29*, 100372.

Alizadeh, M., Krishna, H., Eftekhari, M., Modareskia, M., & Modareskia, M. (2015). Assessment of clonal fidelity in micropropagated horticultural plants. *Journal of Chemical and Pharmaceutical Research*, *7*(12), 977–990.

AL-Kiyyam, M. A., Shibli, R. A., Zatimeh, A., Tahtamouni, R. W., AlQudah, T. S., & Hassan, S. (2024). Evaluation of genetic diversity in Jordanian *Solanum nigrum* plants and genetic stability of *in vitro* grown plant using (AFLP) technique. *Jordan Journal of Biological Sciences*, *17*(1), 45–58.

Álvarez, M. A., Kobak, C. P., Enrico, R. J., Danert, F., Zampini, I. C., & Isla, M. I. (2024). Micropropagation of *Zuccagnia punctata*, an argentine medicinal plant, by means of "in vitro" technology to promove its conservation and the sustainable production of bioactive compounds. *Darwiniana, nueva serie*, *12*(1), 122–135.

Arif, I. A., Bakir, M. A., Khan, H. A., Al Farhan, A. H., Al Homaidan, A. A., Bahkali, A. H., ... & Shobrak, M. (2010). Application of RAPD for molecular characterization of plant species of medicinal value from an arid environment. *Genetics and Molecular Research*, *9*(4), 2191–2198.

Awere, C. O., Rakkammal, K., Ram, P. J., Kumar, K. P., Ragavan, K., Kumari, R. A., ... & Ramesh, M. (2024). Micropropagation, genetic fidelity and chromatographic analysis in *Evolvulus alsinoides* (L.): a potent multipurpose medicinal plant. *Industrial Crops and Products*, *213*, 118444.

Badhepuri, M. K., Aileni, M., Beeravelli, P. R., Durgam, S. K., Sandhya, D., Shekhawat, M. S., & Singisala, N. R. (2025). Production of non-embryogenic synthetic seeds for short-term cold storage and ISSR marker-assisted genetic fidelity analysis of *Coleus forskohlii* (Willd) Briq. plants. *Plant Cell, Tissue and Organ Culture (PCTOC)*, *160*(2), 1–10.

Behera, S., Chauhan, V. B. S., Monalisa, K., Meher, R. K., Kar, S. K., Pati, K., ... & Naik, S. K. (2024). In vitro plant regeneration, genetic fidelity, biochemical analysis, and anticancer activity of anthocyanin-rich purple flesh sweet potato var.'Bhu Krishna'. *South African Journal of Botany*, *166*, 332–343.

Benelli, C. (2021). Plant cryopreservation: a look at the present and the future. *Plants*, *10*(12), 2744.

Bhardwaj, A. K., Singh, B., Kaur, K., Roshan, P., Sharma, A., Dolker, D., ... & Chaurasia, O. P. (2018). In vitro propagation, clonal fidelity and phytochemical analysis of *Rhodiola imbricata* Edgew: a rare trans-Himalayan medicinal plant. *Plant Cell, Tissue and Organ Culture (PCTOC)*, *135*, 499–513.

Bidabadi, S. S., & Jain, S. M. (2020). Cellular, molecular, and physiological aspects of in vitro plant regeneration. *Plants*, *9*(6), 702.

Biswas, P., & Kumar, N. (2023). Application of molecular markers for the assessment of genetic fidelity of in vitro raised plants: current status and future prospects. In Singh, A., & Verma, P. eds. *Molecular Marker Techniques: A Potential Approach of Crop Improvement* (pp. 233–256). Singapore: Springer.

Biswas, P., Kumari, A., & Kumar, N. (2024). Impact of salt strength on in vitro propagation and rebaudioside A content in *Stevia rebaudiana* under semi-solid and liquid MS media. *Scientific Reports*, *14*(1), 22148.

Bose, B., Kumaria, S., Choudhury, H., & Tandon, P. (2016). Assessment of genetic homogeneity and analysis of phytomedicinal potential in micropropagated plants of *Nardostachys jatamansi*, a critically endangered, medicinal plant of alpine Himalayas. *Plant Cell, Tissue and Organ Culture (PCTOC)*, *124*, 331–349.

Botstein, D., White, R. L., Skolnick, M., & Davis, R. W. (1980). Construction of a genetic linkage map in man using restriction fragment length polymorphisms. *American Journal of Human Genetics*, *32*(3), 314–331.

Byrne, M., MacDonald, B., Broadhurst, L., & Brand, J. (2003). Regional genetic differentiation in Western Australian sandalwood (*Santalum spicatum*) as revealed by nuclear RFLP analysis. *Theoretical and Applied Genetics*, *107*(7), 1208–1214.

Cardoso, J. C., Sheng Gerald, L. T., & Teixeira da Silva, J. A. (2018). Micropropagation in the twenty-first century. In Teixeira da Silva, J. A. ed. *Plant Cell Culture Protocols* (pp. 17–46). New York: Springer.

Cavallaro, V., Pellegrino, A., Muleo, R., & Forgione, I. (2022). Light and plant growth regulators on in vitro proliferation. *Plants*, *11*(7), 844.

Chuang, S. J., Chen, C. L., Chen, J. J., & Sung, J. M. (2010). Using bulked AFLP analysis to assess genetic diversity in Echinacea species. *Scientia Horticulturae*, *124*(3), 400–404.

De la Rosa, R., James, C. M., & Tobutt, K. R. (2004). Using microsatellites for paternity testing in olive progenies. *HortScience*, *39*(2), 351–354.

Debnath, M., Malik, C. P., & Bisen, P. S. (2006). Micropropagation: a tool for the production of high quality plant-based medicines. *Current Pharmaceutical Biotechnology*, *7*(1), 33–49.

Dharman, A. K., & Anilkumar, M. (2020). Micropropagation and clonal fidelity testing in *Solanum capsicoides* all. *International Journal of Botany Studies*, *5*, 76–82.

Duta-Cornescu, G., Constantin, N., Pojoga, D. M., Nicuta, D., & Simon-Gruita, A. (2023). Somaclonal variation—advantage or disadvantage in micropropagation of the medicinal plants. *International Journal of Molecular Sciences*, *24*(1), 838.

Fiuk, A., Bednarek, P. T., & Rybczyński, J. J. (2010). Flow cytometry, HPLC-RP, and metAFLP analyses to assess genetic variability in somatic embryo-derived plantlets of *Gentiana pannonica* Scop. *Plant Molecular Biology Reporter*, *28*, 413–420.

Gantait, S., Debnath, S., & Nasim Ali, M. (2014). Genomic profile of the plants with pharmaceutical value. *3 Biotech*, *4*, 563–578.

Gnavi, G., Bertea, C. M., & Maffei, M. E. (2010). PCR, sequencing and PCR–RFLP of the 5S-rRNA-NTS region as a tool for the DNA fingerprinting of medicinal and aromatic plants. *Flavour and fragrance journal*, *25*(3), 132–137.

Grover, A., & Sharma, P. C. (2016). Development and use of molecular markers: past and present. *Critical reviews in biotechnology*, *36*(2), 290–302.

Handy, D. E., Castro, R., & Loscalzo, J. (2011). Epigenetic modifications: basic mechanisms and role in cardiovascular disease. *Circulation*, *123*(19), 2145–2156.

Husen, A. (2022). *Herbs, Shrubs and Trees of Potential Medicinal Benefits*. Boca Raton, FL: CRC Press. https://doi.org/10.1201/9781003205067

Husen, A. (2023). *Exploring Poisonous Plants: Medicinal Values, Toxicity Response and Other Potential Uses*. Boca Raton, FL: CRC Press. https://doi.org/10.1201/b23017

Husen, A. (2024a). *Antimalarial Medicinal Plants*. Boca Raton, FL: Taylor & Francis Group. https://doi.org/10.1201/9781003378396

Husen, A. (2024b). *Medicinal Spices and Condiment Crops*. Boca Raton, FL: Taylor & Francis Group. https://doi.org/10.1201/9781003387046

Husen, H., & Pant, M. (2024a). *Exploring Medicinal Orchids*. Boca Raton, FL: Taylor & Francis Group. https://doi.org/10.1201/b23375

Husen, H., & Pant, M. (2024b). *Tissue Culture Techniques and Medicinal Plants: Enhancing Propagation and Production*. Boca Raton, FL: Taylor & Francis Group. https://doi.org/10.1201/b23374

Hussain, A., Qarshi, I. A., Nazir, H., & Ullah, I. (2012). Plant tissue culture: current status and opportunities. *Recent Advances in Plant in Vitro Culture*, *6*(10), 1–28.

Jha, S. R., Naz, R., Asif, A., Okla, M. K., Soufan, W., Al-Ghamdi, A. A., & Ahmad, A. (2020). Development of an in vitro propagation protocol and a sequence characterized amplified region (SCAR) marker of *Viola serpens* Wall. ex Ging. *Plants*, *9*(2), 246.

Junior, C. A. D. K., Manechini, J. R. V., Corrêa, R. X., Pinto, A. C. R., da Costa, J. B., Favero, T. M., & Pinto, L. R. (2020). Genetic structure analysis in sugarcane (*Saccharum spp.*) using target region amplification polymorphism (TRAP) markers based on sugar-and lignin-related genes and potential application in core collection development. *Sugar Tech*, *22*, 641–654.

Kanthaliya, B., Joshi, A., Verma, K. K., & Arora, J. (2024). Efficient micropropagation protocol and assessment of genetic fidelity of micropropagated plantlets of *Pueraria tuberosa*. *Plant Biosystems: An International Journal Dealing with all Aspects of Plant Biology*, *159*(1), 26–35.

Kherwar, D., Usha, K., Mithra, S. A., & Singh, B. (2018). Microsatellite (SSR) marker assisted assessment of population structure and genetic diversity for morpho-physiological traits in guava (*Psidium guajava* L.). *Journal of Plant Biochemistry and Biotechnology*, *27*, 284–292.

Koppula, T., Sandhya, D., Rohela, G. K., Kommidi, S., & Mohammed, M. (2023). Micropropagation and assessment of genetic homogeneity of regenerants by ISSR and SCoT markers in *Solena amplexicaulis* (Lam.) Gandhi—a threatened medicinal cucurbit. *In Vitro Cellular & Developmental Biology-Plant*, *59*(6), 724–733.

Krishna, H., & Singh, S. K. (2007). Biotechnological advances in mango (*Mangifera indica* L.) and their future implication in crop improvement—a review. *Biotechnology Advances*, *25*(3), 223–243.

Kumari, A., Biswas, P., & Kumar, N. (2024). Molecular alterations for the enhancement of natural products in plants. In Sharma, A., & Verma, P. eds. *Biosynthesis of Natural Products in Plants: Bioengineering in Post-genomics Era* (pp. 39–57). Singapore: Springer Nature Singapore.

Kurepa, J., & Smalle, J. A. (2022). Auxin/cytokinin antagonistic control of the shoot/root growth ratio and its relevance for adaptation to drought and nutrient deficiency stresses. *International Journal of Molecular Sciences*, *23*(4), 1933.

Kwasniewska, J., & Bara, A. W. (2022). Plant cytogenetics in the micronuclei investigation—the past, current status, and perspectives. *International Journal of Molecular Sciences*, *23*(3), 1306.

Lemma, D. T., Banjaw, D. T., & Megersa, H. G. (2020). Micropropagation of medicinal plants. *Journal of Plant Breeding and Crop Science*, *7*, 796–802.

Liu, S., Liu, H. W., Wu, A., Hou, Y., An, Y. L., & Wei, C. L. (2017). Construction of fingerprinting for tea plant (*Camellia sinensis*) accessions using new genomic SSR markers. *Mol Breeding*, *37*, 93.

Luz, G. C., Strioto, D. K., Mangolin, C. A., & Maria, D. F. (2020). ISSR markers to assess genetic diversity of cultivated populations from artificial selection of *Stevia rebaudiana* (Bert.) Bertoni. *Breeding Science*, *70*(4), 508–514.

Moraes, R. M., Cerdeira, A. L., & Lourenço, M. V. (2021). Using micropropagation to develop medicinal plants into crops. *Molecules*, *26*(6), 1752–1768.

Oliya, B. K., Chand, K., Thakuri, L. S., Baniya, M. K., Sah, A. K., & Pant, B. (2021). Assessment of genetic stability of micropropagated plants of *Rhynchostylis retusa* (L.) using RAPD markers. *Scientia Horticulturae*, *281*, 110008.

Pant, M., & Husen, A. (2025). *Exploring Medicinal Bamboos.* Boca Raton, FL: Taylor & Francis Group, LLC. https://doi.org/10.1201/9781003485933

Pant, M., Sohrab, S. S., & Husen, A. (2025). *Genetic Improvement and Conservation Practice of Medicinal Plants.* Boca Raton, FL: Taylor & Francis Group. https://doi.org/10.1201/9781003482659

Permadi, N., Akbari, S. I., Prismantoro, D., Indriyani, N. N., Nurzaman, M., Alhasnawi, A. N., ... & Julaeha, E. (2024). Traditional and next-generation methods for browning control in plant tissue culture: current insights and future directions. *Current Plant Biology*, *31*, 100339.

Pharmawati, M., Yan, G., Sedgley, R., & Finnegan, P. M. (2004). Chloroplast DNA inheritance and variation in Leucadendron species (Proteaceae) as revealed by PCR-RFLP. *Theoretical and Applied Genetics*, *109*, 1694–1701.

Ramawat, K. G., & Arora, J. (2021). Medicinal plants domestication, cultivation, improvement, and alternative technologies for the production of high value therapeutics: an overview. In Ramawat, K. G., & Arora, J. eds. *Medicinal Plants: Domestication, Biotechnology and Regional Importance* (pp. 1–29). Cham: Springer.

Ramesh, M., Vijayakumar, K. P., Karthikeyan, A., & Pandian, S. K. (2011). RAPD based genetic stability analysis among micropropagated, synthetic seed derived and hardened plants of *Bacopa monnieri* (L.): a threatened Indian medicinal herb. *Acta Physiologiae Plantarum*, *33*, 163–171.

Rani, M., Miah, M. A. H., Hasan, M. T., Rashid, M. H. O., Yasmin, S., & Haque, M. S. (2024). Clonal propagation of turmeric (*Curcuma longa*) and confirmation of genetic fidelity of the micropropagated shoots by RAPD markers. *Plant Tissue Culture and Biotechnology*, *34*(1), 55–69.

Rasool Hassan, B. A. (2012). Medicinal plants (importance and uses). *Pharmaceutical Analytical Acta*, *3*(10), 2153–435.

Ravi, D., Siril, E. A., & Nair, B. R. (2021). SCAR marker development for the identification of elite germplasm of *Moringa oleifera* Lam.-a never die plant. *Plant Molecular Biology Reporter*, *39*(4), 850–861.

Rieseberg, L. H. (1996). Homology among RAPD fragments in interspecific comparisons. *Molecular Ecology*, *5*(1), 99–105.

Rohela, G. K., Jogam, P., Shabnam, A. A., Shukla, P., Abbagani, S., & Ghosh, M. K. (2018). In vitro regeneration and assessment of genetic fidelity of acclimated plantlets by using ISSR markers in PPR-1 (*Morus sp.*): an economically important plant. *Scientia Horticulturae*, *241*, 313–321.

Rosli, N. M., Yunus, M. F., Sundram, T. C. M., Zainuddin, Z., Midin, M. R., & Ponari, M. H. A. (2025). In vitro callus induction and identification of DNA variation in *Etlingera elatior* using Inter-simple sequence repeat (ISSR) markers. *Ornamental Horticulture*, *31*, 1–9.

Sarwat, M., Nabi, G., Das, S., & Srivastava, P. S. (2012). Molecular markers in medicinal plant biotechnology: past and present. *Critical Reviews in Biotechnology*, *32*(1), 74–92.

Savitikadi, P., Jogam, P., Rohela, G. K., Ellendula, R., Sandhya, D., Allini, V. R., & Abbagani, S. (2020). Direct regeneration and genetic fidelity analysis of regenerated plants of *Andrographis echioides* (L.): an important medicinal plant. *Industrial Crops and Products*, *155*, 112766.

Sharma, M. M., Verma, R. N., Singh, A., & Batra, A. (2014). Assessment of clonal fidelity of *Tylophora indica* (Burm. f.) Merrill "in vitro" plantlets by ISSR molecular markers. *SpringerPlus*, *3*, 1–9.

Sliwinska, E., Loureiro, J., Leitch, I. J., Šmarda, P., Bainard, J., Bureš, P., ... & Galbraith, D. W. (2022). Application-based guidelines for best practices in plant flow cytometry. *Cytometry Part A*, *101*(9), 749–781.

Sosnowski, J., Truba, M., & Vasileva, V. (2023). The impact of auxin and cytokinin on the growth and development of selected crops. *Agriculture*, *13*(3), 724.

Sultana, K. W., Das, S., Chandra, I., & Roy, A. (2022). Efficient micropropagation of *Thunbergia coccinea* Wall. and genetic homogeneity assessment through RAPD and ISSR markers. *Scientific Reports*, *12*(1), 1683.

Sun, S., Zhong, J., Li, S., & Wang, X. (2013). Tissue culture-induced somaclonal variation of decreased pollen viability in torenia (*Torenia fournieri* Lind.). *Botanical Studies*, *54*, 1–7.

Tikendra, L., Sushma, O., Amom, T., Devi, N. A., Paonam, S., Bidyananda, N., ... & Nongdam, P. (2022). Genetic clonal fidelity assessment of rhizome-derived micropropagated *Acorus calamus* L.–A medicinally important plant by random amplified polymorphic DNA and inter-simple sequence repeat markers. *Pharmacognosy Magazine*, *18*(77).

Tilwari, A., & Sharma, R. (2021). Random amplified polymorphic DNA and inter simple sequence repeat markers reveals genetic diversity between micro propagated, wild and field cultivated genotypes of *Gloriosa superba*: an endangered medicinal plant. *Molecular Biology Reports*, *48*(3), 2437–2452.

Vir, R., Jehan, T., Bhat, K.V., & Lakhanpaul, S. (2024). Assessment of genetic diversity among species mungo-radiata group of subgenus Ceratotropis of genus Vigna Savi. using amplified fragment length polymorphism (AFLP). *Vegetos*, *37*(4), 1443–1456.

Volenzo, T., & Odiyo, J. (2020). Integrating endemic medicinal plants into the global value chains: the ecological degradation challenges and opportunities. *Heliyon*, *6*(9), e04899.

Williams, J. G., Kubelik, A. R., Livak, K. J., Rafalski, J. A., & Tingey, S. V. (1990). DNA polymorphisms amplified by arbitrary primers are useful as genetic markers. *Nucleic Acids Research*, *18*(22), 6531–6535.

Xie, X., Rigor, P., & Baldi, P. (2009). MotifMap: a human genome-wide map of candidate regulatory motif sites. *Bioinformatics*, *25*(2), 167–174.

Zhou, L. G., & Wu, J. Y. (2006). Development and application of medicinal plant tissue cultures for production of drugs and herbal medicinals in China. *Natural Product Reports*, *23*(5), 789–810.

10 Role of Transcriptional Markers during *in vitro* Organogenesis

Gareema Pandey and Arpan Modi

10.1 INTRODUCTION

Unlike somatic cells of animals, plant cells possess the characteristic of regeneration which allow them to convert into an entire plant if specific growing conditions are provided. The phenomenon is called totipotency. In a simpler term, plant cells can undergo developmental plasticity and/or elasticity. The former refers to the permanent changes in the structure of the plants and is often an irreversible, whereas the latter shows the ability to come back to the original state of the plant cell (Gaillochet and Lohmann, 2015). Developmental plasticity often refers to the physiological changes or adaptations of the plants such as photomorphogenesis and shade avoidance mechanisms in model plants (Jong and Leyser, 2012). On the other hand, a physiological process such as guard cell-mediated opening and closing of stomata is a temporary change highlighting plant elasticity (Yi et al., 2019). Using this characteristic of totipotency, millions of plants can be placed in a production line and is a routine practice for plant tissue culture industries across the globe. Specific conditions refer to the medium on which the cells are proliferated, axenic conditions, and other physical conditions such as temperature, humidity, and photoperiod associated with plant growth (Haque et al., 2022). In plant science research, plant tissue culture serves as versatile platform for the starting from the study of plant growth under various experimental conditions such as plant stress (Wijerathna-Yapa and Hiti-Bandaralage, 2023; Pérez-Clemente and Gómez-Cadenas, 2012) to the development of transgenic and genome-edited plants (Loyola-Vargas and Avilez-Montalvo, 2018; Bekalu et al., 2023). One also has to know the difference between two similar but development-wise distinct terms, namely, plant regeneration and organogenesis. Organogenesis is the part of the regeneration process where whole plant formation is targeted. Classical example of both the processes is transgenic plant development and hairy root culture, respectively (Xu, Chang, et al., 2024). One of the techniques of plant tissue culture, called micropropagation, involves several step starting from explant (tiny part of the plant from which the production of million plants is desired) selection; establishment of the axenic culture (free from microbes); proliferation of cells; shoot multiplication; and rooting (Figure 10.1). All these steps are performed under aseptic conditions (*in vitro*). The growth medium plays a vital role in plant development under *in vitro* environments. It constitutes macro- and micronutrients, vitamins, a carbon source, plant growth regulators, and a solidifying agent like agar (in most cases). Growth of the plant under *in vitro* conditions can be manipulated by changing the type and concentrations of one or more of the media constituents. Based on the plant growth regulator supplied, the plant cell can turn into callus (unorganized, undifferentiated, friable/compact mass of cells), embryogenic callus, shoot, or root (Sudheer et al., 2022). *In vitro* flowering

DOI: 10.1201/9781003542681-10

FIGURE 10.1 Micropropagation of recalcitrant plant species *Ficus microcarpa* covering all the stages, namely, surface sterilization (a); initiation (b); proliferation (c); multiplication (d); rooting (e); primary (f); and secondary (g) hardening.

is also induced in orchids (Teixeira et al., 2014). Other types of organogenesis include hairy root cultures (Hanafy et al., 2016); tuber formation in potato (Mohamed et al., 2023); and *in vitro* fruiting (Xiong et al., 2021).

In a plant breeding strategy, molecular markers are used to characterize plants, select hybrids, and identify quantitative trait loci (QTLs). All these exercises are done using information contained in DNA of plants (Collard and Mackill, 2008). However, the genes responsible for particular traits, their associated transcription factors and microRNAs are also equally important to identify the level of expression and thereby the extent to which the traits can be quantified. Such a database is generated using transcriptome sequencing, suppression subtractive hybridization (SSH), expressed sequence tags (ESTs), serial analysis of gene expression (SAGE), cap analysis of gene expression (CAGE), and rapid analysis of cDNA ends (RACE). These platforms are used based on the experimental conditions and desired output (Panahi et al., 2025). Researchers have identified transcription factors, and microRNAs using *in silico* analyses, and have already used the information to develop transgenic lines capable of withstanding adverse environmental conditions, exhibiting good plant vigour, and showing enhanced regenerative capabilities (Pan et al., 2018; Yu et al., 2018; Kong et al., 2020; Boutilier et al., 2002). Transcription factor genes such as *Growth Regulating Factor-5* (*GRF5*), *Baby Boom* (*BBM*), *Wuschel* (*WUS*), *Wound Induced Differentiation 1* (*WIND1*), and *Elongation Of Shoot Regeneration 1* (*ESR1*) were shown to enhance shoot regeneration, somatic embryogenesis, and callus formation in a variety of plant species (Xu, Chang, et al., 2024; Heidmann et al., 2011; Bouchabke-Coussa et al., 2013; Bull et al., 2023; Iwase et al., 2013; Banno et al., 2001).

Organogenesis in tissue culture is defined as the formation of a new organ from either a differentiated or undifferentiated mass of cells under a controlled environment (*in vitro*) where all the conditions for the plant cell to grow are met (Long et al., 2022). Organogenesis can be a separate process, as desired or a part of the regeneration protocols. Shooting and rooting, in most cases, during plant tissue culture operations, are regarded as organogenesis. *De novo* shoot formation takes place from the pre-existing meristematic cells available in the explant, and the process is known as direct organogenesis. Such phenomena are observed in nature when the shoot tips of most plant species are cut, which results in the formation of a shoot from the lateral meristems. However, under tissue culture environment, multiple shoots may also be induced, especially when large numbers of plantlet formations are targeted. In a woody plant species like Indian sandalwood, rosewood, and teak, *in vitro* procedures for the induction of callus, shoot multiplication, and rooting showed poor and inconsistent results due to the nature and stage of the plant (Mexudhan et al., 2023). Woody plant species are highly recalcitrant in terms of *in vitro* manipulations. They take more than 2 years in micropropagation to develop into complete plantlets. Apart from the biology of the plant cells, the external environment, such as nutrient availability, plant growth regulators' treatment, carbon source, etc., will also be responsible for the plant growth under *in vitro* conditions. All the factors together point towards the fate of an explant, whether it would turn into an organ or just callusing would go on. Therefore, while optimizing the plant growth condition in tissue culture, it is extremely important to predict the next structure of the explant, and based on that, further manipulation towards the organogenesis can be scheduled (Patel et al., 2022). Tissue culture manipulations are challenging in certain plant species, and molecular insights into the process of organogenesis can help biotechnologists to employ distinct transcription factors and associated genes to overcome explant recalcitrance.

10.2 CROSSTALK OF ORGANOGENESIS AND GENE EXPRESSION

For the development of new shoots either under *in vitro* or *in vivo* conditions, auxin and cytokinins play a key role. They are also very important during embryogenesis under both forms, namely, the natural way of embryogenesis and somatic embryogenesis. Auxins are involved in cell polarity, cell patterning, embryo development, formation of vascular tissue, phyllotaxis (leaf arrangement), and root formation. Cytokinin, on the other hand, promotes cell division as its name suggests. Not only auxin and cytokinin levels but also their ratio significantly contributes to plant cell development. Auxin to cytokinin ratio modulates the organogenesis pathway through cellular, metabolic, proteomic, and transcriptomic levels (Pernisová et al., 2011). With the thousands of articles published in different regions across the world, it is now well-documented that high auxin to lower cytokinins promotes root growth and high cytokinins to lower auxin promotes shoots (Nath et al., 2025). Once the dormancy is broken during embryogenesis, formation of apical meristems begins. Primary growth regulator for the initiation and proliferation of apical meristem is auxin, and an asymmetric distribution of the auxins directs the organogenesis. Auxin efflux carrier proteins called PIN- FORMED (PIN) are responsible for the directional auxin transport in the plants (Zhou and Luo, 2018). They are also involved in the environmental responses (Su et al., 2011). For the establishment of shoot apical meristem (SAM), radial pattern of the embryo formation takes place first, which is regulated by two transcription factors, namely, the class III homeodomain-leucine zipper (HD-ZIPIII) and KANADI proteins (KAN). HD-ZIPIII promotes the adaxial cell fate, maintains the identity of SAMs, and regulates vascular development, whereas KAN, being an antagonist of HD-ZIPIII, promotes abaxial cell fate and regulates lateral root formation, embryo patterning, and leaf margin development (Emery et al., 2003). At the later stage of embryo development, bilateral symmetry is established in which the interplay of the expression of two genes, namely, *SHOOT MERISTEMLESS* (*STM*) and *CUP SHAPED COTYLEDON* (*CUP*) plays a significant role. The *STM* gene regulates the formation of shoots in developing embryos and has interactions with two other genes *WUSCHEL* (*WUS*) and *ZWILLE* to fine-tune the tissue development (Endrizzi et al., 1996; Lechon et al., 2025). In the later stages of embryogenesis, *WUS* plays a leading role in the development of meristematic stem cells (MSCs). *WUS* expression takes place in a bunch of cells from which MSCs emerge. Experimental evidence suggested that the transport of WUS protein is mediated by plasmodesmata (Mayer et al., 1998; Yadav et al., 2011). *CUC* defines the boundaries between different organs during plant development. It is also an activator of *STM* (Hake, 2019). Further in the development of an MSC, *WUS* activates the production of CLAVATA (CLV) ligand receptor system, which has a feedback inhibition of WUS (Somssich et al., 2016). This controls the size of the stem cells in the developing tissues. During plant regeneration, these transcription factors are governed by the endogenous levels of plant growth regulators, exogenously supplied and the ratio of different plant growth regulators (mainly auxin and cytokinin). However, other factors like light- and enzyme inhibitor-induced changes in the expression of these genes also play a significant role in organ and whole plant development (Zhang et al., 2024; Liu et al., 2025; Park et al., 2025).

In plant tissue culture, organogenesis is achieved using a combination of hormones (auxin and cytokinins). *De novo* organogenesis during the plant regeneration process passes through various developmental stages, namely, acquisition of pluripotency; formation of shoot pro-meristem; induction of shoot progenitor; and shoot outgrowth (Shin et al., 2020). Table 10.1 shows the differentially expressed genes during distinct stages of organogenesis or regeneration in a variety of plant species.

10.3 SCOPE IN IMPROVEMENT OF MEDICINAL PLANTS

Medicinal plants are rich in secondary metabolites and due to which they have been utilized since ages. During the course of their cultivation and utilization, new metabolites were found and using the conventional breeding and modern biotechnological methods, improvement in quality and quantity of such metabolites was observed (Suri et al., 2024; Husen and Pant, 2024; Pant et al., 2025). One of the crop improvement strategies in biotechnology is the over-expression of transcription factors or genes involved in enhancing the concentration of secondary metabolites. This can be achieved through cutting-edge transformation technologies like transgenic plant development, hairy-root culture, cell culture, and genome editing tools (Rabeh et al., 2025). Tissue culture operations are a prerequisite for the successful establishment of genetically engineered or gene-edited plants (Luo et al., 2025). At the same time, several species of medicinal plants are also found to be recalcitrant to tissue culture manipulations (Bennur et al., 2025). Due to this recalcitrance, regeneration or direct organogenesis from differentiated or undifferentiated tissues becomes a real bottleneck in transgenic plant production. Several strategies have been reviewed by Patel et al. (2022) where they explained that explant recalcitrance could be overcome by manipulating media composition, method of micropropagation, and pre-treatment of explants. Experimental evidence suggests that epigenetic modifications are also responsible for the explant recalcitrance (Hesami et al., 2024). Many transcription factor genes are engaged in the regeneration process, and as the database for the transcription factors responsible for the morphogenesis is widely available, these genes may be utilized to overcome the explant recalcitrance and transformation efficiency. These factors are isolated from the source plant and over-expressed ectopically to enhance the organogenesis process required for the successful establishment of transformation process (Kong et al., 2020). Table 10.2 shows the list of transcription factors or genes associated with the regeneration process in plants and their scope in the improvement of medicinal and non-medicinal plant species. In some cases, the ectopic expression of morphogenesis-related genes such as *WUS* showed negative effects on transformed plants (Victorathisayam and Sridevi, 2020).

TABLE 10.1
Expression of Genes during *in vitro* Organogenesis or Regeneration and Their Physiological, Molecular, or Biochemical Effects in Different Plant Species

Plant Species	Stage of Organogenesis/ Plant's Response	Gene(s)/TF(s) Studied	Effect(s)/Function(s)	References
Eucalyptus species	Callus development (mature callus)	LHW (TF); F-box protein; Protein TIFY 10a; Protein EXORDIUM; ß – fructofuranosidase	qRT-PCR results revealed that in a mature callus stage *LHW*; *F-box* protein; Protein *TIFY 10a*; and *β – fructofuranosidase* were up-regulated and Protein *EXORDIUM* was down-regulated.	Zhang et al. (2022)
Arabidopsis thaliana	Pluripotent callus induction	*WRKY23* and *bHLH041*	For the auxin-induced pluripotent callus induction during *in vitro* organogenesis, *WRKY23* and *bHLH041* were found to be transcriptional activator and repressor, respectively.	Xu, Zhong, et al. (2024)
Arachis hypogaea L.	Embryogenic callus induction	*Peroxisomal ABC Transporter 1, PPR superfamily protein, SAUR-like auxin responsive protein, MYB transcription factor, Homeodomain-like transcriptional regulator*	These five genes were associated with embryogenic callus induction over the subculture cycles.	Luo et al. (2024)
Cucumis sativus L.	Direct organogenesis from leaf	*CsARF6 and CsWOX9*	The major-effect QTLs, responsible for the direct organogenesis (shoot regeneration), are under the influence of the genes studied which indicated that *CsARF6* and *ScWOX9* are directly associated with *in vitro* shoot regeneration.	Słomnicka et al. (2025)
Capsicum annuum L.	*In vitro* regeneration	Transcriptome	*A total of 505 differentially expressed genes were found belonging to 55 transcription factor families. The expression pattern of these genes in initial stages may be utilized to develop the in vitro regeneration system in pepper.*	Li et al. (2024)
Santalum album L.	Direct and indirect organogenesis	*Alternative oxidase, LEA, Cytochrome P-450, ABC Transporter, Serine-Threonine Phosphatase*	*LEA* expressions were found to increased up to 70-fold during direct organogenesis pathway and down-regulated during indirect pathway. The expression pattern can be monitored and used to screen the medium when the tissue culture protocol development process is time consuming.	Pandey et al. (2022)
Moringa oleifera Lam.	Embryogenic callus formation	miRNAs-lncRNAs	Expression of mes-miR319b and ptc-miR319a were increased up to 7-fold during initial phases of EC formation.	Yang et al. (2024)
Lactuca sativa L.	*In vitro* morphogenesis	*LsWUS1L, LsWUS2L, LsSTM1L*	Expression of *LsWUS2L, LsSTM1L* increased extremely after 4 days of inoculation.	Farina et al. (2021)

TABLE 10.2
Transcription Factors Used for the Transformation in Recalcitrant Medicinal- and Non-medicinal Plants, Their Effect on Regeneration

Transcription Factor (Gene)	Plant Species (Medicinal Uses)	Transformation Feature	Effect	References
SERK1	*Coffea canephora* (Neuroprotective actions, analgesic)	Overexpression	The rate of somatic embryogenesis was enhanced. The over-expressing lines were also found to show increased expression of other morphogenic genes such as *CcAGL15*, *CcWUS*, *CcBBM*, *CcPKL*, *CcYUC1*, *CcPIN1*, and *CcPIN4*.	Pérez-Pascual et al. (2018)
WUS	*Betula platyphylla* (Diuretic, antirheumatic, skincare)	Overexpression	Transgenic line exhibiting higher expression of *AtWUS* in Birtch showed increased rate of somatic embryogenesis and expression of SE-related genes.	Lou et al. (2022)
WUS	*Medicago truncatula Gaertn* (Digestive disorders)	Overexpression	Increased callogenesis, embryogenesis under hormone-free medium. Enhanced the embryogenic potential of a hairy root fragment.	Kadri et al. (2021)
WUS, SBH, IPT, GRF, GIF *PDS*	*Cannabis sativa* L. (Anti-inflammatory, pain relief, skincare, neuroprotective activities)	*Overexpression of morphogenesis related genes and gene editing of CsPDS1*	Significantly higher shoot induction rate was achieved in transformed line. Albino phenotype was observed in *CsPDS1* line.	Zhang et al. (2021)
BBM	*Malus domestica* L.	Overexpression	Shoot regeneration efficiency during *in vitro* cultures was increased up to 31% as compared to control plants (5–10%).	Chen et al. (2022)
BBM	*Zea maize* L.	Overexpression; Inducible; *Cre/LoxP* system	Significantly increased transformation efficiency; more numbers of usable transformants; drought-induced promoter removed the transgene to alleviate negative effects on plant growth in regenerated shoots.	Kumar et al. (2025)

10.4 CONCLUSIONS AND FUTURE PROSPECTS

Modern techniques of plant improvement comprise of genetic engineering and gene editing which have been utilized since last decades and a number of transgenic and gene-edited plants have been generated. At one end, the techniques are promising; the other side shows the bottlenecks of the protocols. *In vitro* organogenesis is required for both plant production and trait improvement. Transcriptional markers are the signature of expressed genes that can be utilized to predict the developmental stage of the plant. Such techniques can be used to optimize the tissue culture conditions to escape from the lengthy protocols. Moreover, these transcriptional markers are genes associated with morphogenesis and, therefore, they are used to overcome explant recalcitrance by plant transformation methods in a variety of plants, including medicinal species. To date, over-expression of morphogenesis-related genes has been carried out to boost plant growth; however, down-regulation of genes involved in inhibition mechanisms has not been done and, therefore, the future strategies include CRISPR/Cas9-mediated genome editing and RNP (Ribo Nucleo-Protein)-mediated genome editing to release the gene-edited plants that face minimal regulatory constraints.

REFERENCES

Banno, H., Ikeda, Y., Niu, Q.W. and Chua, N.H., 2001. Overexpression of Arabidopsis ESR1 induces initiation of shoot regeneration. *The Plant Cell, 13*(12), pp.2609–2618.

Bekalu, Z.E., Panting, M., Bæksted Holme, I. and Brinch-Pedersen, H., 2023. Opportunities and challenges of in vitro tissue culture systems in the era of crop genome editing. *International Journal of Molecular Sciences, 24*(15), 11920.

Bennur, P.L., O'Brien, M., Fernando, S.C. and Doblin, M.S., 2025. Improving transformation and regeneration efficiency in medicinal plants: insights from other recalcitrant species. *Journal of Experimental Botany, 76*(1), pp.52–75.

Bouchabke-Coussa, O., Obellianne, M., Linderme, D., Montes, E., Maia-Grondard, A., Vilaine, F. and Pannetier, C., 2013. Wuschel overexpression promotes somatic embryogenesis and induces organogenesis in cotton (*Gossypium hirsutum* L.) tissues cultured in vitro. *Plant Cell Reports, 32*(5), pp.675–686.

Boutilier, K., Offringa, R., Sharma, V.K., Kieft, H., Ouellet, T., Zhang, L., Hattori, J., Liu, C.M., van Lammeren, A.A., Miki, B.L. and Custers, J.B., 2002. Ectopic expression of BABY BOOM triggers a conversion from vegetative to embryonic growth. *The Plant Cell, 14*(8), pp.1737–1749.

Bull, T., Debernardi, J., Reeves, M., Hill, T., Bertier, L., Van Deynze, A. and Michelmore, R., 2023. GRF–GIF chimeric proteins enhance in vitro regeneration and Agrobacterium-mediated transformation efficiencies of lettuce (*Lactuca spp.*). *Plant Cell Reports, 42*(3), pp.629–643.

Chen, J., Tomes, S., Gleave, A.P., Hall, W., Luo, Z., Xu, J. and Yao, J.L., 2022. Significant improvement of apple (*Malus domestica* Borkh.) transgenic plant production by pre-transformation with a Baby boom transcription factor. *Horticulture Research, 9*, uhab014.

Collard, B.C. and Mackill, D.J., 2008. Marker-assisted selection: an approach for precision plant breeding in the twenty-first century. *Philosophical Transactions of the Royal Society B: Biological Sciences, 363*(1491), pp.557–572.

De Jong, M. and Leyser, O., 2012. Developmental plasticity in plants. In Bartel, R. and Lawrence, P. eds. *Cold Spring Harbor Symposia on Quantitative Biology* (Vol. 77, pp. 63–73). Cold Spring Harbor Laboratory Press.

Emery, J.F., Floyd, S.K., Alvarez, J., Eshed, Y., Hawker, N.P., Izhaki, A., Baum, S.F. and Bowman, J.L., 2003. Radial patterning of Arabidopsis shoots by class III HD-ZIP and KANADI genes. *Current Biology, 13*(20), pp.1768–1774.

Endrizzi, K., Moussian, B., Haecker, A., Levin, J.Z. and Laux, T., 1996. The SHOOT MERISTEMLESS gene is required for maintenance of undifferentiated cells in Arabidopsis shoot and floral meristems and acts at a different regulatory level than the meristem genes WUSCHEL and ZWILLE. *The Plant Journal, 10*(6), pp.967–979.

Farina, P., Fambrini, M., Pugliesi, C. and Viviani, A., 2021. Expression of homeobox genes during in vitro culture of Lactuca sativa. *Plant Biosystems: An International Journal Dealing with all Aspects of Plant Biology, 155*(3), pp.609–621.

Gaillochet, C. and Lohmann, J.U., 2015. The never-ending story: from pluripotency to plant developmental plasticity. *Development, 142*(13), pp.2237–2249.

Hake, S., 2019. Identification of cup-shaped cotyledon: new ways to think about organ initiation. *The Plant Cell, 31*(6), pp.1202–1203.

Hanafy, M.S., Matter, M.A., Asker, M.S. and Rady, M.R., 2016. Production of indole alkaloids in hairy root cultures of *Catharanthus roseus* L. and their antimicrobial activity. *South African Journal of Botany, 105*, pp.9–18.

Haque, M.I., Singh, P.K., Ghuge, S., Kumar, A., Rai, A.C., Kumar, A. and Modi, A., 2022. A general introduction to and background of plant tissue culture: Past, current, and future aspects. In Rai, A.C. Kumar, A. Modi, A. and Singh, M. eds. *Advances in Plant Tissue Culture* (pp. 1–30). Academic Press.

Heidmann, I., De Lange, B., Lambalk, J., Angenent, G.C. and Boutilier, K., 2011. Efficient sweet pepper transformation mediated by the BABY BOOM transcription factor. *Plant Cell Reports, 30*(6), pp.1107–1115.

Hesami, M., Pepe, M., Spitzer-Rimon, B., Eskandari, M. and Jones, A.M.P., 2024. Epigenetic factors related to recalcitrance in plant biotechnology. *Genome, 68*, pp.1–11.

Husen, H. and Pant, M., 2024. *Tissue Culture Techniques and Medicinal Plants: Enhancing Propagation and Production.* Taylor & Francis Group. https://doi.org/10.1201/b23374

Iwase, A., Mitsuda, N., Ikeuchi, M., Ohnuma, M., Koizuka, C., Kawamoto, K., Imamura, J., Ezura, H. and Sugimoto, K., 2013. Arabidopsis WIND1 induces callus formation in rapeseed, tomato, and tobacco. *Plant Signaling & Behavior*, *8*(12), e27432.

Kadri, A., Grenier De March, G., Guerineau, F., Cosson, V. and Ratet, P., 2021. WUSCHEL overexpression promotes callogenesis and somatic embryogenesis in Medicago truncatula Gaertn. *Plants*, *10*(4), p.715.

Kong, J., Martin-Ortigosa, S., Finer, J., Orchard, N., Gunadi, A., Batts, L.A., Thakare, D., Rush, B., Schmitz, O., Stuiver, M. and Olhoft, P., 2020. Overexpression of the transcription factor GROWTH-REGULATING FACTOR5 improves transformation of dicot and monocot species. *Frontiers in Plant Science*, *11*, 572319.

Kumar, R., Azanu, M.K., Lee, K., Kang, M., Zobrist, J., Bùi, J.T., Wang, N., Gordon-Kamm, W., Van Eck, J. and Wang, K., 2025. Comparative analysis of maize leaf whorl-based transformation using ABA-induced Bbm/Wus2 removal in two Agrobacterium ternary vector systems. *In Vitro Cellular & Developmental Biology-Plant*, *61*, pp.1–14.

Lechon, T., Kent, N.A., Murray, J.A. and Scofield, S., 2025. Regulation of meristem and hormone function revealed through analysis of directly-regulated SHOOT MERISTEMLESS target genes. *Scientific Reports*, *15*(1), p.240.

Li, X., Mushtaq, N., Xing, N., Wu, S., Liu, J. and Wang, Z., 2024. Efficient In vitro regeneration system and comparative transcriptome analysis offer insight into the early development characteristics of explants from cotyledon with partial petiole in small-fruited pepper (*Capsicum annuum*). *International Journal of Molecular Sciences*, *25*(14), 7547.

Liu, C., Jia, S., Wei, Q., Zhang, Z., Xie, L. and Zeng, R., 2025. Mechanism of light induced shoot regeneration from rhizome in *Cymbidium aloifolium*. *Scientia Horticulturae*, *350*, 114324.

Long, Y., Yang, Y., Pan, G. and Shen, Y., 2022. New insights into tissue culture plant-regeneration mechanisms. *Frontiers in Plant Science*, *13*, 926752.

Lou, H., Huang, Y., Wang, W., Cai, Z., Cai, H., Liu, Z., Sun, L. and Xu, Q., 2022. Overexpression of the AtWUSCHEL gene promotes somatic embryogenesis and lateral branch formation in birch (*Betula platyphylla* Suk.). *Plant Cell, Tissue and Organ Culture (PCTOC)*, *150*(2), pp.371–383.

Loyola-Vargas, V.M. and Avilez-Montalvo, R.N., 2018. Plant tissue culture: a battle horse in the genome editing using CRISPR/Cas9. In Teixeira da Silva, J. A. ed. *Plant Cell Culture Protocols* (pp.131–148). Springer.

Luo, D., Shi, L., Sun, Z., Qi, F., Liu, H., Xue, L., Li, X., Liu, H., Qu, P., Zhao, H. and Dai, X., 2024. Genome-wide association studies of embryogenic callus induction rate in peanut (*Arachis hypogaea* L.). *Genes*, *15*(2), p.160.

Luo, G., Trinh, M.D.L., Falkenberg, M.K.D., Chiurazzi, M.J., Najafi, J., Nørrevang, A.F., Correia, P.M.P. and Palmgren, M., 2025. Unlocking in vitro transformation of recalcitrant plants. *Trends in Plant Science*, *30*, 123–136.

Mayer, K.F., Schoof, H., Haecker, A., Lenhard, M., Jürgens, G. and Laux, T., 1998. Role of WUSCHEL in regulating stem cell fate in the Arabidopsis shoot meristem. *Cell*, *95*(6), pp.805–815.

Mexudhan, Jha, Z., Singh, L., Lal, J. and Tirkey, J., 2023. Problem and its remedy of micropropagation in woody trees: a review. *Environment and Ecology*, *41*(1C), pp.633–639.

Mohamed, A.E.S. and Girgis, N.D., 2023. Factors affecting in vitro tuberization of potato. *Bulletin of the National Research Centre*, *47*(1), p.80.

Nath, J., Kumari, A., Joshi, S., Gusain, S., Kumari, K., Yadav, S.K. and Joshi, R., 2025. Understanding the molecular landscape of de novo organogenesis: insights into cell fate reprogramming. *Discover Biotechnology*, *2*(1), pp.1–19.

Pan, J., Huang, D., Guo, Z., Kuang, Z., Zhang, H., Xie, X., Ma, Z., Gao, S., Lerdau, M.T., Chu, C. and Li, L., 2018. Overexpression of microRNA408 enhances photosynthesis, growth, and seed yield in diverse plants. *Journal of Integrative Plant Biology*, *60*(4), pp.323–340.

Panahi, B., Hamid, R. and Jalaly, H.M.Z., 2025. Deciphering plant transcriptomes: leveraging machine learning for deeper insights. *Current Plant Biology*, *32*, 100432.

Pandey, G., Patil, G., Modi, A., Desai, S., Desai, P. and Narayanan, S., 2022. Expression pattern of in vitro organogenesis associated genes as transcriptional marker in sandalwood (*Santalum album* L.) micropropagation. *Plant Tissue Culture and Biotechnology*, *32*(2), pp.103–113.

Pant, M., Sohrab, S.S. and Husen, A. (2025). *Genetic Improvement and Conservation Practice of Medicinal Plants.* Taylor & Francis Group. https://doi.org/10.1201/9781003482659

Park, S.H., Jeong, Y.J., Kim, S., Lee, J., Kim, C.Y. and Jeong, J.C., 2025. Trichostatin A promotes de novo shoot regeneration from Arabidopsis root explants via a cytokinin related pathway. *Scientific Reports*, *15*(1), p.978.

Patel, P., Sarswat, S.K. and Modi, A., 2022. Strategies to overcome explant recalcitrance under in vitro conditions. In Rai, A.C. Kumar, A. Modi, A. and Singh, M. eds. *Advances in Plant Tissue Culture* (pp. 283–294). Academic Press.

Pérez-Clemente, R.M. and Gómez-Cadenas, A., 2012. In vitro tissue culture, a tool for the study and breeding of plants subjected. In Pérez-Clemente, R.M. and Gómez-Cadenas, A. eds. *Recent Advances in Plant in Vitro Culture* (pp. 91–110). Springer.

Pérez-Pascual, D., Jiménez-Guillen, D., Villanueva-Alonzo, H., Souza-Perera, R., Godoy-Hernández, G. and Zúñiga-Aguilar, J.J., 2018. Ectopic expression of the Coffea canephora SERK1 homolog-induced differential transcription of genes involved in auxin metabolism and in the developmental control of embryogenesis. *Physiologia Plantarum*, *163*(4), pp.530–551.

Pernisová, M., Kuderová, A. and Hejátko, J., 2011. Cytokinin and auxin interactions in plant development: metabolism, signalling, transport and gene expression. *Current Protein and Peptide Science*, *12*(2), pp.137–147.

Rabeh, K., Hnini, M. and Oubohssaine, M., 2025. A comprehensive review of transcription factor-mediated regulation of secondary metabolites in plants under environmental stress. *Stress Biology*, *5*(1), p.15.

Shin, J., Bae, S. and Seo, P.J., 2020. De novo shoot organogenesis during plant regeneration. *Journal of Experimental Botany*, *71*(1), pp.63–72.

Słomnicka, R., Cieplak, M., Antosiewicz, M., Sadłos, A., Galczak, A., Kaźmińska, K. and Bartoszewski, G., 2025. Identification of quantitative trait loci for in vitro plant regeneration from leaf microexplants in cucumber (*Cucumis sativus* L.). *Journal of Applied Genetics*, *66*(3), pp.545–555.

Somssich, M., Je, B.I., Simon, R. and Jackson, D., 2016. CLAVATA-WUSCHEL signaling in the shoot meristem. *Development*, *143*(18), pp.3238–3248.

Su, Y.H., Liu, Y.B. and Zhang, X.S., 2011. Auxin–cytokinin interaction regulates meristem development. *Molecular Plant*, *4*(4), pp.616–625.

Sudheer, W.N., Praveen, N., Al-Khayri, J.M. and Jain, S.M., 2022. Role of plant tissue culture medium components. In Rai, A.C. Kumar, A. Modi, A. and Singh, M. eds. *Advances in Plant Tissue Culture* (pp. 51–83). Academic Press.

Suri, A., Bhardwaj, P. and Sharma, T., 2024. Cultivating nature's pharmacy: strategies for medicinal plants improvement. *South African Journal of Botany*, *169*, pp.219–230.

Teixeira da Silva, J.A., Kerbauy, G.B., Zeng, S., Chen, Z. and Duan, J., 2014. In vitro flowering of orchids. *Critical Reviews in Biotechnology*, *34*(1), pp.56–76.

Victorathisayam, T. and Sridevi, G., 2020. Ectopic expression of WUSCHEL (AtWUS) gene alters plant growth and development in rice. *Plant*, *8*(3), p.43.

Wijerathna-Yapa, A. and Hiti-Bandaralage, J., 2023. Tissue culture—A sustainable approach to explore plant stresses. *Life*, *13*(3), p.780.

Xiong, Y., Chen, S., Wei, Z., Yu, X., Pang, J., Zhang, T., Wu, K., Ren, H., Jian, S., da Silva, J.A.T. and Ma, G., 2021. In vitro flowering and fruiting in *Portulaca pilosa* L. *South African Journal of Botany*, *140*, pp.1–3.

Xu, C., Chang, P., Guo, S., Yang, X., Liu, X., Sui, B., Yu, D., Xin, W. and Hu, Y., 2024. Transcriptional activation by *WRKY23* and derepression by removal of *bHLH041* coordinately establish callus pluripotency in Arabidopsis regeneration. *The Plant Cell*, *36*(1), pp.158–173.

Xu, P., Zhong, Y., Xu, A., Liu, B., Zhang, Y., Zhao, A., Yang, X., Ming, M., Cao, F. and Fu, F., 2024. Application of developmental regulators for enhancing plant regeneration and genetic transformation. *Plants*, *13*(9), p.1272.

Yadav, R.K., Perales, M., Gruel, J., Girke, T., Jönsson, H. and Reddy, G.V., 2011. WUSCHEL protein movement mediates stem cell homeostasis in the Arabidopsis shoot apex. *Genes & Development*, *25*(19), pp.2025–2030.

Yang, E., Zheng, M., Zhang, L., Chen, X. and Zhang, J., 2024. Integrated analysis of microRNAs and lncRNAs expression profiles reveals regulatory modules during adventitious shoot induction in Moringa oleifera Lam. *BMC Plant Biology*, *24*(1), p.1237.

Yi, H., Chen, Y., Wang, J.Z., Puri, V.M. and Anderson, C.T., 2019. The stomatal flexoskeleton: how the biomechanics of guard cell walls animate an elastic pressure vessel. *Journal of Experimental Botany*, *70*(14), pp.3561–3572.

Yu, C., Chen, Y., Cao, Y., Chen, H., Wang, J., Bi, Y.M., Tian, F., Yang, F., Rothstein, S.J., Zhou, X. and He, C., 2018. Overexpression of miR169o, an overlapping microRNA in response to both nitrogen limitation and bacterial infection, promotes nitrogen use efficiency and susceptibility to bacterial blight in rice. *Plant and Cell Physiology*, *59*(6), pp.1234–1247.

Zhang, T., Zhang, R., Zeng, X.Y., Lee, S., Ye, L.H., Tian, S.L., Zhang, Y.J., Busch, W., Zhou, W.B., Zhu, X.G. and Wang, P., 2024. GLK transcription factors accompany ELONGATED HYPOCOTYL5 to orchestrate light-induced seedling development in Arabidopsis. *Plant Physiology*, 194(4), pp.2400–2421.

Zhang, X., Xu, G., Cheng, C., Lei, L., Sun, J., Xu, Y., Deng, C., Dai, Z., Yang, Z., Chen, X. and Liu, C., 2021. Establishment of an Agrobacterium-mediated genetic transformation and CRISPR/Cas9-mediated targeted mutagenesis in Hemp (*Cannabis Sativa* L.). *Plant Biotechnology Journal*, *19*(10), pp.1979–1987.

Zhang, Y., Li, J., Li, C., Chen, S., Tang, Q., Xiao, Y., Zhong, L., Chen, Y. and Chen, B., 2022. Gene expression programs during callus development in tissue culture of two Eucalyptus species. *BMC Plant Biology*, *22*(1), p.1.

Zhou, J.J. and Luo, J., 2018. The PIN-FORMED auxin efflux carriers in plants. *International Journal of Molecular Sciences*, *19*(9), p.2759.

11 Genome-Editing Tools for Plant-Based Improved Medicines

Reut Peer, Arpan Modi, and Megha Barot

11.1 INTRODUCTION

In animal kingdom, where movement from one place to the other is possible, the machinery for escape from an adverse situation is mostly physical. However, in the plant kingdom, being a sessile organism, they have to cope with the metabolic machinery in the same situation. In the metabolic context, essential metabolites like carbohydrates, proteins, lipids, nucleic acids, vitamins, and minerals are regarded as primary metabolites. These metabolites are produced in large amounts in plants, and they impart a structural basis and energy production for the survival of plants (Bocso and Butnariu, 2022). There is also a different class of metabolites, which are synthesized from primary metabolites and in very minute quantities when required. These are secondary metabolites. They are also vital, but their synthesis is species-specific, condition-specific, organ-specific, and time-specific. They help the plants under adverse environmental conditions, such as biotic and abiotic stresses, to avoid, tolerate, or impart resistance (Palazon and Alcalde, 2025; Pagare et al., 2015). Common secondary metabolites found in plants are alkaloids, terpenoids, phenolics, flavonoids, tannins, saponins, phytosterols, and glycosides (Mera et al., 2019; Thirumurugan et al., 2018). These secondary chemicals are the plant's natural defence system, which often helps a plant to survive under biotic and abiotic conditions. Interestingly, these secondary chemicals are also helpful for humans. Various chemical-based industries like textile, healthcare, skincare, pharmaceuticals, agrochemicals, paint, and dyes use these metabolites to produce commercially viable products. Not only terrestrial plants, but also aquatic plants have great diversity and provide useful chemicals for the industries (Barot et al., 2019). To date, in the pharmaceutical industry, thousands of metabolites have been produced, formulated, and modified. Many of them are plant-based and, therefore, in the present era, a branch of pharmacology called ethnopharmacology has emerged as a promising alternative in the healthcare industry (Pirintsos et al., 2022). Apart from Indian classical literature, the knowledge on the use of traditional medicinal plants has been recognized in the Chinese book Shannong's Herbal Atlas, classical Greek antiquity, Hippocrates of Kos, Galen, and Pedanios Dioscurides (Vaou et al., 2025).

Strategies for the extraction of natural products from medicinal plants have also been modified over the period of time. Technological interventions have made the extraction of secondary metabolites a simple, cost-effective, fast, and efficient process (Roopashree and Naik, 2019). Conventional methods of extraction comprise maceration, digestion, infusion, percolation, decoction, and Soxhlet extraction. Disadvantages associated with these methods are time-consuming protocols, use of hazardous chemicals in high quantities, and energy consumption. Modern techniques, in contrast, are

DOI: 10.1201/9781003542681-11

accelerated solvent extraction, microwave-assisted extraction, ultrasound-assisted extraction, supercritical fluid extraction, enzyme-assisted extraction, and pressurized hot water extraction. Using modern tools and techniques, a wide range of secondary metabolites from different groups have been isolated from different plant species (Bitwell et al., 2023). Where extraction protocols of these chemicals are well established and feasible, understanding the biosynthesis pathway, enzymes, and genes involved in the production is necessary to employ the biotechnological interventions. Biosynthesis of alkaloids, terpenoids, and flavonoids is synthesized in plants through amino acid metabolism, mevalonic acid or methyl erythritol phosphate, and phenylpropanoid pathways, respectively (Huang et al., 2022). Online resources like Secondary Metabolism Collaboratory (SMC) and Secondary Metabolites Bioinformatics Portal (SMBP) are available from which information on the genes and gene clusters engaged in the synthesis of these metabolites across the plant kingdom can be collected for further genetic engineering-related work (Weber and Kim, 2016; Udwary et al., 2025). Using ectopic expression of secondary metabolite-associated genes or transcription factors, various attempts have already been made to increase the production of valuable compounds from plant species. It is called metabolic engineering, where the corresponding gene is overexpressed in the cells, followed by plant regeneration. The transformants are expected to produce a larger amount of the targeted molecule than the non-transformed lines. Using such techniques, production of various secondary metabolites has been targeted (Mora-Vásquez et al., 2022; Guo et al., 2025). However, the secondary metabolite production is tightly regulated in the plant cells, which includes transcription factors, miRNAs, siRNAs, lncRNAs, circRNAs, DNA methylation, and histone modifications. Researchers have to understand the regulatory network to fine-tune the production or inhibition of target molecules (Zhao et al., 2023; Datta et al., 2024).

Other methods of enhancement of secondary metabolites in medicinal plants are treatment- or elicitor-based (Modi et al., 2011; Modi et al., 2014; Modi and Kumar, 2018), callus or cell culture-based (Indu et al., 2025; Abdulhafiz et al., 2022), and hairy root culture (Singh et al., 2018; Awere et al., 2024). Plant-based herbal formulations have been widely used in countries like India and China for ages (Husen, 2022; 2023a; 2023b; 2024a; 2024b; Husen and Pant, 2024a; 2024b; Pant and Husen, 2025). Traditionally used medicines are also formulated with the WHO and other pharmacopoeial guidelines. Nano-formulations are also one of the promising strategies to enhance the solubility and bioavailability of active ingredients (Patwa et al., 2024). The present chapter highlights the cutting-edge genome-editing tools; the current status of plant-based medicines; and how the gene editing tool serves as a platform for the production of next-generation herbal medicines.

11.2 GENOME-EDITING TECHNIQUES

The genetic modification of crops followed by commercialization is subjected to strict regulatory procedures in many countries and many such crops are still in a queue to be approved. This is due to the transgene (foreign gene from bacteria or virus) introduction in an organism of interest. The alternative to this situation like repairing of mutated gene already exists in target organism. Recent advances in targeted mutagenesis comprise of introduction of double-stranded break (DSB) counteracted by cell's own error-prone DNA repair mechanism. Such repair in plants is made possible by non-homologous end joining (NHEJ) pathway in which exonuclease can often cause nibbling of the ends which results in deletion of base-pairs ranging from few to thousands (Gorbunova and Levy, 1997). Nowadays, custom-designed and engineered nucleases are available, which can cleave the DNA at specific targets. A major advance was made with the development of zinc finger nucleases in humans (Urnov et al., 2005) and later on in plants (Townsend et al., 2009). ZFNs were followed by a new generation of DSB-inducing enzymes engineered, such as Xanthomonas-derived transcription activator-like effector nuclease (TALEN) enzymes (Čermák et al., 2015). Two types of TALENs are available, namely, standard TALEN, which has two DNA recognition regions and is on a dimeric restriction enzyme FokI, and compact TALEN, which is a monomeric DNA-cleaving protein. The success in generating a much shorter endonuclease is mainly based on the substitution of FokI with I-TevI homing endonuclease (Beurdeley et al., 2013). The major bottleneck in engineering ZFNs is that all the regions of a genome cannot be targeted. Genome editing using Clustered Regularly Interspaced Short Palindromic Repeats (CRISPR) approach consists of short guide RNA (sgRNA) and sgRNA-driven Cas9 (CRISPR-associated) proteins, which bind at a specific site (as guided by sgRNA) and introduce a double-stranded break in the functional part of the targeted gene. The break in the genome is repaired by the cell's own repair mechanism which is also error prone. This results in the insertion and deletion (Indels) of the nucleotide with shift in the frame which ultimately results in gene silencing (Figure 11.1). The enzyme Cas9 specifically cuts near NGG sequence known as PAM (Protospacer Adjacent Motif). The system can be designed in such a way that after modification, the entire CRISPR cassette can be excised out to make the transformed plant foreign DNA-free and therefore it faces fewer regulatory constraints than genetically modified crops (Jaganathan et al., 2018). A number of diverse group of plants have been edited for the improvement of traits with this technology, which includes cereals (Svitashev et al., 2015; Shi et al., 2017; Smedly et al., 2021; Yang and Wang, 2025), oilseeds (Jiang et al., 2017; Braatz et al., 2017; Khan et al., 2025), vegetables (Kushwaha et al., 2025; Soyk et al., 2017), fruits (Tripathi et al., 2019; Osakabe et al., 2018; Malnoy et al., 2016), and medicinal plants (Gao et al., 2025; Borah et al., 2024). In the present chapter, engineering and improvement of plant-based medicine using the CRISPR/Cas9 tool have been explored with relevant case studies.

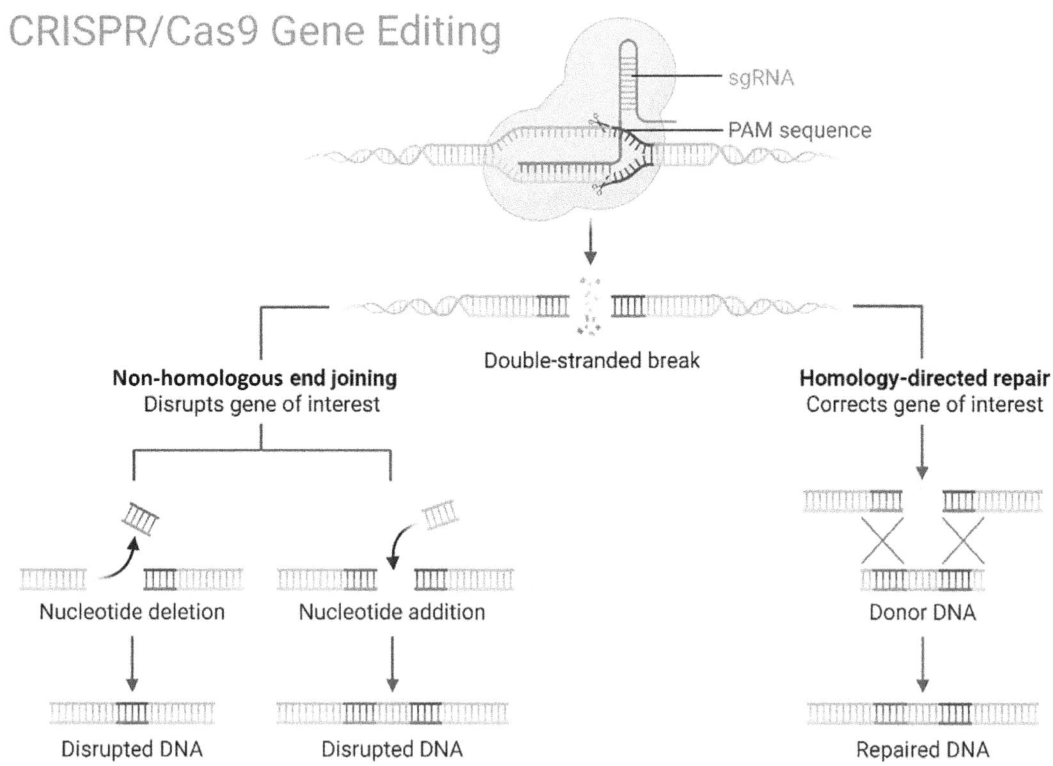

FIGURE 11.1 CRISPR/Cas9-based gene editing using either NHEJ (left) or HDR (right) pathways, results in permanent changes of the gene sequence and altered gene function or mutation.

11.3 CURRENT STATUS OF PLANT-BASED MEDICINE

Morphine was the first plant-based medicine from *Papaver somniferum* in 1803, after which the era of plant-derived chemicals began. Several other important active ingredients were also identified from the plant origin. This includes digitoxin, cocaine, pilocarpine, codeine, and quinine (Vaou et al., 2025). A search for the natural compounds that could be extracted from the cultivable medicinal plants began afterwards. Today medicinal plants comprises of 10% of the total available plant species from which power of the natural medicines is harvested (Nasim et al., 2022; Salmerón-Manzano et al., 2020). Some of the commercially available herbal formulations from India are Ashwagandha (*Withania somnifera*), Giloy (*Tinospora cordifolia*), Arjuna (*Terminalia arjuna*), Ashoka (*Saraca asoca*), Guggulu (*Commiphora wightii*), Brahmi (*Bacopa monnieri*), Shankhpushpi (*Convolvulus pluricaulis*), ginger powder (*Zingiber officinale*), and Shilajit (*Asphaltum* spp.). Complementary and alternative medicines (herbal) are being consumed and prescribed in many parts of the world. Most allopathic practitioners in Japan prescribe 'Kampo' medicines (herbal products) and in Germany 70% doctors advise to use plant-based products. Herbal medicines are consumed by 40%–70% population in the developed and developing countries. Such countries include Chile (71% consumption), Colombia (40% consumption), Australia (46% consumption), France (49% consumption), and Canada (70% consumption). Around 158 million Americans use herbal or natural medicines (Hossain et al., 2022). Although the plant-based medicines have been in use since ancient times, after the COVID-19 pandemic, the use of plant-based medicines increased radically (Panda et al., 2022). Ancient medicinal practices such as Ayurveda (India), Unani (Iran), Traditional Chinese Medicines (China), Kampo (Japan), Traditional Thai Medicine (Thailand), Hippocratic Medicine (Greece), Ancient Egyptian Medicine (Egypt), and Andean Medicine (Peru) used whole plant, plant part, powder or extract which are composed of many metabolites. After the advancement in extraction and formulation procedures, specific metabolites could be isolated from the medicinal plant source. This provides an option to treat the medical conditions with the corresponding active ingredient. As the production of pharmaceutical ingredients is sourced from the plant, process optimization of the extraction, purification, characterization, efficacy, and safety studies is a crucial part. The process is thoroughly reviewed by Chaachouay and Zidane (2024). Currently, many plant-derived natural products with their medicinal applications are available. These include cardiotonic drugs such as Digoxin and Digitoxin from *Digitalis* spp., Convallatoxin from *Convallaria majalis* L., Ouabain from *Strophanthus gratus* Baill.; anticancer drugs such as Taxol from *Taxus brevifolia* Nutt., Vinblastine and Vincristine from *Catharanthus roseus* L., Camptothecin from *Camptotheca acuminata* L., Colchicine from *Colchicum autumnale* L.; central nervous system modulators like Morphine from *P. somniferum* L., Salicin from *Salix alba* L., Caffeine from *Camellia sinensis* L., Cocaine from *Erythroxylum coca* Lam., Scopolamine from *Datura metel* L.; antihypertensive drug Reserpine (from *Rauvolfia serpentina* L.); antimalarial drug Quinine (from *Cinchona officinalis* L.); diuretic drug Theophylline (from *Theobroma cacao* L.); and antifungal like Thymol (from *Thymus vulgaris* L.). Some of the active compounds like vitamin E, β-carotene, ellagic acid, and organosulfur from the plant origin are also being researched (Chaachouay and Zidane, 2024). These medicines are manufactured through the process mentioned. These compounds are present in plants or plant parts and extracted in their original form and concentrations. Biosynthesis of these compounds can be modulated using genetic engineering or genome-editing tools. In the next section, we study how genome-editing tool (CRISPR/Cas9) provides a platform or scope for the quantitative improvement of herbal medicines.

11.4 SCOPE FOR THE IMPROVEMENT USING THE GENOME-EDITING TOOL

It is well known that the classical transgenic plant development system comprises of ectopic expression of trans(cis)gene in plants which makes them overexpress the transferred gene and as a result production of corresponding protein, enzyme, or metabolites increases drastically. Compared to that, CRISPR/Cas9 system or any other gene-editing tool targets the 'faulty' or undesired gene to undergo mutation that results in the knockout or loss-of-function. It is desirable that the mutation is induced near the start codon, followed by mutant screening where stop codon has been introduced through editing machinery of CRISPR/Cas9 so that the protein is not fully formed and the corresponding function is lost. This is in contrast to conventional transgenic plant development. However, gene-editing techniques are still preferred because the accumulation of the biomolecule is not only governed by the expression of corresponding gene but also by the repression of the inhibitory genes in the pathway. In addition, the over-production of competitive products of the pathways hinders the synthesis of desired metabolites (Das et al., 2024). For instance, Shi et al. (2021) created a loss-of-function mutation of an ABA-responsive transcription factor like bZIP2 in *Salvia miltiorrhiza* using CRISPR/Cas9 method, where the edited lines showed increased production of phenolic acid because *bZIP2* is a positive regulator of the competitive pathway of phenolic acids. One of the most common techniques to check the proof of concept of CRISPR system on any plant species is to target phytoene desaturase gene which is present in all plants and mutation of which impart no significant changes in metabolism or physiology except to develop albino phenotypes which is a visible marker and therefore testing of the concept requires no destructive sampling and gene editing experiments can be monitored in a real time as cell grows. Many plant species have been evaluated first for the protocol optimisation for CRISPR/Cas9-based genome editing (Siddappa et al., 2023; Thakur and Meru, 2023; Bulle et al., 2024; Brewer and Chambers, 2022). Not only *PDS* gene, but also the other genes of secondary metabolism such as squalene synthesis, genes involved in lignin synthesis etc., have also been targeted to develop the protocol and establish the proof of concept of genome editing in medicinal plants (Feng et al., 2018; Kui et al., 2017).

Atropa belladonna L. is known to produce anisodamine and scopolamine in the aerial and underground parts of the plants. The synthesis of these compounds is highly prevalent and masks the synthesis of other important metabolite atropine (hyoscyamine). From the non-protein amino acids through tropane indole alkaloid biosynthesis pathway, first hyoscyamine is synthesized which undergoes enzymatic changes (hydroxylation) quickly and produces anisodamine and scopolamine. Hyoscyamine is an anticholinergic drug and used in a variety of disease conditions, such as nausea, bradycardia, Parkinson's, and overactive bladder. As the metabolite is intermediate in the pathway, its production is found significantly less in all parts of the plant. Zeng et al. (2021) developed a CRISPR line of *A. belladonna* L. by targeting the gene responsible for the hydroxylation of hyoscyamine (*AbH6H*). They designed the guide RNA from the second exon of the *AbH6H* gene and placed between under U6 promoter from *Arabidopsis thaliana* and scaffold sequence followed by *SpCas9* driven and terminated by Ubiquitin promoter and terminator, respectively. The recombinant plasmid was transferred to *Agrobacterium* strain EHA105 and transformation was carried out. The resultant plants were screened using polymerase chain reaction (PCR) followed by sequencing to check the editing in the sequence near PAM region. Both the roots and leaves of CRISPR line H9 showed no peaks of scopolamine and anisodamine as compared to non-transformed plants and significantly higher production of hyoscyamine in the same samples (Zeng et al., 2021). *P. somniferum* L. is known for its oil content and medicinally important benzylisoquinoline alkaloids (BIAs) such as codeine, noscapine, papaverine, and morphine. These metabolites are synthesized from L-tyrosine and the key regulatory enzyme for their production is 3′-hydroxy-N-methylcoclaurine 4′-O-methyltransferase (4OMT). The *Ps4OMT* gene was mutated using CRISPR/Cas9 system with synthetic and viral backbone (construct) by Alagoz et al. (2016). The resultant CRISPR line in both the constructs showed reduced levels of all the BIAs suggesting the importance of the enzyme. Surprisingly, the CRISPRed plants showed a novel peak when metabolites were assessed through HPLC-TOF/MS (Alagoz et al., 2016). The result opened up the new possibilities for the development of novel metabolites in other medicinal plants. One of the most important medicinal plants on which research has been conducted in different parts of the world is Cannabis. It contains two major compounds, namely delta-9-tetrahydrocannabidiol and cannabidiol. Cannabis can be found in nature with two distinct types: drug type (high tetrahydrocannabinol (THC) and low cannabidiol (CBD)) and hemp (high CBD and low THC). Because of its psychoactive properties, there is a strict regulation across the globe for the usage of Cannabis, its products, and the research on it. One of the important parameters for the usage of Cannabis is that the THC content should be less than 0.3%, and under any circumstances, the level should not rise (Wheeldon and Heidt, 2023). Shiels et al. (2022) reviewed and explained the possibilities of use of CRISPR technology to reduce the level of THC in the hemp, thereby to increase the concentration of CBD to generate the hemp line regulatory-friendly (Shiels et al., 2022). In a *S. miltiorrhiza* plant, tanshinone production is prevalent and shares the same pathway, which also leads to taxol (a potent anti-cancer drug) production originating from geranyl geranyl diphosphate. The enzyme responsible for the tanshinone production is diterpene synthase governed by the *SmCPS1* gene in the plant. Li et al. (2017) generated a CRISPR line through hairy root culture which showed significantly reduced levels of tanshinones, and they proposed to enhance the taxol production in the roots of *S. miltiorrhiza* using metabolic engineering approaches.

Plant-derived metabolites are not only important medicinally but also used in other chemical-based industries. CRISPR technology provides both knockout and over-expression strategies to enhance the secondary metabolites in the plant. There is a small variation in the structure of the key protein Cas9, where one amino acid substitution makes the protein deactivate (dCas9) in such a way that it can bind to the region of DNA but cannot introduce a double-stranded break. So, if the guide RNA is designed from the promoter region of the gene, expression is temporarily halted. This strategy is used in specific conditions, such as where stress or tissue-specific expression or repression is desired (Karlson et al., 2021).

11.5 REGULATIONS AND GUIDELINES

Regulations of herbal medicines across developed and developing nations consist of different protocols for approval, clinical trials, QC data, and risk assessment. Regulatory authorities for EU, USA, India, Canada, Russia, and China are the European Medicines Agency (EMA), Food and Drug Administration (FDA), CDSCO and AYUSH, Health Canada (HC), Minzdrav, and National Medical Products Administration (NMPA), respectively. Approval pathways in most countries require traditional uses followed by clinical data and good manufacturing practices. Quality controls involve stability studies, good manufacturing practices (GMP), and chemical profiling in all the regions mentioned. However, some countries have specific criteria for quality control like biological assays (USA), alignment with International Council for Harmonisation (ICH) standard (Canada), safety and efficacy validation (China), etc. Risk assessment associated with these medicines is mislabelling, standardization of active ingredient, and genetic and environmental variability (Jadhav et al., 2025).

11.6 CONCLUSION

The demand for the plant-derived natural products is increasing continuously. With the technological advancement, it is now possible to harness the metabolite power of the plants. Due to the same reason, the scope for the improvement also broadens. This includes metabolic engineering, cell culture, tissue and organ culture, and the cutting-edge technology of genome editing. Among various techniques available in genome editing, CRISPR/Cas9 provides promising, reliable, and relatively simple solutions for the enhancement of secondary metabolites and 'tunable' synthesis of metabolites of interest from the pathway. Of course, there is a need to work on some of the demerits like regulatory constraints and off-target mutations, but overall, for the improvement of secondary metabolite-based medicines, genome-editing techniques are reliable.

REFERENCES

Abdulhafiz, F., Mohammed, A., Reduan, M.F.H., Kari, Z.A., Wei, L.S. and Goh, K.W., 2022. Plant cell culture technologies: A promising alternatives to produce high-value secondary metabolites. *Arabian Journal of Chemistry*, *15*(11), 104161.

Alagoz, Y., Gurkok, T., Zhang, B. and Unver, T., 2016. Manipulating the biosynthesis of bioactive compound alkaloids for next-generation metabolic engineering in opium poppy using CRISPR-Cas 9 genome editing technology. *Scientific Reports*, *6*(1), 30910.

Awere, C.O., Rakkammal, K., Mwaura, M.M., Anadebe, V.C. and Ramesh, M., 2024. Hairy-root technology: A metabolic engineering tool and specialized metabolite pathway elucidation and production of secondary metabolites. A review. *Results in Engineering*, *23*, 102697.

Barot, M., Nirmal Kumar, J.I. and Kumar, R.N., 2019. An evaluation of the nutritional composition of seaweeds as potential source of food and feed. *National Academy Science Letters*, *42*(6), pp.459–464.

Beurdeley, M., Bietz, F., Li, J., Thomas, S., Stoddard, T., Juillerat, A., Zhang, F., Voytas, D.F., Duchateau, P. and Silva, G.H., 2013. Compact designer TALENs for efficient genome engineering. *Nature Communications*, *4*(1), pp.1–8.

Bitwell, C., Indra, S.S., Luke, C. and Kakoma, M.K., 2023. A review of modern and conventional extraction techniques and their applications for extracting phytochemicals from plants. *Scientific African*, *19*, e01585.

Bocso, N.S. and Butnariu, M., 2022. The biological role of primary and secondary plants metabolites. *Journal of Nutrition and Food Processing*, *5*(3), pp.1–7.

Borah, A., Singh, S., Chattopadhyay, R., Kaur, J. and Bari, V.K., 2024. Integration of CRISPR/Cas9 with multi-omics technologies to engineer secondary metabolite productions in medicinal plant: Challenges and Prospects. *Functional & Integrative Genomics*, *24*(6), p.207.

Braatz, J., Harloff, H.J., Mascher, M., Stein, N., Himmelbach, A. and Jung, C., 2017. CRISPR-Cas9 targeted mutagenesis leads to simultaneous modification of different homoeologous gene copies in polyploid oilseed rape (Brassica napus). *Plant Physiology*, *174*(2), pp.935–942.

Brewer, S.E. and Chambers, A.H., 2022. CRISPR/Cas9-mediated genome editing of phytoene desaturase in *Carica papaya* L. *The Journal of Horticultural Science and Biotechnology*, *97*(5), pp.580–592.

Bulle, M., Venkatapuram, A.K., Abbagani, S. and Kirti, P.B., 2024. CRISPR/Cas9 based genome editing of phytoene desaturase (PDS) gene in chilli pepper (*Capsicum annuum* L.). *Journal of Genetic Engineering and Biotechnology*, *22*(2), 100380.

Čermák, T., Baltes, N.J., Čegan, R., Zhang, Y. and Voytas, D.F., 2015. High-frequency, precise modification of the tomato genome. *Genome Biology*, *16*(1), p.232.

Chaachouay, N. and Zidane, L. 2024. Plant-derived natural products: A source for drug discovery and development. *Drugs and Drugs Candidate, 3*(1), pp. 184–207. https://doi.org/10.3390/ddc3010011

Das, S., Kwon, M. and Kim, J.Y., 2024. Enhancement of specialized metabolites using CRISPR/Cas gene editing technology in medicinal plants. *Frontiers in Plant Science*, *15*, 1279738.

Datta, R., Kapadia, C., Danish, S. and Singh, S. eds., 2024. *Non-coding RNA in Plants: Modulation and Stress Responses*. Amsterdam: Elsevier.

Feng, S., Song, W., Fu, R., Zhang, H., Xu, A. and Li, J., 2018. Application of the CRISPR/Cas9 system in *Dioscorea zingiberensis. Plant Cell, Tissue and Organ Culture (PCTOC)*, *135*(1), pp.133–141.

Gao, H., Pei, X., Song, X., Wang, S., Yang, Z., Zhu, J., Lin, Q., Zhu, Q. and Yang, X., 2025. Application and development of CRISPR technology in the secondary metabolic pathway of the active ingredients of phytopharmaceuticals. *Frontiers in Plant Science*, *15*, 1477894.

Gorbunova, V. and Levy, A.A., 1997. Non-homologous DNA end joining in plant cells is associated with deletions and filler DNA insertions. *Nucleic Acids Research*, *25*(22), pp.4650–4657.

Guo, C., Xu, S. and Guo, X., 2025. Metabolic engineering of terpenoid biosynthesis in medicinal plants: From genomic insights to biotechnological applications. *Current Issues in Molecular Biology*, *47*(9), p.723.

Hossain, C.M., Gera, M.E.E.T.A. and Ali, K.A., 2022. Current status and challenges of herbal drug development and regulatory aspect: A global perspective. *Asian Journal of Pharmaceutical and Clinical Research*, *15*(2), pp.31–41.

Huang, W., Wang, Y., Tian, W., Cui, X., Tu, P., Li, J., Shi, S. and Liu, X., 2022. Biosynthesis investigations of terpenoid, alkaloid, and flavonoid antimicrobial agents derived from medicinal plants. *Antibiotics*, *11*(10), p.1380.

Husen, A., 2022. *Herbs, Shrubs and Trees of Potential Medicinal Benefits*. Boca Raton, FL: CRC Press. https://doi.org/10.1201/9781003205067

Husen, A., 2023a. *Exploring Poisonous Plants: Medicinal Values, Toxicity Response and Other Potential Uses.* Boca Raton, FL: CRC Press. https://doi.org/10.1201/b23017

Husen, A., 2023b. *Antidiabetic Medicinal Plants and Herbal Treatments.* Boca Raton, FL: CRC Press. https://doi.org/10.1201/b23347

Husen, A., 2024a. *Antimalarial Medicinal Plants.* Boca Raton, FL: Taylor & Francis Group. https://doi.org/10.1201/9781003378396

Husen, A., 2024b. *Medicinal Spices and Condiment Crops*. Boca Raton, FL: Taylor & Francis Group. https://doi.org/10.1201/9781003387046

Husen, H. and Pant, M., 2024a. Exploring Medicinal Orchids. Taylor & Francis Group, LLC, 6000 Broken Sound Pkwy. NW https://doi.org/10.1201/b23375

Husen, H. and Pant, M., 2024b. *Tissue Culture Techniques and Medicinal Plants: Enhancing Propagation and Production*. Boca Raton, FL: Taylor & Francis Group. https://doi.org/10.1201/b23374

Indu, B.K., Balasubramanya, S., Anuradha, M. and Shilpa, P., 2025. Callus and cell suspension cultures for secondary metabolite production. In Anuradha, M. and Balasubramanya, S. eds. *In Vitro Production of Plant Secondary Metabolites: Theory and Practice* (pp. 71–88). Singapore: Springer Nature Singapore.

Jadhav, R., Das, J., Rajyaguru, S., Kyada, S., Kale, V., Teja, P.K., Bhutia, D.O. and Chauthe, S.K., 2025. Regulations, current development, and future prospects of phytopharmaceuticals, a new class of herbal medicines in India. *Discover Pharmaceutical Sciences*, *1*(1), p.6.

Jaganathan, D., Ramasamy, K., Sellamuthu, G., Jayabalan, S. and Venkataraman, G., 2018. CRISPR for crop improvement: An update review. *Frontiers in Plant Science*, *9*, p.985.

Jiang, W.Z., Henry, I.M., Lynagh, P.G., Comai, L., Cahoon, E.B. and Weeks, D.P., 2017. Significant enhancement of fatty acid composition in seeds of the allohexaploid, *Camelina sativa*, using CRISPR/Cas9 gene editing. *Plant Biotechnology Journal*, *15*(5), pp.648–657.

Karlson, C.K.S., Mohd-Noor, S.N., Nolte, N. and Tan, B.C., 2021. CRISPR/dCas9-based systems: Mechanisms and applications in plant sciences. *Plants*, *10*(10), p.2055.

Khan, S.U., Qian, M., Li, S., Fan, Y., Wang, H., Chang, W., Alam, O., Saeed, S. and Lu, K., 2025. Unveiling CRISPR/Cas in rapeseed: Triumphs, trials, and tomorrow. *New Crops*, *2*, 100045.

Kui, L., Chen, H., Zhang, W., He, S., Xiong, Z., Zhang, Y., Yan, L., Zhong, C., He, F., Chen, J. and Zeng, P., 2017. Building a genetic manipulation tool box for orchid biology: identification of constitutive promoters and application of CRISPR/Cas9 in the orchid, Dendrobium officinale. *Frontiers in Plant Science*, *7*, p.2036.

Kushwaha, S.B., Nagesh, C.R., Lele, S.S., Viswanathan, C., Prashat, G.R., Goswami, S., Kumar, R.R., Kunchge, N., Gokhale, J.S. and Vinutha, T., 2025. CRISPR/Cas technology in vegetable crops for improving biotic, abiotic stress and quality traits: Challenges and opportunities. *Scientia Horticulturae*, *341*, 113957.

Li, B., Cui, G., Shen, G., Zhan, Z., Huang, L., Chen, J. and Qi, X., 2017. Targeted mutagenesis in the medicinal plant *Salvia miltiorrhiza*. *Scientific Reports*, *7*(1), 43320.

Malnoy, M., Viola, R., Jung, M.H., Koo, O.J., Kim, S., Kim, J.S., Velasco, R. and Nagamangala Kanchiswamy, C., 2016. DNA-free genetically edited grapevine and apple protoplast using CRISPR/Cas9 ribonucleoproteins. *Frontiers in Plant Science*, *7*, p.1904.

Mera, I.G., Falconí, D.G. and Córdova, V.M., 2019. Secondary metabolites in plants: Main classes, phytochemical analysis and pharmacological activities. *Rev. Bionatura*, *4*(4), 1000–1009.

Modi, A. and Kumar, N., 2018. Conventional and biotechnological approaches to enhance steviol glycosides (SGs) in *Stevia rebaudiana* Bertoni. In Kumar, N. eds. *Biotechnological Approaches for Medicinal and Aromatic Plants: Conservation, Genetic Improvement and Utilization* (pp. 53–62). Singapore: Springer Singapore.

Modi, A.R., Raj, S., Kanani, P., Patel, A. and Narayanan, S., 2014. Analysis of differentially expressed genes involved in stevioside biosynthesis in cultures of *Stevia rebaudiana* Bertoni treated with steviol as an immediate precursor. *Journal of Plant Growth Regulation*, *33*(3), pp.481–488.

Modi, A.R., Shukla, Y.M., Litoriya, N.S., Patel, N.J. and Narayan, S., 2011. Effect of gibberellic acid foliar spray on growth parameters and stevioside content of ex vitro grown plants of *Stevia rebaudiana* Bertoni. *Medicinal Plants-International Journal of Phytomedicines and Related Industries*, *3*(2), pp.157–160.

Mora-Vásquez, S., Wells-Abascal, G.G., Espinosa-Leal, C., Cardineau, G.A. and García-Lara, S., 2022. Application of metabolic engineering to enhance the content of alkaloids in medicinal plants. *Metabolic Engineering Communications*, *14*, e00194.

Nasim, N., Sandeep, I.S. and Mohanty, S., 2022. Plant-derived natural products for drug discovery: Current approaches and prospects. *The Nucleus*, *65*(3), pp. 399–411.

Osakabe, Y., Liang, Z., Ren, C., Nishitani, C., Osakabe, K., Wada, M., Komori, S., Malnoy, M., Velasco, R., Poli, M. and Jung, M.H., 2018. CRISPR–Cas9-mediated genome editing in apple and grapevine. *Nature Protocols*, *13*(12), pp.2844–2863.

Pagare, S., Bhatia, M., Tripathi, N., Pagare, S. and Bansal, Y.K., 2015. Secondary metabolites of plants and their role: Overview. *Current Trends in Biotechnology and Pharmacy*, *9*(3), pp.293–304.

Palazon, J. and Alcalde, M.A., 2025. Secondary metabolites in plants. *Plants*, *14*(14), p.2146.

Panda, A.K., Kar, S., Rai, A.K., Rao, B.C.S. and Srikanth, N., 2022. AYUSH-64: A potential therapeutic agent in COVID-19. *Journal of Ayurveda and Integrative Medicine*, *13*(2), 100538.

Patwa, S., Rasve, V., Sayyad, G. and Garje, S., 2024. Advancements in herbal medicine: A review of current research and applications. *International Journal of Pharmaceutical Sciences,* *2*(11), pp. 1286–1294. https://doi.org/10.5281/zenodo.14217116

Pant, M. and Husen, A., 2025. *Exploring Medicinal Bamboos.* Boca Raton, FL: Taylor & Francis Group. https://doi.org/10.1201/9781003485933

Pirintsos, S., Panagiotopoulos, A., Bariotakis, M., Daskalakis, V., Lionis, C., Sourvinos, G., Karakasiliotis, I., Kampa, M. and Castanas, E., 2022. From traditional ethnopharmacology to modern natural drug discovery: A methodology discussion and specific examples. *Molecules*, *27*(13), p.4060.

Roopashree, K.M. and Naik, D., 2019. Advanced method of secondary metabolite extraction and quality analysis. *Journal of Pharmacognosy and Phytochemistry*, *8*(3), pp.1829–1842.

Salmerón-Manzano, E., Garrido-Cardenas, J.A. and Manzano-Agugliaro, F., 2020. Worldwide research trends on medicinal plants. *International Journal of Environmental Research and Public Health*, *17*(10), p.3376.

Shi, J., Gao, H., Wang, H., Lafitte, H.R., Archibald, R.L., Yang, M., Hakimi, S.M., Mo, H. and Habben, J.E., 2017. ARGOS 8 variants generated by CRISPR-Cas9 improve maize grain yield under field drought stress conditions. *Plant Biotechnology Journal*, *15*(2), pp.207–216.

Shi, M., Du, Z., Hua, Q. and Kai, G., 2021. CRISPR/Cas9-mediated targeted mutagenesis of bZIP2 in Salvia miltiorrhiza leads to promoted phenolic acid biosynthesis. *Industrial Crops and Products*, *167*, 113560.

Shiels, D., Prestwich, B.D., Koo, O., Kanchiswamy, C.N., O'Halloran, R. and Badmi, R., 2022. Hemp genome editing—Challenges and opportunities. *Frontiers in Genome Editing*, *4*, 823486.

Siddappa, S., Sharma, N., Salaria, N., Thakur, K., Pathania, S., Singh, B., Sharma, H., Sood, S., Bhardwaj, V., Thakur, A.K. and Mangal, V., 2023. CRISPR/Cas9-mediated editing of phytoene desaturase (PDS) gene in an important staple crop, potato. *3 Biotech*, *13*(5), p.129.

Singh, R.S., Chattopadhyay, T., Thakur, D., Kumar, N., Kumar, T. and Singh, P.K., 2018. Hairy root culture for in vitro production of secondary metabolites: a promising biotechnological approach. In Kumar, N. eds. *Biotechnological Approaches for Medicinal and Aromatic Plants: Conservation, Genetic Improvement and Utilization* (pp. 235–250). Singapore: Springer Singapore.

Smedley, M.A., Hayta, S., Clarke, M. and Harwood, W.A., 2021. CRISPR-Cas9 based genome editing in wheat. *Current Protocols*, *1*(3), e65.

Soyk, S., Müller, N.A., Park, S.J., Schmalenbach, I., Jiang, K., Hayama, R., Zhang, L., Van Eck, J., Jiménez-Gómez, J.M. and Lippman, Z.B., 2017. Variation in the flowering gene SELF PRUNING 5G promotes day-neutrality and early yield in tomato. Nature Genetics, 49(1), pp.162–168.

Svitashev, S., Young, J.K., Schwartz, C., Gao, H., Falco, S.C. and Cigan, A.M., 2015. Targeted mutagenesis, precise gene editing, and site-specific gene insertion in maize using Cas9 and guide RNA. Plant physiology, 169(2), pp.931–945.

Thakur, S. and Meru, G., 2023. CRISPR/Cas9 mediated editing of phytoene desaturase gene in squash. *Journal of Plant Biochemistry and Biotechnology*, *32*(4), pp.862–869.

Thirumurugan, D., Cholarajan, A., Raja, S.S. and Vijayakumar, R., 2018. An introductory chapter: Secondary metabolites. *Secondary Metabolites-Sources and Applications*, *1*, p.13.

Townsend, J.A., Wright, D.A., Winfrey, R.J., Fu, F., Maeder, M.L., Joung, J.K. and Voytas, D.F., 2009. High-frequency modification of plant genes using engineered zinc-finger nucleases. Nature, 459(7245), pp.442–445.

Tripathi, J.N., Ntui, V.O., Ron, M., Muiruri, S.K., Britt, A. and Tripathi, L., 2019. CRISPR/Cas9 editing of endogenous banana streak virus in the B genome of *Musa spp.* overcomes a major challenge in banana breeding. Communications biology, 2(1), pp.1–11.

Udwary, D.W., Doering, D.T., Foster, B., Smirnova, T., Kautsar, S.A. and Mouncey, N.J., 2025. The secondary metabolism collaboratory: A database and web discussion portal for secondary metabolite biosynthetic gene clusters. *Nucleic Acids Research*, *53*(D1), pp.D717–D723.

Urnov, F.D., Miller, J.C., Lee, Y.L., Beausejour, C.M., Rock, J.M., Augustus, S., Jamieson, A.C., Porteus, M.H., Gregory, P.D. and Holmes, M.C., 2005. Highly efficient endogenous human gene correction using designed zinc-finger nucleases. *Nature*, *435*(7042), pp.646–651.

Vaou, N., Voidarou, C., Rozos, G., Saldari, C., Stavropoulou, E., Vrioni, G. and Tsakris, A., 2025. Unraveling nature's pharmacy: Transforming medicinal plants into modern therapeutic agents. *Pharmaceutics*, *17*(6), p.754.

Weber, T. and Kim, H.U., 2016. The secondary metabolite bioinformatics portal: computational tools to facilitate synthetic biology of secondary metabolite production. *Synthetic and Systems Biotechnology*, *1*(2), pp.69–79.

Wheeldon, J. and Heidt, J., 2023. Cannabis and criminology: A history of race, addiction, and inconvenient research. *Journal of Criminal Justice*, *85*, 101991.

Yang, B. and Wang, K., 2025. CRISPR–Cas-directed genome editing in maize. *Cold Spring Harbor Protocols*. Epub ahead of print 7 October. https://doi.org/10.1101/pdb.top108448

Zeng, L., Zhang, Q., Jiang, C., Zheng, Y., Zuo, Y., Qin, J., Liao, Z. and Deng, H., 2021. Development of *Atropa belladonna* L. plants with high-yield hyoscyamine and without its derivatives using the CRISPR/Cas9 system. *International Journal of Molecular Sciences*, *22*(4), p.1731.

Zhao, Y., Liu, G., Yang, F., Liang, Y., Gao, Q., Xiang, C., Li, X., Yang, R., Zhang, G., Jiang, H. and Yu, L., 2023. Multilayered regulation of secondary metabolism in medicinal plants. *Molecular Horticulture*, *3*(1), p.11.

12 Molecular Insights of Hairy Root Culture

Mahak Sachdev

12.1 INTRODUCTION

A revolutionary method in plant biotechnology, hairy root culture has developed from an intriguing biological phenomenon into a potent instrument for both industry and study. The discovery of *Agrobacterium rhizogenes*, a soil bacterium that can cause plants to grow roots rapidly through horizontal gene transfer, is where it all began. Researchers discovered this bacterium, which is closely linked to *Agrobacterium tumefaciens*, in the middle of the 20th century after noticing unusual root growth at infection locations. The transfer of root-inducing (Ri) plasmid DNA into the plant genome, more especially the T-DNA region encoding rol genes, was eventually identified as the cause of the phenomena. According to Mirmazloum et al. (2024), these genes—rolA, rolB, rolC, and occasionally rolD—reprogram plant cells to start growing quickly, hormone-independent roots with recognizable shape and branching patterns. The potential of these hairy roots for *in vitro* culture was promptly recognized by the scientific community. Hairy roots have exceptional genetic and biochemical stability, which makes them perfect for long-term research and metabolite production in contrast to conventional callus or suspension cultures. Hairy root culture was a mainstay in plant tissue culture labs during the 1980s and 1990s, especially for aromatic and medicinal plants. Secondary metabolites produced by these roots were found to be on par with or even higher than those produced by entire plants. This discovery made it possible to produce useful substances including terpenoids, alkaloids, flavonoids, and phenolics, without harvesting slow-growing or endangered species. Today, the American Type Culture Collection (ATCC) has more than 15,000 strains (variants) of *A. rhizogenes* of the original strain 15,834 (Gantait and Mukharjee, 2021; Kiryushkin et al., 2021).

Hairy root cultivation is significant because of its many uses. It provides a dependable platform for the production of nutraceuticals, medicines, and cosmetic components obtained from plants. The roots lower the difficulty and expense of culture care as they develop quickly in hormone-free media. They are appealing for commercial applications because of their capacity to release metabolites into the surrounding media, which streamlines downstream processing. Additionally, because hairy roots may be genetically modified, scientists can test gene functions, build biosynthetic pathways, and examine promoter activity in a controlled setting. Because of this, they are now essential in metabolic engineering, synthetic biology, and functional genomics (Liu et al., 2025). Additionally, hairy root systems have shown value in agricultural and environmental settings. According to Bharagava (2019), they can be employed for phytoremediation, which involves removing and purifying contaminants from soil and water. They can be used to research plant defence mechanisms and create natural insecticides because of their sensitivity to stress and elicitors. Hairy root cultures aid in crop development efforts in agriculture by clarifying root

DOI: 10.1201/9781003542681-12

physiology and nutrient uptake. They also play a role in the expression of recombinant proteins, which have uses in the creation of therapeutic proteins, enzymes, and vaccines. Hairy roots plant-based glycosylation machinery makes it possible to produce complex proteins with the right post-translational modifications, which is essential for the effectiveness of pharmaceuticals (Zhu et al., 2024; Trujillo and Angulo, 2025).

Transgenic plants in 89 distinct taxa, spanning 79 species from 55 genera and 27 families, have been produced through transformation mediated by *A. rhizogenes* (Christey, 2001). Hairy root cultures have been used in cotton (Gao et al., 2013; Cui et al., 2020), soybeans (Kereszt et al., 2007; Cai et al., 2015), tomatoes (Ron et al., 2014), groundnuts (Huai et al., 2023), and other plants. Binary vectors, such as *GFP/NPTII*, can be transferred by infection with *A. rhizogenes* to create altered hairy roots for crop improvement and plant genetic transformation. *A. rhizogenes* can also be utilized to incorporate gene silencing or RNA interference constructs for the rapid over-expression or knockout of genes in roots. CRISPR/Cas9-mediated gene alteration or editing has also been integrated with hairy root transformation in soybean. This rendered the assessment of editing efficiency rapid in regenerated transgenic hairy roots and established a powerful tool for root-specific functional studies (Cai et al., 2015; Srivastava et al., 2018). *A. rhizogenes*-mediated hairy root transformation can also be used for tissue-culture-mediated generation of composite plants containing a transgenic root and wild-type shoot. They are important for studies related to root biology and root pathogen interaction. There are many similarities in infection patterns of *A. tumefaciens* and *A. rhizogenes*. It has been suggested that the most important oncogenes, the so-called rol genes, encode proteins involved in the regulation of plant hormone metabolism, but rol genes have a function different from creating mere alterations in plant hormone concentration (Nilsson and Olsson, 2006).

12.2 MOLECULAR INSIGHTS: THE ROLE OF *ROL* GENES

The plasmid carrying *rol* genes is a versatile tool and prepares the host plant cell to perform multiple functions once integrated stably. A dissimilar strain of *A. tumefaciens*, *A. rhizogenes* (also called *Rhizobium rhizogenes*), carries a root-inducing Ri plasmid having the ability to cause infection and then integrate T-DNA into the host plant genome, which is later expressed as polyadenylated mRNA within the host plant system (Cui et al., 2020; Nemoto et al., 2009; Datta et al., 2024). This induces proliferation of genetically similar hairy roots from the infection site (Tepfer, 1990; Christey, 2001; Bahramnejad et al., 2019). These hairy roots are induced rapidly and efficiently from explants and are genetically and biochemically stable (Morey and Peebles, 2022; Sun et al., 2017); moreover, they can grow in phytohormone-free media. In addition, the regenerated roots are expected to be non-chimeric as they originate individually from single cells (Aggarwal et al., 2018). Most common Agropine type *A. rhizogenes* have two T-DNA regions, TL-DNA and TR-DNA, capable of independent transfer to the plant genome. The T-DNA is region-metered by left-border and right-border sequences, 25-bp direct repeat regions (Nemoto et al., 2009; Bahramnejad et al., 2019). The TL-DNA region carries around 18 *rol* genes; 4 of the *rol* genes, A, B, C, and D, play an important role in hairy root generation in transformed or infected plant cells. The Ri TR-DNA carries genes for opine biosynthesis and two genes for auxin biosynthesis, aux1 and aux2. The TR-DNA also contains a homologue of the *rolB* gene, the function of which is unknown. It has been demonstrated by various experiments that rol genes are necessary and enough to induce hairy roots, while aux genes do not play an important role in the hairy root culture (Spena et al., 1987; Cardarelli et al., 1987; Boulanger et al., 1986). Plants regenerated after infection with *A. rhizogenes* show phenotypic alterations like dwarfism, reduced apical dominance, increased rooting, increased branching, wrinkled leaves, and altered flowering, called Ri syndrome. The disarmed *A. rhizogenes* strain, with *rol* genes removed from the Ri plasmid, is also being used to reduce the extent of undesirable hairy root formation. Functions and regulations of these genes are mentioned in Table 12.1.

TABLE 12.1
Summary of the Function and Regulation of *rol* Genes from the Ri Plasmid of *Agrobacterium rhizogenes*

Gene	Function	Regulation
rolA	• Alters leaf morphology (wrinkling, reduced size) • Modulates plant growth and architecture • May influence cytokinin sensitivity	• The expression is domain-dependent: active in leaves when domain A is present. • Subcellular localization regulate its function
rolB	• Strong inducer of hairy root formation • Enhances auxin sensitivity • Stimulates secondary metabolite production (e.g., alkaloids, flavonoids) • May possess tyrosine phosphatase activity	• Expression regulated by promoter elements and tissue specificity • Enzymatic activity linked to auxin-glucoside hydrolysis • Subcellular localization influences activity
rolC	• Modifies plant stature and internode length • Influences cytokinin metabolism • Enhances stress tolerance and antioxidant capacity • Promotes secondary metabolite biosynthesis	• Expression varies across tissues and developmental stages • Protein may act on cytokinin glucosides • Regulation linked to hormonal balance and environmental cues
rolD	• Less characterized • May contribute to root morphogenesis and metabolic shifts	• Regulation not well understood • Likely co-expressed with other rol genes in T-DNA context

Source: Mauro et al. (2017).

12.3 FUNCTIONAL IMPROVEMENT: 'THE GENOME ENGINEER'

For the crop improvement programme, various biotechnological interventions such as genetic engineering, genome editing, a combination of engineering and editing, and hairy root culture have been widely used methods since the last three decades. Transformation of the gene of interest or the machinery which edits the gene of interest in the host plant cells has been carried out using various methods, and *Agrobacterium*-mediated transformation (AMT) is one of them. The method requires the engineering or editing cassette to be armed on a specific vector (Altpeter et al., 2016; Bélanger et al., 2024). Conventionally, a few strains of *Agrobacterium* were used for the genetic transformation experiments. However, many plant species were still found to be resistant to transformation and showed low transformation efficiencies. Nowadays, sequence information of various plasmids used in AMT enables researchers to engineer further for the functional improvement of the strain. The technique is called strain engineering. It has also been reported that during AMT, the host plant cell perceives the signal as a pathogen attack and initiates the defence mechanism (Pitzschke, 2013). Therefore, various strategies have been developed to mitigate the situation. These include supplemental *vir* genes, induction of plant hormone-modulating genes, temporary silencing of plant-host immunity, and the use of improved *Agrobacterium* strains. From the engineering perspective and strain improvement, various protocols have been developed, such as CRISPR/Cas base editing, RNA-guided sequence insertion, and integration of a recombinase system called recombineering (Goralogia et al., 2025).

12.4 CHALLENGES, PROSPECTS, AND CONCLUSION

Despite their advantages, hairy root cultures are not without challenges. Species-specific transformation efficiency, scalability issues, and regulatory concerns regarding transgenic products can limit their broader adoption. However, ongoing research aims to overcome these hurdles through improved transformation protocols, non-transgenic alternatives, and advanced bioreactor designs.

The integration of CRISPR/Cas9 and other genome-editing tools has further enhanced the precision and versatility of hairy root systems, enabling targeted modifications and pathway optimizations.

In summary, hairy root culture represents a convergence of natural biology and technological innovation. Its historical roots in microbial–plant interactions have blossomed into a robust platform for sustainable production, scientific exploration, and industrial application. As the demand for plant-based solutions grows in medicine, agriculture, and environmental management, hairy root culture stands out as a scalable, efficient, and ecologically responsible method. Its continued evolution promises to deepen our understanding of plant biology while offering tangible benefits across multiple sectors. Associated genes (*rolA*, *rolB*, *rolC*, and *rolD*) from the *A. rhizogenes* Ri plasmid modulate plant growth and development by altering hormone sensitivity, root architecture, and secondary metabolite biosynthesis—particularly through enhanced auxin and cytokinin responsiveness. Future directions involve leveraging rol gene functions for targeted metabolic engineering, stress-resilient crop development, and scalable biofactories using hairy root systems integrated with CRISPR and synthetic biology platforms.

REFERENCES

Aggarwal, P.R., Nag, P., Choudhary, P., Chakraborty, N. and Chakraborty, S., 2018. Genotype-independent Agrobacterium rhizogenes-mediated root transformation of chickpea: A rapid and efficient method for reverse genetics studies. Plant Methods. 14, 55.

Altpeter, F., Springer, N.M., Bartley, L.E., Blechl, A.E., Brutnell, T.P., Citovsky, V., Conrad, L.J., Gelvin, S.B., Jackson, D.P., Kausch, A.P. and Lemaux, P.G., 2016. Advancing crop transformation in the era of genome editing. Plant Cell. 28(7), 1510–1520.

Bahramnejad, B., Naji, M., Bose, R. and Jha, S., 2019. A critical review on use of Agrobacterium rhizogenes and their associated binary vectors for plant transformation. Biotechnol. Adv. 37, 107405.

Bélanger, J.G., Copley, T.R., Hoyos-Villegas, V., Charron, J.B. and O'Donoughue, L., 2024. A comprehensive review of in planta stable transformation strategies. Plant Methods. 20(1), 79.

Bharagava, R. N. (Ed.). (2019). *Recent advances in environmental management.* CRC Press, Taylor & Francis Group. https://doi.org/10.1201/9781351011259

Boulanger, F., Berkaloff, A. and Richaud, F., 1986. Identification of hairy root loci in the T-regions of Agrobacterium rhizogenes Ri plasmids. Plant Mol. Biol. 6, 271–279. doi: 10.1007/BF00015233.

Cai, Y., Chen, L., Liu, X., Sun, S., Wu, C., Jiang, B., Han, T., and Hou, W., 2015. CRISPR/Cas9-mediated genome editing in soybean hairy roots. PLoS ONE, 10,e0136064. https://doi.org/10.1371/journal.pone.0136064

Cardarelli, M., Mariotti, D., Pomponi, M., Spanò, L., Capone, I. and Costantino, P., 1987. Agrobacterium rhizogenes T-DNA genes capable of inducing hairy root phenotype. Mol. Gen. Genet. 209, 475–480. doi: 10.1007/BF00331152.

Christey, M.C., 2001. Use of ri-mediated transformation for production of transgenic plants. In Vitro Cell. Dev. Biol. -Plant. 37, 687–700. https://doi.org/10.1007/s11627-001-0120-0

Cui, M.-L., Liu, C., Piao, C.-L. and Liu, C.-L., 2020. A stable Agrobacterium rhizogenes-mediated transformation of cotton (Gossypium hirsutum L.) and plant regeneration from transformed hairy root via embryogenesis. Front. Plant Sci. 11, 604255.

Datta, R., Kapadia, C., Danish, S. and Singh, S. eds., 2024. Non-coding RNA in Plants: Modulation and Stress Responses. Elsevier, Amsterdam.

Gantait, S. and Mukherjee, E., 2021. Hairy root culture technology: Applications, constraints and prospect. Appl. Microbiol. Biotechnol. 105(1), 35–53.

Gao, Z., Liu, C., Zhang, Y., Li, Y., Yi, K., Zhao, X. and Cui, M.L., 2013. The promoter structure differentiation of a MYB transcription factor RLC1 causes red leaf coloration in Empire Red Leaf Cotton under light. PLoS One. 8(10), e77891. doi: 10.1371/journal.pone.0077891.

Goralogia, G.S., Willig, C. and Strauss, S.H., 2025. Engineering Agrobacterium for improved plant transformation. Plant J. 121(5), e70015.

Huai, D., et al., 2023. Red fluorescence protein (DsRed2) promotes the screening efficiency in peanut genetic transformation. Front. Plant Sci. 14, 1123644.

Kereszt, A., et al. 2007. Agrobacterium rhizogenes-mediated transformation of soybean to study root biology. Nat. Protoc. 2, 948–952.

Kiryushkin, A.S., Ilina, E.L., Guseva, E.D., Pawlowski, K. and Demchenko, K.N., 2021. Hairy CRISPR: genome editing in plants using hairy root transformation. Plants. 11(1), 51.

Liu, C., Ahmad, N., Tao, Y., Hussain, H., Chang, Y., Umar, A.W. and Liu, X., 2025. Reprogramming hairy root cultures: A synthetic biology framework for precision metabolite biosynthesis. Plants. 14(13), 1928.

Mauro, M.L., Costantino, P. and Bettini, P.P., 2017. The never ending story of rol genes: A century after. Plant Cell, Tissue Organ Cult. 131(2), 201–212.

Mirmazloum, I., Slavov, A.K. and Marchev, A.S., 2024. The untapped potential of hairy root cultures and their multiple applications. Int. J. Mol. Sci. 25(23), 12682.

Morey, K. J. and Peebles, C.A.M., 2022. Hairy roots: An untapped potential for production of plant products. Front. Plant Sci. 13, 9370–9395.

Nemoto, K. et al., 2009. Function of the aux and rol genes of the Ri plasmid in plant cell division in vitro. Plant Signal. Behav. 4, 1145–1147.

Nilsson, O. and Olsson, O., 1997. Getting to the root: The role of the Agrobacterium rhizogenes rol genes in the formation of hairy roots. Physiol. Plant. 100, 463–473.

Pitzschke, A., 2013. Agrobacterium infection and plant defense—Transformation success hangs by a thread. Front. Plant Sci. 4, 519.

Ron, M., Kajala, K., Pauluzzi, G., Wang, D., Reynoso, M.A., Zumstein, K., et al., 2014. Hairy root transformation using Agrobacterium rhizogenes as a tool for exploring cell type-specific gene expression and function using tomato as a model. Plant Physiol. 166, 455–469. doi: 10.1104/pp.114.239392.

Spena, A., Schmulling, T., Koncz, C. and Schell, J., 1987. Independent and synergistic activity of the rol A, B and C loci in stimulating abnormal growth in plants. EMBO J. 6, 3891–3899. doi: 10.1002/j.1460-2075.1987.tb02729.x.

Srivastava, V., Mehrotra, S. and Mishra, S., eds, 2018. *Hairy Roots: An Effective Tool of Plant Biotechnology*. Springer Singapore. https://doi.org/10.1007/978-981-13-2562-5.

Sun, J., Ma, L., San, K. and Peebles, C.A.M., 2017. Still stable after 11 years: A Catharanthus roseus Hairy root line maintains inducible expression of anthranilate synthase. Biotechnol. Prog. 33, 66–69.

Tepfer, D., 1990. Genetic transformation using Agrobacterium rhizogenes. Physiol. Plant. 79, 140–146. https://doi.org/10.1111/j.1399-3054.1990.tb05876.x

Trujillo, E. and Angulo, C., 2025. Perspectives on the use of the CRISPR system in plants to improve recombinant therapeutic protein production. J. Biotechnol. 405, 111–123.

Zhu, Y., Zhu, X., Wen, Y., Wang, L., Wang, Y., Liao, C., Zhao, M., Li, T., Liu, D., Li, B. and Zhu, T., 2024. Plant hairy roots: Induction, applications, limitations and prospects. Ind. Crops Prod. 219, 119104.

13 Scope of Molecular Markers in Cannabis Research

Anna Zilov and Arpan Modi

13.1 INTRODUCTION

The annual herbaceous species *Cannabis sativa* L. primarily displays a dioecious phenotype, where male and female flowers appear on separate plants, but it also occasionally displays a monoecious (hermaphrodite) phenotype, where male and female flowers emerge on the same plant. The species is a member of the family Cannabaceae. Its amazing chemical diversity and broad spectrum of biological activity have garnered a lot of scientific and public attention in recent years. For the majority of the twentieth century, there was little scientific research on cannabis, despite the fact that it has been used as a medicinal herb since ancient times and is prized for its analgesic, anti-inflammatory, and neuroactive qualities. Despite centuries of traditional use, legal constraints in many parts of the world precluded comprehensive research on its pharmacology, chemistry, and genetics (Thomas & ElSohly, 2016; Zuardi, 2006; Gonçalves et al., 2019). The last 10 years have seen a significant change in this circumstance. Cannabis has resurfaced as a scientific and economic focus as more nations legalize its usage for medical and, in some cases, recreational purposes. A new age of multidisciplinary study integrating pharmacology, agronomy, molecular biology, and genomics has been made possible by this change (Monthony et al., 2021a; Sirangelo et al., 2022). These days, *C. sativa* is acknowledged as a versatile crop in addition to a source of medicinal cannabinoids. Its seeds and inflorescences are utilized as raw materials for dietary supplements, cosmetics, oils, and biofuels; its fibrous stalks are used to make textiles and paper; and its secondary metabolites are being investigated for a variety of pharmacological uses (Zuardi, 2006; Gonçalves et al., 2019; Ren et al., 2021). Cannabis is one of the most sophisticated and adaptable plant species in terms of chemistry. Approximately 125 cannabinoids, 120 terpenoids, 34 flavonoids, 42 phenolics, 11 steroids, and 2 alkaloids are among the more than 545 chemicals that have been discovered thus far (Radwan et al., 2021). The glandular trichomes, which are specialized secretory structures found on female flowers, are where these metabolites are produced and stored. Because they have the largest quantities of bioactive chemicals, female inflorescences are the primary focus for both medical and recreational production (Livingston et al., 2020).

Nonetheless, the plant's remarkable metabolic variety is a reflection of its wide genetic diversity. The morphology, flowering behaviour, cannabinoid makeup, and environmental tolerance of populations and cultivars vary greatly. The creation of superior female lines with consistent and reliable metabolic profiles is one of the main obstacles facing contemporary cannabis research. The purpose of breeding efforts is to create plants with consistent chemotypes, such as low-delta-9-tetrahydrocannabidiol (THC) and high-cannabidiol (CBD) variants, although this is challenging due to the species' complicated DNA. To overcome these obstacles, scientists are using molecular markers more and more as instruments to measure genetic diversity, direct selection, and speed up breeding.

 DOI: 10.1201/9781003542681-13

13.2 MOLECULAR MARKERS

Molecular markers are distinct and recognizable DNA or RNA sequences that are useful instruments for genetic study. By serving as landmarks within the genome, these markers let scientists identify genetic differences, examine evolutionary trends, and investigate trait inheritance at the molecular level. Traditional phenotypic markers, which are frequently impacted by environmental factors and might not accurately reflect genetic differences, are not as accurate or reliable as molecular markers (Chamberlin & Lawit, 2017). In several disciplines, such as genetics, evolutionary biology, conservation, and breeding programmes, molecular markers have become essential, especially when it comes to medicinal plants.

The idea of genetic variation is the basis of molecular markers. The term 'genetic variation' describes the variations in nucleotide sequences between species or within a group. These natural variances serve as the foundation for species variety and evolutionary adaptation. These differences, which appear as insertions or deletions (Indels), microsatellites, and single-nucleotide polymorphisms (SNPs), are used by molecular markers. SNPs, for instance, are extremely prevalent and helpful for accurate genetic mapping and diversification research as they represent single base alterations in the DNA sequence. Simple sequence repeats (SSRs), also known as microsatellites, are short tandemly repeated sequences that are perfect for analysing population genetics and evolutionary relationships due to their high mutation rates (Joshi & Nguyen, 1993).

Molecular markers function at the DNA level, in contrast to conventional phenotypic markers, which depend on observable traits like plant height, flower colour, or leaf form. Because observable features can be modified by developmental phases, interactions with other species, and environmental factors, traditional phenotypic markers are prone to errors. It might be difficult to identify the genetic basis of characteristics like flower size or secondary metabolite synthesis because their expression can vary greatly depending on the environment. Molecular markers, on the other hand, offer a reliable and environment-independent way to detect genetic variation. Because of their accuracy, molecular markers can be employed extensively in marker-assisted selection (MAS) to improve crops and medicinal plants (Winter & Kahl, 1995).

Technological advancements have further revolutionized the utility of molecular markers. The advent of polymerase chain reaction (PCR) technology significantly increased the efficiency and accessibility of molecular marker techniques, such as random amplified polymorphic DNA (RAPD), amplified fragment length polymorphism (AFLP), and SSRs. High-throughput sequencing technologies, such as next-generation sequencing (NGS), have further expanded the scope of molecular marker research by enabling genome-wide marker discovery and genotyping at an unprecedented scale. For example, genome-wide association studies (GWAS) utilize molecular markers such as SNPs to identify genes associated with economically important traits in crops and medicinal plants. These advancements have facilitated rapid and cost-effective genetic studies, accelerating research in genetic diversity and trait mapping (Cheng et al., 2025).

In conservation genetics, the function of molecular markers is especially significant. Genetic diversity in uncommon and endangered medicinal plant species has been evaluated using molecular markers. To keep populations resilient and adaptable, genetic variety is essential. For example, research on endangered medicinal plant species utilizing SSR markers has helped to design successful conservation strategies by offering important insights about genetic bottlenecks and population dynamics. The safety and effectiveness of traditional medicine have also been guaranteed by the application of molecular markers to identify adulteration in herbal products and authenticate medicinal plant species (Semagn et al., 2006).

13.3 THE SIGNIFICANCE OF MOLECULAR MARKERS IN CANNABIS RESEARCH

The study of molecular markers in *C. sativa* has rapidly progressed over the past 20 years. Early studies demonstrated the first link between genotype and phenotype, but progress accelerated with the availability of genome-scale data. By producing one of the first high-quality morphological and genetic maps of *C. sativa*, Laverty et al. (2019) made a substantial contribution. These maps revealed notable rearrangements at the THC/CBD synthase loci. This revealed information about the structural arrangement of the cannabis synthase gene cluster, which was used to develop reliable markers. Using these genetic resources, Zhang et al. (2020) created SSR markers for *C. sativa* on a large scale for the first time. After discovering over 92,000 SSR motifs and creating over 63,000 primers, their analysis yielded one of the species' largest SSR databases. These markers enabled robust evaluations of genetic diversity and population structure, offering a strong basis for germplasm management and cultivar validation. Then, Toth et al. (2020) developed functional genetic markers for both plant sex and cannabis chemotype through PCR Allele Competitive Extension (PACE) tests. By using their study to identify male plants and distinguish THC-dominant from CBD-dominant chemotypes at early embryonic stages, breeders were able to boost efficiency and ensure conformity to THC laws. Borin et al. (2021) expanded the use of SSR markers by creating and validating a panel of highly polymorphic loci examined across several hemp varieties. By demonstrating genetic grouping matching to dioecious and monoecious breeding systems, their study advanced our understanding of population structure and offered resources for MAS and variety authentication. As the discipline evolved, researchers began using multi-omics tools to examine how genetic variability fosters biochemical diversity. In their summary of developments in cannabis metabolomics, transcriptomics, and genomes, Sirangelo et al. (2022) emphasized the potential of multi-layered data to enhance our comprehension of cannabinoid biosynthesis and gene-metabolite connections. Additionally, Watts et al. (2021) demonstrated that variation in terpene synthase (TPS) genes accounts for a significant amount of the chemical variety observed in commercial cannabis cultivars. Using whole-genome resequencing and terpene research, they showed that traditional 'Sativa–Indica–Hybrid' classifications do not accurately reflect true genetic and chemical structure. Instead, variations in TPS loci are strongly associated with distinct terpene profiles, making these genes excellent candidates for the development of future markers in breeding for medicinal, flavour, and aroma properties. Lastly, de Ronne and Torkamaneh (2025) used GWAS to identify 33 important quantitative trait loci (QTLs) associated with cannabis production, including a 60-Mb haplotype on chromosome 7 associated with THC-dominant genotypes. Together, these advances reflect the transition from marker discovery to quantitative, predictive genomics in cannabis genetics.

Collectively, these studies demonstrate the rapid evolution of *C. sativa* genetics, from the development of early markers to more sophisticated, genome-informed breeding methods. The development of SSR, SNP, and functional marker resources, along with multi-omics and GWAS studies, has significantly enhanced the ability to study, classify, and alter cannabis cultivars. Although the field is still relatively new compared to major crops, these advancements are gradually constructing the molecular infrastructure required for reliable genetic research. Because of molecular technologies, breeders and researchers may now more readily determine sex and chemotype, trace lineage, and begin mapping characteristics related to plant design, environmental adaptation, and THC content. Marker-based methods are already helpful for germplasm characterization, cultivar legitimacy, and early sex determination. Studies like Watts et al. (2021) show that TPS gene variation broadens the range of marker usage beyond cannabinoids alone. However, the widespread use of these technologies in predictive breeding and genomic selection is still in its infancy. The intricacy, dioeciousness, high heterozygosity, sporadic polyploidy, and historically limited breeding efforts of the cannabis genome continue to be problems. However, development cannot be denied. A more consistent relationship between genotype and phenotype is being made possible by standardized genotyping and the quick development of molecular resources. In addition to breeding, these tools are essential for traceability, product authenticity, and regulatory compliance, especially when chemical profiles need to be verified for legal or medical purposes. They also contribute to biodiversity conservation by documenting genetic variation between landraces and industrial hemp lines before more domestication reduces variety.

Like many dioecious plant species, Cannabis also possess male, female, and hermaphrodite sex types. Some cultivated varieties of hemp also exhibit monoecy (Malabadi et al., 2024). Molecular markers, targeting male or female specific regions, for the early detection in dioecious plants, have wider scopes in scientific and industrial research communities (Modi et al., 2018; Li et al, 2025; Leite et al., 2021). Two major compounds namely CBD and THC found in Cannabis are also regulated with sex types along with many other environmental factors. Flowers of unpollinated female plants tend to have more CBD and THC level as compared to male flowers, whereas, male plants are rich in fibre content. On the sex-specific markers have already been developed and used for the early detection of gender in Cannabis (Baek and Vergara, 2025). Sex reversion from female to male and vice versa using silver thiosulphate and ethephon are also well-characterized and widely used practice among research communities (Garcia-de Heer et al., 2025; Malabadi et al., 2024). Considering these facts and resources, researchers and industrialist use molecular markers to identify gender at the seedling stage and propagate the specific gender population according to the demand of the industry or research.

13.4 A FUTURE DIRECTION USING BIOTECHNOLOGY

Although research on cannabis has been carried out in all the area of life sciences (Monthony et al., 2021b), there is scope for improvement in terms of development of a new genotype, transgenic lines, and genome-edited lines (Shujat et al., 2024). Cannabinoid production in non-native hosts like bacteria, other plant cell lines, yeasts, and fungi is one of the promising alternatives in which physical involvement of the plant or its cultivation is not required. This enables the producers to skip the strict regulations imposed on the cultivation of the plant by many governments (Li et al., 2022). Using transgenic technology and genome-editing tools such as CRISPR/Cas9, researchers have already established a proof-of-concept by overexpressing *GUS* and *GFP* or targeting the phytoene desaturase (*PDS*) gene in Cannabis (Mohammad et al., 2024; Zhang et al., 2021). These works opened up the gates for the researchers to establish transgenic or genome-edited lines overexpressing the genes involved in CBD synthesis or suppressing the genes of THC biosynthesis, respectively.

13.5 CONCLUSION

To sum up, the application of molecular markers in *C. sativa* is moving beyond testing to broader utilization. SSR, SNP, TPS, and QTL research developments are working together to lay the foundation for a more transparent, data-driven, and long-term approach to cannabis enhancement. The way forward is clear, despite the fact that there are still legal and technological obstacles to be addressed: molecular insights will allow for dependable production, prudent breeding, and a deeper understanding of this incredibly complex plant.

REFERENCES

Baek, Y. and Vergara, D., 2025. A review of sexual strategies in *Cannabis sativa* L. under genomic and environmental controls. *Agrosystems, Geosciences & Environment*, *8*(1), e70050.

Borin, M., Palumbo, F., Vannozzi, A., Scariolo, F., Sacilotto, G.B., Gazzola, M. and Barcaccia, G., 2021. Developing and testing molecular markers in *Cannabis sativa* (Hemp) for their use in variety and dioecy assessments. *Plants*, *10*(10), p.2174.

Chamberlin, M.A. and Lawit, S.J., 2017. Development and observation of mature megagametophyte cell-specific fluorescent markers. In M.A. Germana and M. Lambardi (Eds.), *In vitro embryogenesis in higher plants* (pp. 237–248). Humana Press. https://pubmed.ncbi.nlm.nih.gov/28936649/

Cheng, L.T., Wang, Z.L., Zhu, Q.H., Ye, M. and Ye, C.Y., 2025. A long road ahead to reliable and complete medicinal plant genomes. *Nature Communications*, *16*(1), p.2150.

de Ronne, M. and Torkamaneh, D., 2025. Discovery of major QTL and a massive haplotype associated with cannabinoid biosynthesis in drug-type Cannabis. *The Plant Genome*, *18*(2), e70031.

Garcia-de Heer, L., Guo, Q., Mieog, J., Nolan, M., Liu, L., Dimopoulos, N., Melzer, R. and Kretzschmar, T., 2025. Transcriptomic analysis of ethephon-induced sex reversion of male *Cannabis sativa* reveals changes in expression of floral homeotic genes and a distinct trichome morphology. *Journal of Experimental Botany*, *76*(18), pp.5401–5418.

Gonçalves, J., Rosado, T., Soares, S., Simão, A.Y., Caramelo, D., Luís, Â., Fernández, N., Barroso, M., Gallardo, E. and Duarte, A.P., 2019. Cannabis and its secondary metabolites: their use as therapeutic drugs, toxicological aspects, and analytical determination. *Medicines*, *6*(1), p.31.

Joshi, C.P. and Nguyen, H.T., 1993. Application of molecular markers in plant biology. *Current Science*, *64*(8), pp.616–622.

Laverty, K.U., Stout, J.M., Sullivan, M.J., Shah, H., Gill, N., Holbrook, L., Deikus, G., Sebra, R., Hughes, T.R., Page, J.E. and Van Bakel, H., 2019. A physical and genetic map of *Cannabis sativa* identifies extensive rearrangements at the THC/CBD acid synthase loci. *Genome Research*, *29*(1), pp.146–156.

Leite Montalvão, A.P., Kersten, B., Fladung, M. and Müller, N.A., 2021. The diversity and dynamics of sex determination in dioecious plants. *Frontiers in Plant Science*, *11*, 580488.

Li, L., Yu, S., Chen, J., Cheng, C., Sun, J., Xu, Y., Deng, C., Dai, Z., Yang, Z., Chen, X. and Tang, Q., 2022. Releasing the full potential of cannabis through biotechnology. *Agronomy*, *12*(10), p.2439.

Li, W.L., Lu, Y.J., Ren, D.S., Li, J.H., Ma, Y.B., Li, X.Y., Li, M.J., Li, C.X., Wang, Y. and Ma, X.R., 2025. The sex-determining region and early-stage gender identification in the dioecious industrial plant *Herpetospermum pedunculosum*. *Industrial Crops and Products*, *236*, 122038.

Livingston, S.J., Quilichini, T.D., Booth, J.K., Wong, D.C., Rensing, K.H., Laflamme-Yonkman, J., Castellarin, S.D., Bohlmann, J., Page, J.E. and Samuels, A.L., 2020. Cannabis glandular trichomes alter morphology and metabolite content during flower maturation. *The Plant Journal*, *101*(1), pp.37–56.

Malabadi, R.B., Kolkar, K.P., Chalannavar, R.K. and Baijnath, H., 2024. *Cannabis sativa*: Monoecious species and Hermaphroditism: Feminized seed production-A breeding effort. *World Journal of Biology Pharmacy and Health Sciences*, *20*(3), pp.169–183.

Modi, A., Suthar, K., Thakkar, P., Mankad, M.C., Kumari, S., Narayanan, S., Singh, A.S. and Kumar, N., 2018. Evaluation of sex specific RAPD and SCAR markers linked to papaya (*Carica papaya* L.). *Biocatalysis and Agricultural Biotechnology*, *16*, pp.271–276.

Mohammad, T., Ghogare, R., Morton, L.B., Dhingra, A., Potlakayala, S., Rudrabhatla, S. and Dhir, S.K., 2024. Evaluation of parameters affecting Agrobacterium-mediated transient gene expression in industrial hemp (*Cannabis sativa* L.). *Plants*, *13*(5), p.664.

Monthony, A.S., Kyne, S.T., Grainger, C.M. and Jones, A.M.P., 2021a. Recalcitrance of *Cannabis sativa* to de novo regeneration; a multi-genotype replication study. *PLoS One*, *16*(8), e0235525.

Monthony, A.S., Page, S.R., Hesami, M. and Jones, A.M.P., 2021b. The past, present and future of *Cannabis sativa* tissue culture. *Plants*, *10*(1), p.185.

Radwan, M.M., Chandra, S., Gul, S. and ElSohly, M.A., 2021. Cannabinoids, phenolics, terpenes and alkaloids of cannabis. *Molecules*, *26*(9), p.2774.

Ren, G., Zhang, X., Li, Y., Ridout, K., Serrano-Serrano, M.L., Yang, Y., Liu, A., Ravikanth, G., Nawaz, M.A., Mumtaz, A.S. and Salamin, N., 2021. Large-scale whole-genome resequencing unravels the domestication history of *Cannabis sativa*. *Science Advances*, *7*(29), eabg2286.

Semagn, K., Bjørnstad, Å. and Ndjiondjop, M.N., 2006. An overview of molecular marker methods for plants. *African Journal of Biotechnology*, *5*(25), pp.2540–2568.

Shujat, S., Robinson, G.I., Norouzkhani, F. and Kovalchuk, I., 2024. Using advanced biotechnological techniques to improve cannabis cultivars. *Biocatalysis and Agricultural Biotechnology*, *60*, 103250.

Sirangelo, T.M., Ludlow, R.A. and Spadafora, N.D., 2022. Multi-omics approaches to study molecular mechanisms in *Cannabis sativa*. *Plants*, *11*(16), p.2182.

Toth, J.A., Stack, G.M., Cala, A.R., Carlson, C.H., Wilk, R.L., Crawford, J.L., Viands, D.R., Philippe, G., Smart, C.D., Rose, J.K. and Smart, L.B., 2020. Development and validation of genetic markers for sex and cannabinoid chemotype in *Cannabis sativa* L. *Gcb Bioenergy*, *12*(3), pp.213–222.

Watts, S., McElroy, M., Migicovsky, Z., Maassen, H., van Velzen, R. and Myles, S., 2021. Cannabis labelling is associated with genetic variation in terpene synthase genes. *Nature Plants*, *7*(10), pp.1330–1334.

Winter, P. and Kahl, G., 1995. Molecular marker technologies for plant improvement. *World Journal of Microbiology and Biotechnology*, *11*(4), pp.438–448. https://doi.org/10.1007/BF00364604

Zhang, J., Yan, J., Huang, S., Pan, G., Chang, L., Li, J., Zhang, C., Tang, H., Chen, A., Peng, D. and Biswas, A., 2020. Genetic diversity and population structure of cannabis based on the genome-wide development of simple sequence repeat markers. *Frontiers in Genetics*, *11*, p.958.

Zhang, X., Xu, G., Cheng, C., Lei, L., Sun, J., Xu, Y., Deng, C., Dai, Z., Yang, Z., Chen, X. and Liu, C., 2021. Establishment of an Agrobacterium-mediated genetic transformation and CRISPR/Cas9-mediated targeted mutagenesis in Hemp (*Cannabis sativa* L.). *Plant Biotechnology Journal*, *19*(10), pp.1979–1987.

Zuardi, A.W., 2006. History of cannabis as a medicine: a review. *Brazilian Journal of Psychiatry*, *28*, pp.153–157.

14 Genetic Characterization of *Artemisia annua* L. Using EST-SSR Markers

Surojeet Das, Divyang Ghanshyambhai Savaliya, Raviraj Bharatbhai Barot, Hiral Sanghavi, and Sachidanand Singh

14.1 INTRODUCTION

Genetic characterization underpins breeding, conservation, and functional genomics in medicinal plants. *Artemisia annua* L. is a diploid annual (2n = 18) in *Asteraceae* that produces artemisinin, a sesquiterpene lactone endoperoxide central to artemisinin-based combination therapies (ACTs) (Tu, 2011; World Health Organization, Regional Office for Africa, 2015). The species shows wide phenotypic variation; plants range roughly 30–200 cm in height and differ in glandular trichome density, the specialized structures that synthesize and store artemisinin, with reported leaf artemisinin from approximately 0.01% to >1% dry weight in elite lines (Ikram & Simonsen, 2017). This diversity is valuable for selection, but together with a large, repetitive, and heterozygous genome, it makes genetic analysis demanding (Liao *et al.*, 2022b; Shen *et al.*, 2018).

To tackle this, co-dominant, polymerase chain reaction (PCR)-amenable simple sequence repeats (SSRs) are widely used because they are polymorphic, reproducible, and informative for heterozygosity and parentage (Kage *et al.*, 2016). Transcriptome-derived SSRs (expressed sequence tag (EST)-SSRs) add a functional anchor to standard SSRs by placing primers in expressed genes, allowing allelic variation to be linked to gene activity; they are straightforward to mine from public RNA-seq/EST datasets and, because primer sites lie in conserved exons, they frequently transfer across related species where genomic resources are sparse (Varshney *et al.*, 2005). The chapter describes the discovery, primer design, optimization, and validation in *A. annua*, and elucidates how they resolve genetic diversity and structure, connect markers to the artemisinin pathway, and support breeding strategies aligned with ACT supply goals.

14.2 *ARTEMISIA ANNUA*: BOTANY AND ECONOMIC CONTEXT

Known as qinghao in traditional Chinese medicine, *A. annua* has been used against fever and is now cultivated on several continents to supply ACTs (Tu, 2011; World Health Organization, Regional Office for Africa, 2015). The crop's value depends on biomass and on trichome-localized artemisinin content, which is controlled by pathway enzymes and transcriptional regulators. Modern breeding programmes combining field selection with molecular tools have delivered lines consistently above approximately 1.0% artemisinin in dried leaves under good agronomy, compared with approximately 0.5% or less in older material (Larson *et al.*, 2013; Townsend *et al.*, 2013). Linking genetic markers to these traits is therefore a direct route to faster genetic gain.

 DOI: 10.1201/9781003542681-14

14.3 GENETIC BASIS OF ARTEMISININ BIOSYNTHESIS

Artemisinin biosynthesis proceeds from farnesyl diphosphate (FPP) via amorpha-4, 11-diene synthase (ADS), followed by sequential oxidations by CYP71AV1 (with its P450 reductase partner) to artemisinic alcohol and artemisinic aldehyde. Artemisinic aldehyde Δ11(13) reductase (DBR2) reduces this intermediate to dihydroartemisinic aldehyde, which aldehyde dehydrogenase 1 (ALDH1) converts to dihydroartemisinic acid. The final conversion to artemisinin occurs non-enzymatically through photo-oxidation and spontaneous rearrangement (Bouwmeester *et al.,* 1999; Brown, 2010; Zhao *et al.,* 2022). Many pathway genes and regulators are preferentially expressed in glandular trichomes. Transcription factors in the AP2/ERF, bHLH, WRKY, and MYB families coordinate pathway expression, often under jasmonate or light cues (Hassani *et al.,* 2020). Structural variation and gene dosage, especially at *ADS,* are associated with high artemisinin chemotypes, supported by genome assemblies that reveal tandem gene expansions (Liao *et al.,* 2022b; Shen *et al.,* 2018).

14.4 SSRS AND EST-SSRS: MOLECULAR BASIS AND PRACTICAL FEATURES

SSRs (microsatellites) are tandem repeats one to six nucleotides long, flanked by unique sequences for primer binding. Repeat-number variation arises mainly from replication slippage, creating length polymorphisms that are easy to score (Ellegren, 2004; Tautz, 1989). Because SSRs are co-dominant, both alleles can be resolved in heterozygotes; standard metrics such as allele number, expected heterozygosity, and polymorphism information content (PIC) describe informativeness (Gupta *et al.,* 2003). Repeat classes are not evenly distributed: trinucleotides are enriched in coding regions (frame-preserving), whereas many dinucleotides occur in introns and intergenic DNA (Katti *et al.,* 2001). A recurring quality issue is null alleles, typically caused by primer-site mutations or introns that obstruct amplification (Dakin & Avise, 2004; Martins *et al.,* 2011).

EST-SSRs are mined from ESTs or RNA-seq assemblies. Tools such as Microsatellite Identification (MISA) scan unigenes for repeat thresholds (for example, ≥6 repeats for dinucleotides and ≥5 for trinucleotides), after which Primer3 is used to design primers with typical annealing temperatures around 55–60 °C and 100–400 bp amplicons (Untergasser *et al.,* 2012; Z. Zhang *et al.,* 2019). Because primer sites lie in conserved exons, EST-SSRs tend to amplify across related species more reliably than anonymous genomic SSRs, which is useful where genomic resources are limited (Table 14.1) (C.-Y. Li *et al.,* 2016; Varshney *et al.,* 2005). The trade-off is that genic loci experience purifying selection and often show moderate, not very high, polymorphism (typical PIC approximately 0.3–0.6) (Morgante *et al.,* 2002).

TABLE 14.1
Practical Comparison of Genomic SSRs and EST-SSRs in *A. annua* and Plants Generally

Aspect	Genomic SSRs	EST-SSRs (Genic)
Origin	Random genomic DNA (often intergenic)	Expressed sequences (cDNA/RNA-sequnigenes)
Link to function	Usually, neutral	In/near genes; may tag functional variation
Polymorphism	Often high (PIC frequently > 0.5)	Moderate (PIC 0.3–0.6 due to constraint)
Cross-species transferability	Lower (flanks diverge faster)	Higher within genus/family (conserved exons)
Development cost/time	Requires genomic libraries or Whole Genome Sequencing (WGS) mining	Leverages public transcriptomes; faster, cheaper
Null-allele risk	Primer mutation; long/complex repeats	Exon–intron complications; junction-spanning primers
Typical *A. annua* use-case	Broad diversity scans, dense maps	Gene-targeted scans, comparative work, MAS anchors

14.5 GENOMIC AND TRANSCRIPTOMIC RESOURCES FOR *A. ANNUA*

High-quality assemblies now enable gene discovery and candidate-gene anchoring of markers in *A. annua*. A 1.74-Gb genome with approximately 63,000 predicted protein-coding genes clarified repeat content and supported pathway analyses (Shen *et al.,* 2018). An allele-aware chromosome-level assembly for contrasting chemotypes later resolved haplotypes, revealing structural variants and tandem expansions, especially at ADS that correlate with artemisinin yield (Liao *et al.,* 2022a). Complementary transcriptome studies across tissues and conditions, together with functional annotation and gene ontology, highlight trichome-biased expression of pathway genes and transcriptional regulators (Huang *et al.,* 2023; W. Wang *et al.,* 2009). These resources provide the raw material for EST-SSR discovery in genes of interest (e.g., *ADS*, *CYP71AV1*, *DBR2*, *ALDH1*, or their regulators) and for aligning legacy marker maps to the physical genome.

14.6 DEVELOPING AND CHARACTERIZING EST-SSR MARKERS

A typical pipeline begins with assembling non-redundant unigenes from RNA-seq, scanning for SSRs (e.g., MISA), and designing primers with Primer3 (Untergasser *et al.,* 2012; Z. Zhang *et al.,* 2019). After in-silico filtering to avoid low-complexity or duplicated regions, primers are tested on genomic DNA. In *A. annua* studies, 60–80% of primer pairs yield clear amplicons, and roughly half of those are polymorphic in diverse panels, with 4–12 alleles per locus and moderate PIC (Z. Li *et al.,* 2022; Y. Wang *et al.,* 2012). Capillary electrophoresis improves allele resolution and repeatability; markers producing multi-band patterns (suggesting paralogous amplification) are dropped or interpreted cautiously. It is good practice to BLAST the source unigene and record functional annotation so that polymorphisms can be linked to candidate pathways or traits.

14.7 VALIDATION AND DATA QUALITY

Robust genotyping depends on tight, single-locus amplification and concordant allele sizing. Expected heterozygosity for polymorphic EST-SSRs in *A. annua* often falls between 0.01 and 0.6, indicating moderate information content (Abbasi Holasou *et al.,* 2023; Z. Li *et al.,* 2022). Deviations from Hardy–Weinberg expectations, especially excess homozygosity, can signal null alleles; designing primers well within exons and avoiding transcript termini reduces this risk (Dakin & Avise, 2004). Cross-lab ring tests and naming conventions help standardize allele bins, enabling meta-analyses.

14.8 GENETIC DIVERSITY IN *A. ANNUA* USING EST-SSRS

Global germplasm panels genotyped with approximately 20–30 EST-SSRs consistently show substantial within-population variation, as expected for an outcrossing, widely distributed plant (Z. Li *et al.,* 2022). Average gene diversity (He) values of approximately 0.6–0.7 and allelic richness of nearly six alleles per locus have been reported in wide panels, with highest allelic variation reported in native Asian accessions and lower diversity in introduced or intensively bred populations, consistent with founder effects (Z. Li *et al.,* 2022). Model-based clustering (STRUCTURE) typically resolves two to four groups reflecting geography and breeding history, often separating modern high-artemisinin cultivars from wild or landrace accessions (Z. Li *et al.,* 2022; Pritchard *et al.,* 2010). These results guide germplasm conservation and parent selection to maximum heterosis and complementarity.

14.9 POPULATION STRUCTURE AND GENE FLOW

Population differentiation metrics based on EST-SSRs generally indicate moderate structure among regions, with F_{ST} (Wright's fixation index) values around 0.12–0.18 in multi-regional comparisons implying most variation resides within populations (D. Chen *et al.,* 2014; Slatkin, 1987; Whitlock

& McCauley, 1999). Isolation by distance patterns in some areas suggest genetic similarity declines with geographic distance, shaped by seed movement and cultivation practices (Smouse & Peakall, 1999; Heywood, 1991). Bottleneck tests often detect reduced diversity in newly introduced or narrowly selected material. Together, these analyses frame sampling strategies and help prioritize in-situ and ex-situ conservation.

14.10 LINKAGE MAPPING, QUANTITATIVE TRAIT LOCI (QTLS), AND CANDIDATE GENES

Early genetic maps resolved nine linkage groups in *A. annua* and identified QTLs for artemisinin yield (Graham *et al.*, 2010). Subsequent studies using SSRs, including EST-SSR, have associated multiple genomic regions with artemisinin content, with consistent signals near *ADS* and *CYP71AV1* (Graham *et al.*, 2010). As high-quality genomes became available, these QTLs could be anchored to physical coordinates, facilitating nomination of underlying genes. With denser single nucleotide polymorphism (SNP) markers, EST-SSRs remain useful anchors to connect historical datasets, enable across-study comparisons, and fingerprint breeding lines.

14.11 COMPARATIVE GENOMICS AND CROSS-SPECIES TRANSFERABILITY

Because EST-SSR primer sites lie in conserved exons, cross-amplification across species is common. In *Artemisia*, many *A. annua* EST-SSRs amplify in related species (e.g., *Artemisia absinthium*, *Artemisia vulgaris*, *Artemisia afra*), and transfer success can extend to other *Asteraceae*, supporting comparative mapping and diversity surveys

> EST-SSR markers are known to exhibit high cross-species transferability due to their origin in conserved expressed regions, and have been successfully applied across Asteraceae genera (e.g., from *Stevia rebaudiana* to *Artemisia annua*), supporting their utility in comparative mapping and diversity surveys—though explicit transfer among *Artemisia* species (e.g. *A. absinthium*, *A. vulgaris*, *A. afra*) remains to be verified in the literature.
>
> (Poonam, 2014; Victoria *et al.*, 2011)

14.12 BREEDING APPLICATIONS

The practical aim is improved cultivars that combine high artemisinin with good agronomics. EST-SSRs have been used for parent selection, hybrid testing, and marker-assisted selection (MAS) when markers sit near artemisinin QTLs (Graham *et al.*, 2010). MAS can shorten cycles because seedlings can be genotyped for favourable alleles before costly field and chemical testing. Reported breeding programmes that combined molecular markers with classical selection produced lines consistently exceeding 1% artemisinin dry weight and improved field stability (Townsend *et al.*, 2013). EST-SSR fingerprints also protect varietal identity and help manage relatedness to avoid inbreeding.

14.13 LIMITATIONS AND TECHNICAL CAUTIONS

EST-SSRs tend to be less polymorphic than genomic SSRs because they are genic and often have shorter repeat tracts. Consequently, more loci may be required to achieve the same discrimination or map density that fewer highly polymorphic genomic SSRs can deliver (Morgante *et al.*, 2002). EST-SSRs also sample only transcribed regions, leaving regulatory and intergenic variation under-represented. Introns can interrupt amplicons, producing null alleles or misleading size shifts if primers straddle exon–intron junctions; careful primer placement and, where possible, consultation of the genome mitigate this issue (Dakin & Avise, 2004; Varshney *et al.*, 2005). In multigene families, paralogous amplification can cause multi-band patterns; such loci should be discarded or interpreted cautiously. Finally, SSR genotyping is sensitive to laboratory practice (e.g., stutter, allele dropout) and requires standardization across labs. Despite these caveats, EST-SSRs in *A. annua* have proven robust and reproducible when best practices are followed (Han *et al.,* 2019; Z. Li *et al.,* 2022).

14.14 ADDITIONAL PRACTICAL CONSIDERATIONS: STUDY DESIGN AND ANALYSIS

Study design benefits from balancing locus number, population size, and geographic coverage. For diversity surveys, choosing 20–40 well-attributed EST-SSR loci that are distributed across linkage groups captures most of the signal in *A. annua*, and adding a few highly polymorphic genomic SSRs increases power where structure is weak (Meyer *et al.,* 2017; Morgante *et al.,* 2002). Panels should include wild accessions, landraces, and elite cultivars to avoid over-estimating differentiation driven only by management history (Morand-Ferron *et al.,* 2016). Sampling at least 30 individuals per putative population and sampling eight to ten populations across the species range stabilize FST and assignment accuracy (Aguirre-Liguori *et al.,* 2020).

Genotyping workflows usually rely on fluorescently labelled primers and multiplexed capillary electrophoresis. Alleles are binned using internal size standards, and automated calls are visually checked to correct stutter-related mis-binning. Replicate genotyping of roughly 10% of samples is a practical quality check; error rates below 1% per locus are achievable. Observed and expected heterozygosity are calculated per locus, and PIC values are computed from allele frequencies to rank informativeness.

Population analysis proceeds from distance matrices to ordination and model-based clustering. Genetic distances such as Nei's DA feed Unweighted Pair Group Method with Arithmetic Mean (UPGMA) or neighbour-joining trees and principal coordinates analysis. STRUCTURE software is commonly run with admixture and correlated allele frequencies, with adequate burn-in and chain lengths across a sensible range of K and multiple replicates; the Evanno ΔK method helps identify the most likely K (Pritchard *et al.,* 2000). Discriminant analysis of principal components is a complementary, assumption light approach. Isolation by distance is tested by correlating genetic and geographic distances, while analysis of molecular variance (AMOVA) partitions variance among and within populations. Reporting both FST and Jost's D provides a more rounded picture of differentiation (Wegmann & Currat, 2024; Whitlock & McCauley, 1999).

In biparental mapping populations, SSRs are scored as co-dominant markers and combined with phenotypic data such as leaf artemisinin percentage to build maps with Join Map or MST map. QTLs are detected by interval or composite interval mapping with per-mutation derived thresholds; confidence intervals then guide candidate gene searches and marker design for MAS (Neto *et al.,* 2012; Sheng *et al.,* 2021). Where EST-SSRs tag pathway genes, co-location with QTL peaks increases confidence (Gailing *et al.,* 2013).

In breeding pipelines, a practical sequence is to identify EST-SSR markers linked to artemisinin QTLs, genotype seedlings from candidate crosses, advance those carrying favourable haplotypes at multiple loci, and confirm gains in replicated field trials before scale-up. Molecular fingerprints safeguard varietal identity through seed multiplication, while background selection maintains genetic diversity and avoids inbreeding depression.

14.15 APPLIED WORKFLOW IN PRACTICE: FROM EST-SSR DISCOVERY TO MAS IN *A. ANNUA*

To make the workflow concrete, consider a global panel of 180 *A. annua* accessions sampled from six regions: China, Vietnam, India, East Africa, Europe, and South America, with 30 individuals per region. Sampling is stratified across altitudes and collection sites to capture local heterogeneity. Plants are raised in a common garden, and young leaves are harvested for DNA extraction using a CTAB protocol adapted for terpenoid-rich tissues. Each accession is also scored for plant height, branch number, days to first flowering, trichome density on the upper leaf surface (microscope counts per mm^2), and artemisinin percentage dry weight at early flowering (HPLC), so that genetic patterns can be interpreted against phenotypes (Czechowski *et al.,* 2018).

For marker discovery, a combined de novo transcriptome is assembled from pooled leaf and glandular trichome RNA-seq data representing high and low artemisinin chemotypes. After quality trimming, contigs are clustered into non-redundant unigenes and annotated by BLAST against UniProt and *Arabidopsis* references. MISA scans identify SSRs meeting standard thresholds (e.g., ≥6 di-nucleotide, ≥5 tri-nucleotide repeats) (The Arabidopsis Initiative, 2000; Beier *et al.,* 2017). We prioritize motifs within long exons of single-copy genes to lower paralog risk, and favour primer pairs with 58–60 °C melting temperatures and 100–280 bp products to support multiplexing (Untergasser *et al.,* 2012). Candidate loci in pathway genes (*ADS, CYP71AV1, DBR2, ALDH1*) and in transcription factors (AP2/ERF, bHLH, WRKY, MYB) are deliberately over-sampled to maximize functional signal (T. Chen *et al.,* 2021; Hassani *et al.,* 2020; Shen *et al.,* 2019).

Primer screening proceeds in three tiers. Tier 1 tests 240 primer pairs on 24 diverse accessions; approximately 65–70% yield single, bright amplicons. Tier 2 assays the clean loci on all 180 accessions; approximately half prove polymorphic with 4–12 alleles and PIC mostly between 0.35 and 0.60, matching expectations for genic SSRs (Keerthi *et al.,* 2018). Tier 3 reduces the set to 36 high-throughput loci that show consistent sizing, low stutter, and minimal missing data, and adds six highly polymorphic genomic SSRs as 'power boosters' for structure inference (Ding *et al.,* 2017; Jo *et al.,* 2017; Morgante *et al.,* 2002).

Genotyping uses two multiplexes with fluorescently labelled forward primers. Amplicons are run on a capillary platform with a LIZ500 size standard. Automated allele calls are binned in 2–3 bp windows and manually curated. Ten per cent of samples are duplicated blind; discordance above 1% triggers locus review. Loci with evidence of null alleles (e.g., excess homozygosity and allele-specific dropout) are either redesigned or excluded (Chung *et al.,* 2007; Dakin & Avise, 2004).

Diversity statistics show mean expected heterozygosity of approximately 0.62 and allelic richness around 6.3 across loci, with China and Vietnam carrying the highest private allele counts and East Africa showing a mixture of Asian and local variants, consistent with seed movement (Meirmans, 2015; Kumar *et al.,* 2014). STRUCTURE software runs (admixture model, correlated allele frequencies, burn-in 100,000, MCMC 200,000) across K = 1–8 with 10 replicates each identify a clear ΔK at K = 3: an East Asian cluster, a South Asian/African cluster with admixture, and a Western introduction cluster that includes Europe and South America (Pritchard *et al.,* 2000). Principal coordinates analysis agrees with this pattern. A Mantel test reveals isolation-by-distance within China and within India (Liu *et al.,* 2023). Global FST is 0.15 (95% CI 0.12–0.18), indicating moderate differentiation, and AMOVA places approximately 12–16% of variation among regions, the rest within regions (Slatkin, 1987; Whitlock & McCauley, 1999; Wright, 1978).

Linkage mapping is integrated by genotyping a biparental F2 derived from a high-artemisinin Chinese line and a vigorous, lower-artemisinin Indian line (n = 240). Phenotyping for artemisinin content occurs at the same developmental stage to control for ontogeny. Composite interval mapping detects QTLs on three linkage groups, with the largest on LG3 co-localizing with *ADS* and explaining approximately 18% of variance; smaller effects sit near *CYP71AV1* and a WRKY candidate (Graham *et al.,* 2010; Teoh *et al.,* 2006; L. Zhang *et al.,* 2009). Support intervals are aligned to the genome; gene models under the peaks include terpene synthase-like sequences, P450s, and regulators that can be prioritized for functional tests.

MAS is then implemented in a recurrent selection programme. Seedlings from diallel crosses among six parents are genotyped for favourable haplotypes at three EST-SSR loci proximal to the LG3 and LG5 QTLs. Only individuals carrying two or more favourable alleles are advanced, cutting the field evaluation set by half. Two cycles of MAS and field testing deliver lines that exceed 1.0% artemisinin dry weight under standard agronomy, while maintaining biomass and synchronous flowering needed for mechanical harvest (Graham *et al.,* 2010; Knudsmark Jessing *et al.,* 2014). SSR fingerprints are used to certify seed lots and to monitor genetic distance among parents so that heterosis is sustained over cycles (Collard & Mackill, 2008; Varshney *et al.,* 2005).

Cross-species transferability is checked by testing the same EST-SSR set in *A. afra* and *A. absinthium.* Around 75–85% of loci amplify, and approximately 40% are polymorphic at useful levels. These data support quick diversity scans in under-resourced species and enable comparative mapping, where shared markers reveal blocks of synteny and help assess whether terpenoid QTLs are conserved across congeners (Han *et al.,* 2019; Gupta *et al.,* 1999; Varshney *et al.,* 2005; Y. Wang *et al.,* 2012).

Finally, the study connects EST-SSRs to other marker systems. A low-density GBS panel of SNPs is added to the 36-locus SSR backbone for 300 accessions. The combined data increase resolution in clustering detects fine-scale relatedness among breeding lines, while EST-SSRs retain their role as robust anchors across datasets and laboratories (Elshire *et al.,* 2011). This hybrid strategy is practical where budgets are limited and where historical SSR datasets exist.

Taken together, this example shows how EST-SSRs can be designed, validated, and applied end-to-end discovery to diversity, mapping, and selection in a way that is realistic for many labs and breeding groups. The same outline can be adapted for regional germplasm surveys, seed-system audits, or conservation planning.

14.16 CONCLUSION AND FUTURE DIRECTIONS

EST-SSRs have given *A. annua* researchers a practical, gene-linked marker system that works across labs and, often, across species. They have helped quantify diversity, clarify population structure, and position QTLs connected to artemisinin biosynthesis. Looking forward, EST-SSRs will continue to serve as bridges between legacy studies and genome-wide approaches. Dense SNP genotyping and genotyping by sequencing will provide higher resolution, while EST-SSRs maintain continuity, support variety fingerprinting, and contribute to multi-omics integration (Elshire *et al.,* 2011). Genomic selection can leverage genome-wide markers to predict breeding values for complex traits like artemisinin yield and stress tolerance, increasing selection intensity and reducing cycle time (Heffner *et al.,* 2009; Meuwissen *et al.,* 2001). Functional genomics will tighten causal inference: transcription factors and pathway genes associated with favourable EST-SSR alleles can be validated via CRISPR-Cas editing or targeted overexpression, with trichome-focused assays to confirm effects (Ikram & Simonsen, 2017; Shen *et al.,* 2016). Single-cell transcriptomics of glandular trichomes, together with pan-genome resources for *Artemisia,* should expose regulatory nodes and structural variants that explain high-artemisinin (Liao *et al.,* 2022a; Zhou *et al.,* 2022). As public RNA-seq and genome resources expand, EST-SSR discovery will become routine, enabling cost-effective, gene-proximal marker sets for new traits and species. In summary, two decades of work show that EST-SSRs remain useful, complementary tools; combined with high-density SNPs, genomic selection, and precise gene editing, they can accelerate breeding of superior *A. annua* cultivars, strengthen global artemisinin supply chains, and deepen understanding of the species' evolutionary genetics (Elshire *et al.,* 2011; Graham *et al.,* 2010; Ikram & Simonsen, 2017; Varshney *et al.,* 2005).

REFERENCES

Abbasi Holasou, H., Valizadeh, N., & Mohammadi, S. A. (2023). Molecular insights into the genetic diversity and population structure of *Artemisia annua* L. as revealed by insertional polymorphisms. *Brazilian Journal of Botany,* 46(1), 51–60. https://doi.org/10.1007/s40415-022-00860-x

Aguirre-Liguori, J. A., Luna-Sánchez, J. A., Gasca-Pineda, J., & Eguiarte, L. E. (2020). Evaluation of the minimum sampling design for population genomic and microsatellite studies: an analysis based on wild maize. *Frontiers in Genetics*, 11, 870. https://doi.org/10.3389/fgene.2020.00870

Beier, S., Thiel, T., Münch, T., Scholz, U., & Mascher, M. (2017). MISA-web: a web server for microsatellite prediction. *Bioinformatics*, 33(16), 2583–2585. https://doi.org/10.1093/bioinformatics/btx198

Bouwmeester, H. J., Wallaart, T. E., Janssen, M. H. A., van Loo, B., Jansen, B. J. M., Posthumus, M. A., Schmidt, C. O., De Kraker, J.-W., König, W. A., & Franssen, M. C. R. (1999). Amorpha-4,11-diene synthase catalyses the first probable step in artemisinin biosynthesis. *Phytochemistry,* 52(5), 843–854. https://doi.org/10.1016/S0031-9422(99)00206-X

Brown, G. D. (2010). The biosynthesis of Artemisinin (Qinghaosu) and the phytochemistry of *Artemisia annua* L. (Qinghao). *Molecules*, 15(11), 7603–7698. https://doi.org/10.3390/molecules15117603

Chen, D., Cui, G., Zhang, X., & Li, L. (2014). [Genetic structure and genetic diversity of *Artemisia annua* varieties (strains) populations based on SCoT markers]. *Zhongguo Zhong Yao Za Zhi = Zhongguo Zhongyao Zazhi = China Journal of Chinese Materia Medica*, 39(17), 3254–3258.

Chen, T., Li, Y., Xie, L., Hao, X., Liu, H., Qin, W., Wang, C., Yan, X., Wu-Zhang, K., Yao, X., Peng, B., Zhang, Y., Fu, X., Li, L., & Tang, K. (2021). AaWRKY17, a positive regulator of artemisinin biosynthesis, is involved in resistance to *Pseudomonas syringae* in *Artemisia annua. Horticulture Research*, 8(1), 217. https://doi.org/10.1038/s41438-021-00652-6

Chung, M. Y., Nason, J. D., & Chung, M. G. (2007). Effects of population succession on demographic and genetic processes: predictions and tests in the daylily *Hemerocallis thunbergii* (Liliaceae). *Molecular Ecology*, 16(13), 2816–2829. https://doi.org/10.1111/j.1365-294X.2007.03361.x

Collard, B. C. Y., & Mackill, D. J. (2008). Marker-assisted selection: an approach for precision plant breeding in the twenty-first century. *Philosophical Transactions of the Royal Society B: Biological Sciences*, 363(1491), 557–572. https://doi.org/10.1098/rstb.2007.2170

Czechowski, T., Larson, T. R., Catania, T. M., Harvey, D., Wei, C., Essome, M., Brown, G. D., & Graham, I. A. (2018). Detailed phytochemical analysis of high- and low Artemisinin-producing chemotypes of *Artemisia annua. Frontiers in Plant Science*, 9, 641. https://doi.org/10.3389/fpls.2018.00641

Dakin, E. E., & Avise, J. C. (2004). Microsatellite null alleles in parentage analysis. *Heredity*, 93(5), 504–509. https://doi.org/10.1038/sj.hdy.6800545

Ding, M., Wang, K., Wang, W., Chen, M., Wu, D., Xu, C., & Chen, K. (2017). Development of high quality EST-SSR markers without stutter bands in peach and their application in cultivar discrimination and hybrid authentication. *HortScience*, 52(1), 24–30. https://doi.org/10.21273/HORTSCI11314-16

Ellegren, H. (2004). Microsatellites: simple sequences with complex evolution. *Nature Reviews Genetics*, 5(6), 435–445. https://doi.org/10.1038/nrg1348

Elshire, R. J., Glaubitz, J. C., Sun, Q., Poland, J. A., Kawamoto, K., Buckler, E. S., & Mitchell, S. E. (2011). A robust, simple genotyping-by-sequencing (GBS) approach for high diversity species. *PLoS ONE*, 6(5), e19379. https://doi.org/10.1371/journal.pone.0019379

Gailing, O., Bodénès, C., Finkeldey, R., Kremer, A., & Plomion, C. (2013). Genetic mapping of EST-derived simple sequence repeats (EST-SSRs) to identify QTL for leaf morphological characters in a Quercus robur full-sib family. *Tree Genetics & Genomes*, 9(5), 1361–1367. https://doi.org/10.1007/s11295-013-0633-9

Graham, I. A., Besser, K., Blumer, S., Branigan, C. A., Czechowski, T., Elias, L., Guterman, I., Harvey, D., Isaac, P. G., Khan, A. M., Larson, T. R., Li, Y., Pawson, T., Penfield, T., Rae, A. M., Rathbone, D. A., Reid, S., Ross, J., Smallwood, M. F., … Bowles, D. (2010). The genetic map of *Artemisia annua* L. identifies loci affecting yield of the antimalarial drug artemisinin. *Science*, 327(5963), 328–331. https://doi.org/10.1126/science.1182612

Gupta, P. K., Rustgi, S., Sharma, S., Singh, R., Kumar, N., & Balyan, H. S. (2003). Transferable EST-SSR markers for the study of polymorphism and genetic diversity in bread wheat. *Molecular Genetics and Genomics*, 270(4), 315–323. https://doi.org/10.1007/s00438-003-0921-4

Gupta, P. K., Varshney, R. K., Sharma, P. C., & Ramesh, B. (1999). Molecular markers and their applications in wheat breeding. *Plant Breeding*, 118(5), 369–390. https://doi.org/10.1046/j.1439-0523.1999.00401.x

Han, X., Xu, R., Zheng, Y., Gao, M., & Sui, L. (2019). Development of EST-SSR markers and genetic diversity analysis among three Artemia species from different geographic populations. *Crustaceana*, 92(7), 841–851. https://doi.org/10.1163/15685403-00003916

Hassani, D., Fu, X., Shen, Q., Khalid, M., Rose, J. K. C., & Tang, K. (2020). Parallel transcriptional regulation of Artemisinin and flavonoid biosynthesis. *Trends in Plant Science*, 25(5), 466–476. https://doi.org/10.1016/j.tplants.2020.01.001

Heffner, E. L., Sorrells, M. E., & Jannink, J. (2009). Genomic selection for crop improvement. *Crop Science*, 49(1), 1–12. https://doi.org/10.2135/cropsci2008.08.0512

Heywood, J. S. (1991). Spatial analysis of genetic variation in plant populations. *Annual Review of Ecology and Systematics*, 22, 335–355. www.jstor.org/stable/2097265

Huang, D., Zhong, G., Zhang, S., Jiang, K., Wang, C., Wu, J., & Wang, B. (2023). Trichome-specific analysis and weighted gene co-expression correlation network analysis (WGCNA) reveal potential regulation mechanism of artemisinin biosynthesis in *Artemisia annua*. *International Journal of Molecular Sciences*, 24(10), 8473. https://doi.org/10.3390/ijms24108473

Ikram, N. K. B. K., & Simonsen, H. T. (2017). A review of biotechnological artemisinin production in plants. *Frontiers in Plant Science*, 8, 303099. https://doi.org/10.3389/FPLS.2017.01966

Jo, W.-S., Kim, H.-Y., & Kim, K.-M. (2017). Development and characterization of polymorphic EST based SSR markers in barley (*Hordeum vulgare*). *3 Biotech*, 7(4), 265. https://doi.org/10.1007/s13205-017-0899-y

Kage, U., Kumar, A., Dhokane, D., Karre, S., & Kushalappa, A. C. (2016). Functional molecular markers for crop improvement. *Critical Reviews in Biotechnology*, 36(5), 917–930. https://doi.org/10.3109/07388551.2015.1062743

Katti, M. V., Ranjekar, P. K., & Gupta, V. S. (2001). Differential distribution of simple sequence repeats in eukaryotic genome sequences. *Molecular Biology and Evolution*, 18(7), 1161–1167. https://doi.org/10.1093/oxfordjournals.molbev.a003903

Keerthi, C. M., Ramesh, S., Byregowda, M., & Vaijayanthi, P. V. (2018). Simple sequence repeat (SSR) marker assay-based genetic diversity among Dolichos bean (*Lablab purpureus* L. Sweet) advanced breeding lines differing for productivity per se traits. *International Journal of Current Microbiology and Applied Sciences*, 7(5), 3736–3744. https://doi.org/10.20546/ijcmas.2018.705.433

Knudsmark Jessing, K., Duke, S. O., & Cedergreeen, N. (2014). Potential ecological roles of artemisinin produced by *Artemisia annua* L. *Journal of Chemical Ecology*, 40(2), 100–117. https://doi.org/10.1007/s10886-014-0384-6

Kumar, J., Bajaj, P., Mishra, G. P., Singh, S. B., Singh, H., & Naik, P. K. (2014). Utilization of EST-derived SSRs in the genetic characterization of *Artemisia annua* L. genotypes from Ladakh, India. *Indian Journal of Biotechnology*, 13, 464–472.

Larson, T. R., Branigan, C., Harvey, D., Penfield, T., Bowles, D., & Graham, I. A. (2013). A survey of artemisinic and dihydroartemisinic acid contents in glasshouse and global field-grown populations of the artemisinin-producing plant *Artemisia annua* L. *Industrial Crops and Products*, 45, 1–6. https://doi.org/10.1016/j.indcrop.2012.12.004

Li, C.-Y., Chiang, T.-Y., Chiang, Y.-C., Hsu, H.-M., Ge, X.-J., Huang, C.-C., Chen, C.-T., & Hung, K.-H. (2016). Cross-species, amplifiable EST-SSR markers for amentotaxus species obtained by next-generation sequencing. *Molecules*, 21(1), 67. https://doi.org/10.3390/molecules21010067

Li, Z., Yun, L., Gao, Z., Wang, T., Ren, X., & Zhao, Y. (2022). EST-SSR primer development and genetic structure analysis of *Psathyrostachys juncea* Nevski. *Frontiers in Plant Science*, 13, 787. https://doi.org/10.3389/fpls.2022.837787

Liao, B., Shen, X., Xiang, L., Guo, S., Chen, S., Meng, Y., Liang, Y., Ding, D., Bai, J., Zhang, D. and Czechowski, T. (2022a). Allele-aware chromosome-level genome assembly of *Artemisia annua* reveals the correlation between ADS expansion and artemisinin yield. *Molecular Plant*, 15(8), 1310–1328.

Liao, B., Shen, X., Xiang, L., Guo, S., Chen, S., Meng, Y., Liang, Y., Ding, D., Bai, J., Zhang, D., Czechowski, T., Li, Y., Yao, H., Ma, T., Howard, C., Sun, C., Liu, H., Liu, J., Pei, J., … Chen, S. (2022b). Allele-aware chromosome-level genome assembly of *Artemisia annua* reveals the correlation between ADS expansion and artemisinin yield. *Molecular Plant*, 15(8), 1310–1328. https://doi.org/10.1016/j.molp.2022.05.013

Liu, H., Wang, Z., Zhang, Y., Li, M., Wang, T., & Su, Y. (2023). Geographic isolation and environmental heterogeneity contribute to genetic differentiation in *Cephalotaxus oliveri*. *Ecology and Evolution*, 13(3), e9869. https://doi.org/10.1002/ece3.9869

Martins, E. M., Vilarinho, L., Esteves, S., Lopes-Marques, M., Amorim, A., & Azevedo, L. (2011). Consequences of primer binding-sites polymorphisms on genotyping practice. *Open Journal of Genetics*, 1, 15–17. https://doi.org/10.4236/ojgen.2011.12004

Meirmans, P. G. (2015). Seven common mistakes in population genetics and how to avoid them. *Molecular Ecology*, 24(13), 3223–3231. https://doi.org/10.1111/mec.13243
Meuwissen, T. H. E., Hayes, B. J., & Goddard, M. E. (2001). Prediction of total genetic value using genome-wide dense marker maps. *Genetics*, 157(4), 1819–1829. https://doi.org/10.1093/genetics/157.4.1819
Meyer, L., Causse, R., Pernin, F., Scalone, R., Bailly, G., Chauvel, B., Délye, C., & Le Corre, V. (2017). New gSSR and EST-SSR markers reveal high genetic diversity in the invasive plant *Ambrosia artemisiifolia* L. and can be transferred to other invasive Ambrosia species. *PLOS ONE*, 12(5), e0176197. https://doi.org/10.1371/journal.pone.0176197
Morand-Ferron, J., Cole, E. F., & Quinn, J. L. (2016). Studying the evolutionary ecology of cognition in the wild: a review of practical and conceptual challenges. *Biological Reviews*, 91(2), 367–389. https://doi.org/10.1111/brv.12174
Morgante, M., Hanafey, M., & Powell, W. (2002). Microsatellites are preferentially associated with nonrepetitive DNA in plant genomes. *Nature Genetics*, 30(2), 194–200. https://doi.org/10.1038/ng822
Neto, E. C., Keller, M. P., Broman, A. F., Attie, A. D., Jansen, R. C., Broman, K. W., & Yandell, B. S. (2012). Quantile-based permutation thresholds for quantitative trait loci hotspots. *Genetics*, 191(4), 1355–1365. https://doi.org/10.1534/genetics.112.139451
Poonam. (2014). Studies on cross-transferability of EST-SSRS of *Stevia rebaudiana* Bertoni in related genera. Retrieved September 1, 2025, from https://krishikosh.egranth.ac.in/items/99c52b87-8d6f-42b7-a869-7e858e5b5bd9
Pritchard, J. K., Stephens, M., & Donnelly, P. (2000). Inference of population structure using multilocus henotype data. *Genetics*, 155(2), 945–959. https://doi.org/10.1093/genetics/155.2.945
Pritchard, J. K., Wen, X., & Falush, D. (2010). Documentation for structure software: Version 2.3. Retrieved from http://pritch.bsd.uchicago.edu/structure.html
Shen, Q., Huang, H., Zhao, Y., Xie, L., He, Q., Zhong, Y., Wang, Y., Wang, Y., & Tang, K. (2019). The transcription factor Aabzip9 positively regulates the biosynthesis of artemisinin in *Artemisia annua*. *Frontiers in Plant Science*, 10, e01294. https://doi.org/10.3389/fpls.2019.01294
Shen, Q., Yan, T., Fu, X., & Tang, K. (2016). Transcriptional regulation of artemisinin biosynthesis in *Artemisia annua* L. *Science Bulletin*, 61(1), 18–25. https://doi.org/10.1007/s11434-015-0983-9
Shen, Q., Zhang, L., Liao, Z., Wang, S., Yan, T., Shi, P., Liu, M., Fu, X., Pan, Q., Wang, Y., Lv, Z., Lu, X., Zhang, F., Jiang, W., Ma, Y., Chen, M., Hao, X., Li, L., Tang, Y., … Tang, K. (2018). The genome of *Artemisia annua* provides insight into the evolution of asteraceae family and artemisinin biosynthesis. *Molecular Plant*, 11(6), 776–788. https://doi.org/10.1016/j.molp.2018.03.015
Sheng, C., Song, S., Zhou, R., Li, D., Gao, Y., Cui, X., Tang, X., Zhang, Y., Tu, J., Zhang, X., & Wang, L. (2021). QTL-seq and transcriptome analysis disclose major QTL and candidate genes controlling leaf size in sesame (*Sesamum indicum* L.). *Frontiers in Plant Science*, 12, 580846. https://doi.org/10.3389/fpls.2021.580846
Slatkin, M. (1987). Gene flow and the geographic structure of natural populations. *Science*, 236(4803), 787–792. https://doi.org/10.1126/science.3576198
Smouse, P. E., & Peakall, R. (1999). Spatial autocorrelation analysis of individual multiallele and multilocus genetic structure. *Heredity*, 82(5), 561–573. https://doi.org/10.1038/sj.hdy.6885180
Tautz, D. (1989). Hypervariability of simple sequences as a general source for polymorphic DNA markers. *Nucleic Acids Research*, 17(16), 6463–6471. https://doi.org/10.1093/nar/17.16.6463
Teoh, K. H., Polichuk, D. R., Reed, D. W., Nowak, G., & Covello, P. S. (2006). *Artemisia annua* L. (Asteraceae) trichome-specific cDNAs reveal CYP71AV1, a cytochrome P450 with a key role in the biosynthesis of the antimalarial sesquiterpene lactone artemisinin. *FEBS Letters*, 580(5), 1411–1416. https://doi.org/10.1016/j.febslet.2006.01.065
The Arabidopsis Initiative. (2000). Analysis of the genome sequence of the flowering plant *Arabidopsis thaliana*. *Nature,* 408(6814), 796–815. https://doi.org/10.1038/35048692
Townsend, T., Segura, V., Chigeza, G., Penfield, T., Rae, A., Harvey, D., Bowles, D., & Graham, I. A. (2013). The use of combining ability analysis to identify elite parents for *Artemisia annua* F1 hybrid production. *PLoS ONE*, 8(4), e61989. https://doi.org/10.1371/journal.pone.0061989
Tu, Y. (2011). The discovery of artemisinin (qinghaosu) and gifts from Chinese medicine. *Nature Medicine*, 17(10), 1217–1220. https://doi.org/10.1038/nm.2471
Untergasser, A., Cutcutache, I., Koressaar, T., Ye, J., Faircloth, B. C., Remm, M., & Rozen, S. G. (2012). Primer3—new capabilities and interfaces. *Nucleic Acids Research*, 40(15), e115–e115. https://doi.org/10.1093/nar/gks596

Varshney, R. K., Graner, A., & Sorrells, M. E. (2005). Genic microsatellite markers in plants: features and applications. *Trends in Biotechnology*, 23(1), 48–55. https://doi.org/10.1016/j.tibtech.2004.11.005
Victoria, F. C., da Maia, L. C., & de Oliveira, A. C. (2011). In silicocomparative analysis of SSR markers in plants. *BMC Plant Biology,* 11(1), 15. https://doi.org/10.1186/1471-2229-11-15
Wang, W., Wang, Y., Zhang, Q., Qi, Y., & Guo, D. (2009). Global characterization of *Artemisia annua* glandular trichome transcriptome using 454 pyrosequencing. *BMC Genomics*, 10(1), 465. https://doi.org/10.1186/1471-2164-10-465
Wang, Y., Cai, H., & Jia, W. (2012). [Analysis of SSR information in EST resource of sweet wormwood (*Artemisia annua*) and development of EST-SSR marker]. *China Journal of Chinese Materia Medica*, 37(5), 570–574.
Wegmann, D., & Currat, M. (2024). A brief overview of analysis of molecular variance. *Molecular Ecology*, 23(1), 16–19. https://doi.org/10.1111/MEC.12604
Whitlock, M. C., & McCauley, D. E. (1999). Indirect measures of gene flow and migration: FST≠1/(4Nm+1). *Heredity*, 82(2), 117–125. https://doi.org/10.1038/sj.hdy.6884960
World Health Organization, Regional Office for Africa. (2015 April). *Guidelines for the treatment of malaria*. Third edition. Retrieved September 1, 2025, from www.afro.who.int/publications/guidelines-treatment-malaria-third-edition
Wright, S. (1978). Evolution and the genetics of populations. v. 4.Variability within and among natural populations. University of Chicago Press. Retrieved from https://books.google.com/books/about/Evolution_and_the_Genetics_of_Population.html?id=q0RDJf3K_aUC
Zhang, L., Jing, F., Li, F., Li, M., Wang, Y., Wang, G., Sun, X., & Tang, K. (2009). Development of transgenic *Artemisia annua* (Chinese wormwood) plants with an enhanced content of artemisinin, an effective anti-malarial drug, by hairpin-RNA-mediated gene silencing. *Biotechnology and Applied Biochemistry*, 52(3), 199–207. https://doi.org/10.1042/BA20080068
Zhang, Z., Xie, W., Zhao, Y., Zhang, J., Wang, N., Ntakirutimana, F., Yan, J., & Wang, Y. (2019). EST-SSR marker development based on RNA-sequencing of *E. sibiricus* and its application for phylogenetic relationships analysis of seventeen Elymus species. *BMC Plant Biology*, 19(1), 235. https://doi.org/10.1186/s12870-019-1825-8
Zhao, L., Zhu, Y., Jia, H., Han, Y., Zheng, X., Wang, M., & Feng, W. (2022). From plant to yeast—advances in biosynthesis of artemisinin. *Molecules*, 27(20), 6888. https://doi.org/10.3390/molecules27206888
Zhou, P., Chen, H., Dang, J., Shi, Z., Shao, Y., Liu, C., Fan, L., & Wu, Q. (2022). Single-cell transcriptome of *Nepeta tenuifolia* leaves reveal differentiation trajectories in glandular trichomes. *Frontiers in Plant Science*, 13, 988594. https://doi.org/10.3389/fpls.2022.988594

15 Past, Present, and Future of Molecular Markers in Herbal Medicines

Amanullakhan Pathan, Tanvi Nayak, Dipika Pandya, and Sohilkhan Chauhan

15.1 INTRODUCTION

Herbal medicines have been utilized by humans for millennia, with roots in ancient traditions of healing and wellness. From the Ayurvedic practices of India to traditional Chinese medicine (TCM), plants have long been revered for their therapeutic properties (Fugh-Berman, 2000; Hu *et al.*, 2005; Husen, 2022; 2023a; 2023b; 2024a; 2024b; Husen and Pant, 2024a; 2024b; Pant and Husen, 2025). In recent decades, the global demand for herbal products has surged, as individuals seek natural alternatives to synthetic drugs and look for remedies with fewer side effects. The World Health Organization (WHO) has reported that approximately 80% of the global population in developing countries relies on plant-based medicines for their primary healthcare needs (WHO, 2008). Despite their widespread use, the authenticity, safety, and efficacy of herbal medicines are often questioned due to challenges such as misidentification, adulteration, contamination, and variations in plant quality (Newmaster *et al.*, 2013).

Traditional methods of plant identification, such as morphological characteristics, anatomical features, and organoleptic evaluation (taste, smell, and texture), have been historically used to authenticate medicinal plants. However, these methods have significant limitations, such as the high potential for misidentification between closely related species and variability due to environmental and developmental factors (Benzie and Wachtel-Galor, 2011). Furthermore, morphological traits can be altered during the drying, storage, or processing of herbal materials, which may obscure key identifying features. These limitations are particularly problematic when assessing the identity and quality of raw plant material in commercial herbal products, where plant species may be substituted, adulterated, or contaminated.

With the advent of molecular biology, researchers have turned to molecular markers as a solution to these challenges. Molecular markers, which include genetic markers (such as DNA barcoding, microsatellites, and single nucleotide polymorphisms) and protein-based markers, provide a precise, reproducible, and objective means of identifying plant species and assessing genetic diversity (Hyder *et al.*, 2024). DNA barcoding, for example, uses short, standardized gene sequences (such as the rbcL (ribulose-1,5-bisphosphate carboxylase/oxygenase large subunit), matK (maturase K), and ITS (internal transcribed spacer) regions) to identify plant species with high accuracy (Kress *et al.*, 2005; Kress and Horvitz, 2005). The DNA barcode approach has revolutionized plant taxonomy and allowed for the identification of species even when only small fragments of plant material are available, a significant advantage for the authentication of herbal medicines (Chen *et al.*, 2010; Zhang *et al.*, 2014). In addition, molecular markers can aid in detecting adulterants or contaminants in herbal products, ensuring that consumers receive the correct medicinal plant and that products are free of harmful substances (Newmaster *et al.*, 2013).

DOI: 10.1201/9781003542681-15

The role of molecular markers in herbal medicine gained momentum in the early 2000s, with the development of techniques such as polymerase chain reaction (PCR) amplification and DNA sequencing, which enabled the identification of plants at the molecular level (Li *et al.*, 2008). This molecular approach was particularly useful in situations where the plant's morphology could not provide definitive identification, such as when herbal products were highly processed, ground into powders, or mixed with other substances. Researchers began applying these techniques not only for species identification but also for assessing genetic variation and studying plant phylogenetics (Yu *et al.*, 2021; Zhou *et al.*, 2018). The introduction of next-generation sequencing (NGS) in the last decade has further expanded the possibilities of molecular markers, allowing for even more detailed and comprehensive analysis of plant genomes (Ray Soham *et al.*, 2014; Yang *et al.*, 2012; Satam *et al.*, 2023; Gao *et al.*, 2023).

Today, molecular markers are widely used in the herbal medicine industry for quality control, species authentication, and the detection of potential contamination. Regulatory bodies such as the US Pharmacopeia (USP) and European Medicines Agency (EMA) have acknowledged the importance of molecular technologies for the standardization and authentication of herbal medicines (Wang *et al.*, 2023). DNA barcoding, for instance, has become an essential tool in ensuring the authenticity of herbal supplements, helping to safeguard consumers from the risks associated with fraudulent or substandard products (Li *et al.*, 2015). These advances are particularly important as the demand for herbal products grows and the risk of adulteration, including the use of cheaper or potentially harmful plant species, increases.

Looking to the future, molecular markers hold tremendous promise for further advancing the field of herbal medicine. The application of gene-editing technologies such as CRISPR-Cas 9 could allow for the development of plants with enhanced medicinal properties, greater consistency in active compounds, and improved resistance to pests and diseases (Kumar *et al.*, 2013). Furthermore, the integration of molecular techniques with artificial intelligence (AI) and machine learning (ML) algorithms has the potential to revolutionize quality control processes, enabling real-time monitoring and analysis of herbal products (Ghimire *et al.*, 2023). As the field of molecular biology continues to evolve, new technologies will further improve the accuracy and efficiency of molecular markers, making them even more accessible and affordable for widespread use in the herbal industry.

In this chapter, we will explore the past, present, and future of molecular markers in herbal medicines, tracing their development from early molecular biology techniques to their current applications in plant authentication, quality control, and genetic research. Additionally, we will discuss emerging technologies and trends that could shape the future of herbal medicine, including the integration of molecular markers with modern digital and analytical tools. Through this exploration, we aim to highlight how molecular markers are transforming herbal medicine, bridging the gap between traditional knowledge and modern scientific innovation.

15.2 PAST: HISTORICAL DEVELOPMENT OF MOLECULAR MARKERS IN HERBAL MEDICINE

15.2.1 Traditional Methods of Herbal Medicine Identification

For thousands of years, traditional herbal medicine practices have relied on empirical and observational methods to identify and utilize plants. Practitioners of traditional medicine systems, such as TCM, Ayurveda, and Indigenous healing practices, primarily employed morphological features – leaf shape, flower colour, bark texture, root structure, and other external characteristics – as the basis for plant identification (Benzie and Wachtel-Galor, 2011; Fugh-Berman, 2000; Hu *et al.*, 2005). Additionally, sensory evaluation (organoleptic testing) through taste, smell, and touch has been integral in determining the medicinal properties of plants. For instance, many plants with bitter tastes were identified as useful for digestive issues and detoxification, while sweet or aromatic plants were often used for calming or respiratory conditions (Koroma *et al.*, 2022).

However, these traditional identification methods had substantial limitations. Morphological traits, while generally dependable, are often influenced by environmental factors such as climate, soil composition, or developmental stage, causing variability that could lead to incorrect identification (Yuan *et al.*, 2016; Maddala *et al.*, 2024; Sao *et al.*, 2022). Additionally, many plant species used in herbal medicine look strikingly similar but may have vastly different chemical profiles and medicinal uses. This morphological similarity between species is especially problematic when plants are processed, dried, or powdered, further complicating accurate identification (Al-Snafi, 2016). Misidentification, adulteration, and contamination became significant issues as the herbal medicine market expanded globally, contributing to the need for more reliable methods of authentication and quality control.

This growing concern over misidentification and the lack of regulatory oversight led to an increasing demand for accurate, standardized identification methods. As herbal medicine became commercialized, it became critical to ensure that the correct species were being used in herbal products and that there was transparency in sourcing, which spurred the introduction of molecular techniques in plant authentication.

15.2.2 Introduction of Molecular Biology to Herbal Medicine

The 20th century ushered in a revolutionary era in plant biology, thanks in large part to advances in molecular biology. Molecular techniques, especially DNA-based methods, began to supplement and, in some cases, replace traditional morphological and sensory identification methods. The emergence of molecular biology provided tools to examine the genetic makeup of plants, allowing for species identification that transcended morphological limitations.

The first molecular marker techniques applied to plant identification were based on restriction fragment length polymorphism (RFLP) and PCR. RFLP, developed in the 1980s, uses restriction enzymes to cut DNA at specific sites, producing distinctive DNA fragment patterns that could be used to identify genetic differences between species (Smith *et al.*, 1986). This technique was a significant breakthrough, but it was labour-intensive, expensive, and required high-quality DNA samples, making it less practical for routine use.

The introduction of PCR technology by Mullis and Faloona (1987) was a watershed moment in molecular biology. PCR enables the amplification of specific DNA sequences, providing a much-needed solution to the limitations of RFLP. PCR is faster, more flexible, and requires smaller amounts of DNA, making it highly suitable for plant species identification. This was particularly useful in the herbal medicine context, where plant samples may be fragmented, dried, or otherwise degraded. PCR allowed for the amplification and subsequent identification of specific genes that could serve as molecular 'fingerprints' for plant species (Zhang *et al.*, 2000). PCR-based techniques, such as random amplified polymorphic DNA (RAPD) and amplified fragment length polymorphism (AFLP), were developed to detect genetic variation in plants and rapidly distinguish between species (Williams *et al.*, 1990).

The application of molecular markers in herbal medicine research dates back to the late 20th century. Early molecular techniques, including RAPD, RFLP, and AFLP, laid the foundation for the subsequent development of more sophisticated genetic tools.

RAPD: RAPD, introduced by Williams *et al.* (1990), was one of the first molecular techniques used to generate DNA fingerprints for plant species. RAPD uses random primers to amplify short segments of DNA, producing polymorphic band patterns that are unique to specific plant species. This technique was initially applied in plant genetics for species identification and the characterization of genetic diversity. RAPD proved to be an effective tool for the authentication of medicinal plants such as *Echinacea purpurea* (Shinde *et al.*, 2007) and Ginseng (*Panax ginseng*) (Um *et al.*, 2001). However, RAPD has limitations, such as its sensitivity to environmental conditions, which affect reproducibility and reliability.

RFLP: RFLP analysis was another early molecular technique widely used for plant identification and genetic diversity studies. This method detects variations in DNA sequence by observing the differences in fragment lengths following digestion by restriction enzymes. Zaharleva *et al.* (2001) employed RFLP markers to assess genetic diversity in medicinal plants, including species of *Coriandrum sativum* (coriander). RFLP was instrumental in understanding the genetic structure of medicinal plants and served as a useful tool for species identification in cases where morphological traits were insufficient.

Applications of RAPD and RFLP in Herbal Medicine: Both RAPD and RFLP were pioneering methods for molecular authentication in herbal medicine. They were pivotal in verifying the authenticity of plant-based materials in the marketplace and detecting fraudulent products. These markers were used extensively to identify and authenticate high-demand herbal products like ginseng and *Echinacea*, addressing issues of substitution and adulteration (Moon *et al.*, 2016). While these technologies contributed significantly to early herbal medicine research, their limitations prompted the search for more precise and reproducible methods.

However, the major breakthrough in molecular identification of plants came with the advent of DNA barcoding in the early 2000s. DNA barcoding uses short, standardized regions of DNA to identify species, similar to how the barcodes on supermarket items uniquely identify products. The introduction of DNA barcoding for plants was championed by Kress *et al.* (2005), who proposed that regions of the chloroplast genome, such as rbcL and matK, could serve as reliable markers for plant identification. This technique was quickly adopted because it offered a standardized, efficient, and robust approach to identifying plant species with high accuracy, even for fragmented or processed plant material (Kress *et al.*, 2005; Hebert *et al.*, 2003). In particular, rbcL and matK sequences became widely used as barcodes for plant species identification, forming the foundation for large, publicly accessible databases of plant genetic sequences, such as the Barcode of Life Data (BOLD) Systems and GenBank (Hebert *et al.*, 2003; Kress *et al.*, 2005). DNA barcoding revolutionized the field of herbal medicine by providing a solution to many of the authentication challenges posed by traditional methods. It enabled researchers, regulatory bodies, and manufacturers to verify the species used in herbal products, ensuring product consistency and safety (Newmaster *et al.*, 2013). Barcoding also played a significant role in detecting adulteration and mislabelling in the herbal supplement market, helping to address issues such as species substitution, contamination, and fraudulent practices.

In addition to DNA barcoding, molecular biology also contributed to the development of other molecular markers. Microsatellites (also known as simple sequence repeats (SSRs)) and single nucleotide polymorphisms (SNPs) became increasingly popular for studying genetic diversity and resolving closely related species (Liberty *et al.* 2025). Microsatellites, which are highly polymorphic, were used to detect genetic variation within populations, while SNPs were employed to investigate finer-scale genetic differences between individuals and species (Ghimire *et al.*, 2023). These markers offered enhanced resolution in species identification and provided tools for studying the genetic structure and population dynamics of medicinal plant species.

15.2.3 Challenges and Limitations of Early Molecular Marker Approaches

While the introduction of molecular markers marked a significant advancement in plant authentication, early applications faced several challenges and limitations that hindered their widespread adoption in the herbal medicine sector.

15.2.3.1 Cost and Technical Complexity

Early molecular techniques, including RFLP and PCR, were expensive and required specialized laboratory equipment, skilled personnel, and substantial time investment. These limitations made these techniques difficult to implement on a large scale, particularly for small-scale herbal producers and practitioners who lacked the necessary resources (Zhang *et al.*, 2000). Furthermore, some PCR-based methods were prone to contamination and false positives, which could lead to inaccurate identification results (Li X., *et al.*, 2017).

15.2.3.2 DNA Quality and Sample Degradation

In herbal medicine, the quality of plant material used for analysis is often compromised, particularly for products that are dried, powdered, or otherwise processed. DNA degradation due to improper storage or processing made it difficult to obtain high-quality DNA samples for analysis. While PCR provided a solution for amplifying small amounts of DNA, some plant species had low or degraded DNA content, which posed a significant challenge for successful molecular identification (Ghimire *et al.*, 2023).

15.2.3.3 Lack of Comprehensive Genetic Databases

One of the most significant challenges in early molecular identification was the absence of comprehensive, standardized genetic reference databases. For DNA barcoding to be effective, a large database of reference sequences for known plant species is essential. In the early 2000s, when DNA barcoding was first introduced, many medicinal plant species lacked adequate representation in genetic databases, complicating the identification process for lesser-known or poorly studied plants (Chen *et al.*, 2010; Li *et al.*, 2017). Furthermore, many herbal products were made from wild-collected plants or from cultivars with limited genomic data available, which meant that these plants could not be easily matched to existing barcode sequences.

15.2.3.4 Marker Specificity and Resolution

Although DNA barcoding represented a breakthrough, it was not without limitations. In some plant families, genetic divergence between species is minimal, making it difficult for barcode markers such as rbcL and matK to distinguish between closely related species (Newmaster *et al.*, 2013). In such cases, higher resolution markers like SSRs or SNPs were necessary, but these markers were more complex and required more sophisticated technology, further limiting their accessibility. Despite these early challenges, molecular marker technologies have evolved significantly over time. Advances in high-throughput sequencing technologies, such as NGS, have made plant species identification more efficient and cost-effective. NGS technologies, which allow for the sequencing of entire plant genomes at a fraction of the cost of earlier methods, have drastically improved the resolution of molecular markers and expanded the potential for detecting adulteration and contamination in herbal products (Ghimire *et al.* 2023).

In conclusion, while early molecular marker techniques had their limitations, they paved the way for the more advanced, cost-effective, and accurate methods available today. Molecular biology has proven to be a valuable tool in addressing the challenges of plant misidentification, adulteration, and contamination in the herbal medicine industry, providing researchers and regulators with the means to authenticate herbal products and ensure their safety and efficacy.

15.3 PRESENT: CURRENT USE OF MOLECULAR MARKERS IN HERBAL MEDICINE

The application of molecular marker technologies has become a cornerstone in the identification, quality control, and authentication of herbal medicine. These tools have revolutionized the field by providing precise, dependable, and standardized methods to verify the species, detect adulteration, and explore the genetic diversity of medicinal plants. Today, molecular markers are critical for ensuring the safety, efficacy, and authenticity of herbal products, particularly in the context of global herbal trade, where misidentification and contamination remain pressing concerns.

15.3.1 Current Molecular Marker Technologies

Molecular markers are identifiable sequences within the genome that can be used to distinguish species, track genetic variation, and assess the authenticity of plant materials. Among the numerous molecular marker technologies in use, DNA-based markers dominate in herbal medicine due to their precision, versatility, and reproducibility. These include DNA barcoding, simple sequence repeats (SSRs), SNPs, and newer techniques such as proteomics and metabolomics.

15.3.1.1 DNA Barcoding

DNA barcoding is a method used to identify species by using a short, standardized region of their genome. For plants, common DNA barcodes include regions of the chloroplast genome, such as rbcL and matK, and the nuclear genome, specifically the ITS regions (Kress *et al.*, 2005; Hebert *et al.*, 2003). The use of DNA barcoding has revolutionized plant species identification, enabling researchers and herbal practitioners to accurately authenticate herbal materials, even when they are dried, powdered, or processed. rbcL is a commonly used gene in plant DNA barcoding because of its conserved nature across plant species, making it a reliable identifier for a wide range of plants. However, it has limited resolution at distinguishing closely related species, which is why it is often used in combination with other markers like matK and ITS (Chen *et al.,* 2010). matK is another chloroplast gene used alongside rbcL for plant barcoding. Its variability between species is higher than rbcL, making it particularly useful for identifying species within closely related genera (Kress *et al.*, 2005). ITS regions of the nuclear genome are highly variable and have become a popular barcode marker for plant identification, particularly in the angiosperms (flowering plants). ITS regions are often used when the goal is to distinguish between species in a genus where other markers (rbcL or matK) may lack sufficient resolution (Kress *et al.*, 2005; Liberty *et al.*, 2025). DNA barcoding is primarily used to verify the species identity of raw herbal materials and products, ensuring that they are correctly labelled and free from contamination or adulteration. For instance, Newmaster *et al.* (2013) used DNA barcoding to examine 16 herbal products sold in the market and found that 20% of them were misidentified or contaminated with species not listed on the label. This demonstrates the potential of DNA barcoding to detect fraudulent labelling in the herbal medicine industry. DNA barcoding also helps herbal practitioners, manufacturers, and consumers ensure that they are using the correct plant species, especially when dealing with high-value medicinal plants such as ginseng (*P. ginseng*), which is often adulterated with similar species (Li *et al.*, 2017). As a result of its ability to verify species identity, DNA barcoding has become integral to the quality control processes in the herbal supplement industry (Newmaster *et al.*, 2013).

15.3.1.2 SSR Markers

SSRs, also known as microsatellites, are short tandem repeats of 1–6 base pairs found throughout the genome. These markers are highly polymorphic, meaning they exhibit a large degree of variation between individuals of the same species, making them valuable for genetic diversity studies, population genetics, and species identification (Goyal *et al.*, 2021).

Applications of SSR Markers in Herbal Medicine: SSR markers are widely used to assess the genetic diversity of medicinal plant species and to confirm the identity of plant cultivars, particularly in cases where morphological features may not be reliable.

Genetic Diversity and Conservation: SSR markers have been used to assess genetic variation within populations of medicinal plants, helping to establish conservation strategies for endangered species used in herbal medicines. For example, SSRs have been used to investigate the genetic diversity of Echinacea purpurea, a popular herb in the US for immune support (Li *et al.*, 2017).

Species and Cultivar Identification: SSRs can identify and differentiate cultivars of a single species, ensuring that the correct variety is used for medicinal purposes. For example, in *Salvia miltiorrhiza* (Danshen), SSR markers were used to differentiate between cultivated and wild populations, which are chemically distinct (Ghimire *et al.*, 2023).

Authentication of Adulterants: SSR markers also help detect adulterants in herbal products by distinguishing between closely related species that might be used as substitutions (Liu *et al.*, 2019).

15.3.1.3 SNP Markers

SNPs are the most common type of genetic variation and involve a single nucleotide change in the DNA sequence. These variations can be found in coding and non-coding regions of the genome and provide high-resolution markers for species identification and phylogenetic studies. SNPs are particularly useful for distinguishing between closely related species or varieties that might be difficult to differentiate using traditional morphological methods.

Applications of SNPs in Herbal Medicine: SNP markers are applied in cases where the species being studied are genetically very similar, as in the case of closely related medicinal plants or different varieties of a single species. The use of SNPs in species identification has been beneficial in distinguishing adulterants from the target species, even in processed or dried plant materials (Li *et al.*, 2017).

Distinguishing Closely Related Species: SNPs are extremely useful for identifying plants that are genetically similar but phenotypically different. For example, in *Panax* species, SNPs can differentiate between *P. ginseng* and its adulterants, such as *Panax quinquefolius* (American ginseng) (Liu *et al.*, 2019).

Fraud Detection and Species Authentication: By using SNP markers, fraudulent claims in the herbal medicine market can be identified, ensuring consumers receive products made from the correct plant species (Li *et al.*, 2017).

Varietal Classification: SNP markers can also be used to distinguish between varieties within a species, such as different varieties of turmeric (*Curcuma longa*), which may differ in their chemical composition and therapeutic efficacy (Jain *et al.*, 2024).

15.3.2 Proteomics and Metabolomics

While DNA-based molecular markers provide insights into the genetic identity of plants, proteomics and metabolomics offer complementary approaches for studying the chemical composition and biological activity of medicinal plants.

Proteomics: Proteomics involves the study of the entire complement of proteins expressed by a plant. In herbal medicine, proteomic analysis can provide information on the presence of specific bioactive compounds and their expression in different plant tissues. By analysing protein expression profiles, researchers can track the quality and consistency of herbal products, identifying proteins associated with therapeutic effects.

Metabolomics: Metabolomics studies the metabolites present in a plant, such as alkaloids, terpenoids, flavonoids, and other secondary metabolites that contribute to the plant's medicinal properties. Metabolomic profiling is particularly useful for quality control and authenticating herbal medicines by detecting specific metabolites associated with the plant's medicinal properties. For instance, the metabolomic fingerprinting of Echinacea species can help differentiate between genuine and adulterated products, ensuring that only plants with the desired medicinal properties are used in formulations (Salem *et al.*, 2020). Both proteomics and metabolomics contribute significantly to ensuring that herbal medicines contain the correct active compounds and meet quality standards. By combining these approaches with genetic techniques, it is possible to verify not only the species of a plant but also the presence and concentration of bioactive compounds.

15.4 APPLICATIONS OF MOLECULAR MARKERS IN QUALITY CONTROL AND AUTHENTICATION

The use of molecular markers in the quality control of herbal medicines is essential for ensuring the integrity and safety of herbal products. Traditional quality control methods, such as organoleptic analysis or chemical assays, have limitations in detecting contamination, adulteration, or species substitution. Molecular markers overcome these limitations by providing objective, reproducible results that can be used to verify the species identity and purity of plant materials (Li *et al.*, 2017).

15.4.1 Authentication and Traceability

DNA barcoding, SSRs, and SNP markers are widely used to authenticate herbal products and trace their origins. For example, the authenticity of herbal teas, powders, and extracts can be verified by matching the DNA sequences from the product with known sequences in reference databases (Newmaster *et al.*, 2013). This ensures that consumers receive products that contain the correct plant species and that there has been no substitution with cheaper or potentially harmful species.

15.4.2 Detection of Adulteration

Molecular markers play a critical role in detecting adulterants in herbal products, especially when the adulterant is a closely related or morphologically similar species. For example, molecular markers have been used to identify the presence of substituted species in ginseng, saw palmetto, and turmeric (Hyder *et al.*, 2024). The ability to detect adulterants and ensure the authenticity of herbal medicine is crucial for protecting consumers and maintaining the credibility of the herbal industry.

15.4.3 Regulatory Compliance

Many countries have established regulations that require herbal products to meet specific quality and safety standards. Molecular markers are often employed to demonstrate compliance with these regulations, particularly for species identification and purity testing. For example, the EMA and USP have incorporated DNA barcoding as a standard method for verifying the authenticity of herbal medicines used in clinical settings (Fugh-Berman, 2000).

15.4.4 Case Studies and Current Trends

Several studies and real-world applications demonstrate the growing importance of molecular markers in herbal medicine:

Case Study 1: Authentication of Ginseng Species: Ginseng, a globally recognized medicinal herb, is often subject to adulteration due to its high market value. DNA barcoding and SNP markers have been successfully used to distinguish between *P. ginseng* (Asian ginseng) and *P. quinquefolius* (American ginseng), ensuring that consumers receive the correct species (Liu *et al.*, 2019). Molecular methods also help identify other related species that may be used as substitutes.

Case Study 2: Authentication of Saw Palmetto: Saw palmetto, used for prostate health, is another herb prone to adulteration. A study by Little *et al.* (2013) used DNA barcoding to authenticate saw palmetto products sold on the market. The results showed that some products contained incorrect species, including adulteration with other palm species. DNA barcoding thus provides an effective tool for ensuring the purity and safety of saw palmetto products.

Case Study 3: Herbal Tea Authentication: A study on herbal teas used molecular markers to authenticate products containing chamomile (*Matricaria chamomilla*). By employing both DNA barcoding and metabolomics, the study successfully verified the authenticity of the herbal teas and detected the presence of contaminants, such as grass species and other plants (Uncu *et al.*, 2015).

Molecular marker technologies have become indispensable tools in the modern herbal medicine industry, playing a pivotal role in the authentication, quality control, and genetic diversity studies of medicinal plants. Technologies such as DNA barcoding, SSR markers, and SNPs are widely used to identify plant species, detect adulteration, and ensure the safety and efficacy of herbal products. As the demand for herbal medicine continues to grow, the integration of proteomics, metabolomics, and advanced sequencing technologies promises to further enhance the precision and reliability of plant authentication methods. As these technologies become more affordable and accessible, they will continue to drive the future of herbal medicine toward greater standardization and consumer safety.

15.5 FUTURE: THE POTENTIAL OF MOLECULAR MARKERS IN THE FUTURE OF HERBAL MEDICINES

Herbal medicine has evolved into a multi-billion-dollar global industry, driven by increasing demand for natural and plant-based therapeutic products. The integration of molecular technologies into the field of herbal medicine holds great promise for improving the accuracy, consistency, and safety of herbal products. As the demand for herbal remedies grows, so does the need for advanced tools to ensure the authenticity, quality, and efficacy of these products. Molecular markers, such as DNA barcoding, SNPs, SSRs, proteomics, and metabolomics, have already made substantial contributions to the field. However, their potential is far from fully realized, and the future of herbal medicine will likely be shaped by significant technological advancements and the integration of diverse molecular techniques. This chapter outlines the future potential of molecular markers in herbal medicine, focusing on advancements in molecular technologies, the integration of molecular markers with other technologies, and the challenges that need to be overcome.

15.5.1 Advancements in Molecular Technologies

The future of molecular markers in herbal medicine is inextricably tied to ongoing advancements in molecular technologies. These advancements are anticipated to enhance the accuracy, accessibility, and scalability of molecular identification, strengthen the reliability of identifying herbal species, and improve their medicinal properties and authentication techniques.

15.5.1.1 Next-Generation Sequencing

NGS, which involves high-throughput sequencing of DNA, is a revolutionary technology that allows for sequencing at a faster, more accurate, and cost-effective scale than traditional sequencing techniques (Mardis, 2008). NGS has already been applied to the field of DNA barcoding for plant identification and will continue to play a crucial role in the future of herbal medicine. NGS technologies provide the ability to sequence large portions of the genome, enabling the generation of comprehensive DNA barcodes that include not only the typical chloroplast markers like rbcL and matK but also regions of the nuclear genome such as the ITS regions, improving species identification (Newmaster *et al.*, 2013; Kress *et al.*, 2005). As NGS continues to develop, more detailed and global genomic databases for medicinal plants will be established, facilitating species identification and phylogenetic analyses on an unprecedented scale (Satam *et al.*, 2023). NGS enables detailed exploration of genetic variation within and between species, which is critical for assessing the genetic diversity of herbal species used in medicine. By understanding genetic diversity, it will be easier to select plants with optimal characteristics such as higher yields of active ingredients or resistance to diseases (Satam *et al.*, 2023).

15.5.1.2 CRISPR-Cas9 and Gene Editing

CRISPR-Cas9 is a ground-breaking gene-editing technology that has the potential to revolutionize plant breeding and genetic improvement in herbal medicine. With CRISPR-Cas9, specific genes within plants can be targeted, modified, or silenced with unprecedented precision. This could enable the development of medicinal plants with enhanced therapeutic properties, such as increased production of bioactive compounds like alkaloids or flavonoids (Ahmad, 2023).

The integration of CRISPR-Cas9 into the production of herbal medicines could lead to

- enhanced yield of active pharmaceutical ingredients in plants (e.g., increasing the concentration of ginsenosides in ginseng).
- resistance to pests and diseases, ensuring consistent quality and production of medicinal plants.
- the ability to modify plants for improved pharmacological profiles, making them more effective or safer for use in herbal formulations (Gan *et al.*, 2022).

15.5.1.3 Single-Cell Genomics

Another emerging technology with vast potential is single-cell genomics, which allows the sequencing of individual cells within a plant tissue. This technology could be used to:

- Identify specific cell types responsible for the synthesis of bioactive compounds in plants, offering insights into plant metabolism and its therapeutic potential.
- Investigate gene expression profiles of medicinal plants at the cellular level, providing a deeper understanding of how plants produce compounds that contribute to their medicinal properties (Cuperus, 2022).
- Trace genetic variation at the cellular level, potentially revealing how different plant tissues contribute to medicinal efficacy.

These advancements promise to deepen our understanding of plant biology and pave the way for more effective and sustainable production of herbal medicines.

15.5.2 Integration of Molecular Markers with Other Technologies

The future of molecular markers in herbal medicine will not only rely on advancements in molecular biology but also on the integration of these technologies with other scientific tools and methodologies. This integration will enable more comprehensive analysis and better management of herbal medicines.

15.5.2.1 Integration with AI and ML

The integration of AI and ML with molecular markers could dramatically improve the accuracy and efficiency of herbal medicine authentication and quality control. AI algorithms, trained on large datasets of molecular, chemical, and pharmacological data, can quickly analyse complex biological data and identify patterns or anomalies that may be difficult for humans to discern.

AI could assist in

- predicting the therapeutic efficacy of plant species based on genetic and metabolic data.
- enhancing authentication processes by automatically analysing DNA sequences, metabolite profiles, and protein expression patterns.
- optimizing herbal product formulations based on genomic and metabolomic data, ensuring maximum therapeutic benefit.

ML models could also be used to predict how plants will react under different growing conditions, which can help optimize cultivation practices for medicinal plants (Patil *et al.*, 2024).

15.5.2.2 Integration with Metabolomics and Proteomics

Metabolomics and proteomics are complementary fields that focus on the analysis of metabolites and proteins, respectively. When combined with molecular markers like DNA barcoding and SNPs, these technologies provide a holistic view of a plant's genetic makeup, biological activity, and chemical composition.

For instance:

- Metabolomics provides insights into the specific compounds that contribute to the therapeutic effects of a plant, such as alkaloids, terpenoids, or flavonoids. By integrating metabolomic data with genetic data, researchers can identify correlations between plant genetics and the presence of bioactive metabolites (Salem *et al.*, 2020).
- Proteomics can provide insights into the plant's protein profile, which is crucial for understanding its biochemical pathways and potential therapeutic mechanisms. This information can be used to optimize the cultivation of plants that produce high levels of beneficial proteins (Al-Amrani *et al.*, 2021).

The integration of these technologies will enable more precise and targeted approaches to herbal medicine, where both the genetic and biochemical profiles of plants can be manipulated to maximize their health benefits.

15.5.2.3 Integration with Smart Sensors for Real-Time Monitoring

As the demand for high-quality herbal medicines continues to rise, there will be increasing pressure to monitor the quality and purity of herbal products throughout their production cycle. The integration of smart sensors with molecular markers will enable real-time monitoring of the authenticity and quality of herbal products from cultivation to final product formulation.

These sensors could

- detect the presence of adulterants or contaminants in herbal products by analysing DNA, proteins, or metabolites.
- monitor environmental factors during cultivation that may affect the quality of bioactive compounds, such as light, temperature, and soil nutrients.
- provide continuous feedback to manufacturers, ensuring that herbal products meet quality and regulatory standards.

15.6 CHALLENGES TO OVERCOME

Despite the promise of molecular markers in herbal medicine, several challenges remain in realizing their full potential. These challenges span technical, regulatory, and cultural domains, and addressing them will require collaboration between researchers, industry stakeholders, and policymakers.

15.6.1 Lack of Standardization in Molecular Marker Use

One significant challenge is the lack of standardization in the application of molecular markers for herbal medicine. While DNA barcoding has become a widely accepted method for plant identification, discrepancies in the choice of molecular markers, sequence analysis, and data interpretation can lead to inconsistent results (Hyder *et al.*, 2024). There is a need for:

- Global standards for molecular markers used in herbal medicine identification and quality control.
- Reference databases containing DNA barcodes, proteomic, and metabolomic data from a wide range of medicinal plants to ensure consistency across regions.

Standardization will ensure that herbal products are consistently identified and meet regulatory requirements.

15.6.2 Cost and Accessibility of Advanced Technologies

While technologies like NGS and CRISPR-Cas9 hold great potential, they can be expensive and technically challenging to implement, particularly in resource-limited settings. The costs associated with high-throughput sequencing and genetic modification may limit their widespread adoption in herbal medicine production. To overcome this challenge,

- governments and industry leaders must collaborate to lower the cost of advanced molecular technologies.
- public-private partnerships can help support the integration of innovative technologies into the herbal medicine industry.
- training and capacity-building programs are necessary to ensure that researchers and practitioners have the skills to implement these advanced technologies effectively.

15.6.3 Regulatory Hurdles

The introduction of molecular technologies into herbal medicine also raises regulatory challenges. There is a need for clear and harmonized regulatory frameworks that account for the use of molecular markers in plant identification, quality control, and genetic modification. The regulatory landscape will need to address the following:

- The validation and acceptance of molecular marker-based methods for herbal product certification.
- The use of genetically modified plants in herbal medicine, which may be subject to varying regulatory standards across different countries (Ghimire *et al.*, 2023).

15.6.4 Cultural and Consumer Acceptance

While molecular technologies offer precision and reliability, their use in herbal medicine may face resistance from consumers and traditional practitioners. Many people still rely on traditional knowledge and practices in herbal medicine, and there may be scepticism about the role of molecular science in these practices. Overcoming this challenge will require

- Education and awareness campaigns to highlight the benefits of using molecular markers in ensuring the safety, quality, and authenticity of herbal products.

- Collaboration between traditional healers and molecular biologists to bridge the gap between ancient wisdom and modern scientific advancements.

The future of herbal medicine lies in the continued integration of molecular markers and advanced technologies. NGS, gene editing, metabolomics, and proteomics will enhance our understanding of the genetic and biochemical properties of medicinal plants, leading to the development of more effective and standardized herbal products. The integration of these technologies with tools like artificial intelligence, smart sensors, and real-time monitoring will further revolutionize the production, authentication, and quality control of herbal medicines.

However, to fully realize the potential of molecular markers, challenges such as standardization, cost, regulatory hurdles, and cultural acceptance must be addressed. Collaboration between scientists, industry leaders, regulators, and traditional practitioners will be essential in overcoming these barriers and ensuring that the future of herbal medicine is safe, effective, and sustainable.

15.7 FUTURE DIRECTIONS FOR HERBAL MEDICINE AND MOLECULAR MARKERS

Herbal medicine continues to gain popularity as a sustainable and effective form of healthcare. As the market for herbal products expands, ensuring the authenticity, quality, and efficacy of these medicines becomes increasingly critical. The application of molecular marker technologies is poised to address many of these challenges. The future of herbal medicine will likely be defined by the integration of advanced molecular tools, including DNA barcoding, genomic editing, metabolomics, and proteomics, with a focus on personalized medicine, sustainability, and quality control. This section explores key future directions in herbal medicine with a specific focus on molecular markers and their integration with emerging technologies.

15.7.1 Advancing Herbal Medicine Authentication with Molecular Markers

Ensuring the authenticity of herbal medicines remains one of the biggest challenges in the industry. Adulteration and misidentification of plant species can undermine both the quality and safety of herbal products. Molecular marker technologies, particularly DNA barcoding, hold immense potential in this area. The future of DNA barcoding in herbal medicine will involve the creation of more comprehensive global reference databases. These databases will serve as repositories for standardized DNA sequences that can be used to authenticate plants across different stages of the supply chain (Yang *et al.*, 2018; Kress *et al.*, 2005). As genomic technologies continue to evolve, databases will increasingly include genomic signatures that not only allow for species identification but also help to track plant genetic variation, which may affect the chemical composition and efficacy of the medicinal product. SNPs and DNA fingerprinting technologies are expected to become even more important for identifying closely related species or genetic variants that impact therapeutic efficacy. These markers can also help detect mislabelling and prevent the use of substitute species (Lee *et al.*, 2012). Recent advances in high-resolution SNP analysis are enabling more precise identification, even in complex herbal mixtures or extracts. Future approaches will likely combine multiple molecular markers such as DNA barcodes, microsatellites (SSRs), and SNPs in multimarker authentication systems to improve the accuracy of herbal medicine verification (Newmaster *et al.*, 2013). These systems will enhance quality control, ensuring that consumers receive the correct plant species in its proper form.

15.7.2 The Role of Genomic Editing in Enhancing Herbal Medicine Production

The advent of CRISPR-Cas9 and other gene-editing technologies offers tremendous potential for enhancing the production of bioactive compounds in medicinal plants. Through genetic modification, plants can be engineered to increase the yield of active ingredients or to improve their therapeutic efficacy.

15.7.2.1 Targeted Biosynthesis of Bioactive Compounds

Advances in genomic editing will enable the manipulation of plant metabolic pathways to optimize the production of important bioactive compounds. For example, the manipulation of the terpenoid pathway in mint species or ginsenoside biosynthesis in ginseng can increase the concentration of specific bioactive compounds, enhancing the plant's medicinal effects (Ghimire *et al.*, 2023).

15.7.2.2 Stress Resistance and Sustainability

By editing genes responsible for stress tolerance, plants can be made more resilient to climate change and pests. This will not only increase the yield of medicinal plants but also reduce the need for harmful pesticides and chemical fertilizers, which can negatively impact the environment and plant quality (Gan *et al.*, 2022). For instance, editing genes in *Cinnamomum camphora* to increase resistance to fungal infections can help safeguard production and meet growing demand.

15.7.3 Integration with Metabolomics and Proteomics for Comprehensive Profiling

To ensure that herbal products meet consistent quality standards, a comprehensive biochemical profile of the plants needs to be established. The integration of metabolomics and proteomics with molecular markers can provide valuable insights into the composition and biological activity of medicinal plants.

15.7.3.1 Metabolomics

The growing field of metabolomics enables detailed analysis of small molecules, such as secondary metabolites that often contribute to the medicinal properties of plants. Technologies like mass spectrometry (MS) and nuclear magnetic resonance (NMR) spectroscopy are allowing for a deeper understanding of metabolic pathways and how they correlate with genetic markers (Salem *et al.*, 2020). These tools will enhance the ability to identify new bioactive compounds and optimize their production through both traditional breeding and genetic engineering.

15.7.3.2 Proteomics

The application of proteomics to herbal medicine involves the study of plant proteins that mediate pharmacological effects. By using techniques like high-performance liquid chromatography (HPLC) and two-dimensional gel electrophoresis (2-DE), proteomics can identify and characterize plant proteins that interact with drug targets, helping to elucidate the mechanism of action of herbal products (Yan *et al.*, 2022). This will further refine the development of targeted herbal therapies with predictable outcomes.

15.7.3.3 Omics Integration

Combining genomic, transcriptomic, proteomic, and metabolomic data into a comprehensive approach will offer a comprehensive blueprint of medicinal plants' biochemical networks. This integration will guide the personalization of herbal treatments, leading to more effective and consistent therapeutic interventions (Li *et al.*, 2017).

15.7.4 Personalized Herbal Medicine Using Molecular Markers

One of the most exciting future directions for herbal medicine is the shift toward personalized therapies. By leveraging genetic markers, herbal treatments can be tailored to an individual's genetic makeup, metabolic pathways, and environmental factors, resulting in more effective and safer therapies.

15.7.4.1 Pharmacogenomics in Herbal Medicine

Molecular markers, particularly SNPs, will allow for pharmacogenomic profiling of individuals to predict how they will metabolize specific herbal medicines. For example, genetic variations in enzymes responsible for drug metabolism, like cytochrome P450, may influence how herbal compounds are processed in the body (Liu *et al.*, 2015). Personalized herbal treatments could thus be tailored to enhance therapeutic efficacy and minimize side effects.

15.7.4.2 Precision Herbal Formulations

Using genomic and metabolomic data, future herbal treatments can be customized for each individual's specific genetic profile, ensuring that herbal formulations are both safe and effective. This will enable more precise dosing of bioactive compounds, minimize adverse reactions, and improve therapeutic outcomes (Ghimire *et al.*, 2023).

15.7.4.3 Consumer Demand for Personalized Medicine

With the rise of consumer-driven healthcare, the demand for personalized herbal medicine will increase. Molecular markers will empower consumers and healthcare providers to make informed decisions about which herbal products are best suited to their genetic profiles and health conditions (Li *et al.*, 2015).

15.7.5 Sustainability and Environmental Impact of Herbal Medicine

The increasing demand for medicinal plants has raised concerns about the sustainability of herbal medicine production. Molecular marker technologies can be used to support sustainable practices in herbal medicine production, including the conservation of endangered species and the optimization of cultivation methods.

15.7.5.1 Conservation of Endangered Species

Many plants used in herbal medicine are threatened by overharvesting and habitat destruction. Molecular markers can be used to identify genetically distinct populations that can be used for conservation and restoration programs. Additionally, the application of clonal propagation techniques and seed banking will help conserve valuable plant species (Tasheva and Kosturkova, 2013).

15.7.5.2 Sustainable Cultivation

The use of molecular breeding techniques and genetic selection will allow for more efficient and resilient herbal plants that require fewer pesticides and fertilizers, contributing to more sustainable production practices (Gan *et al.*, 2022). This will also reduce the environmental footprint of herbal medicine production and improve the long-term viability of medicinal plant species.

15.7.5.3 Reducing the Environmental Footprint

Research into sustainable agriculture methods, including biocontrol and organic farming, supported by molecular tools, will lead to the development of cultivation systems that preserve biodiversity while meeting the growing global demand for herbal products (Ghimire *et al.*, 2023).

15.7.6 Regulatory Harmonization and Global Standardization

As herbal medicine becomes increasingly globalized, regulatory harmonization and standardization of molecular marker methods will be crucial. Establishing universal guidelines and standards for the use of molecular markers in herbal medicine will promote international trade, improve consumer safety, and protect plant biodiversity. Organizations such as the WHO, the International Union for Conservation of Nature (IUCN), and the Food and Agriculture Organization (FAO) will play an instrumental role in establishing global standards for molecular marker applications. These frameworks will ensure that herbal products meet quality assurance and safety standards. Standardization of Analytical Methods: Standardized methodologies for molecular marker use in plant authentication, genetic testing, and quality control will become essential for ensuring the authenticity and quality of herbal products. This includes DNA barcoding protocols, SNP genotyping, and the establishment of global molecular databases that ensure consistent data sharing across borders (Hyder *et al.*, 2024).

In conclusion, the future of herbal medicine will be shaped by the continuous evolution and integration of molecular marker technologies, allowing for more precise identification, improved quality control, sustainable production, and personalized herbal therapies. As genetic tools such as CRISPR-Cas9, metabolomics, and proteomics continue to develop, they will enable the herbal medicine industry to meet the growing global demand for high-quality, effective, and safe plant-based therapies. By overcoming challenges related to standardization, cost, regulation, and cultural acceptance, the use of molecular markers will revolutionize the way herbal products are cultivated, developed, and consumed worldwide.

15.8 CONCLUSION

The journey of molecular markers in herbal medicine has witnessed significant advancements, from their early exploration to their current integration into various facets of plant science and therapeutic applications. The past of molecular markers in herbal medicine was characterized by the foundational development of techniques like DNA barcoding and genetic fingerprinting, which provided a reliable method for the identification and authentication of plant species. Early research focused on overcoming the challenges of adulteration and misidentification, which posed significant risks to the safety, efficacy, and integrity of herbal products. The establishment of DNA databases and the use of molecular markers to discern species variation laid the groundwork for a more standardized and scientifically validated approach to herbal medicine.

In the present, molecular markers have become essential tools for addressing critical issues in the herbal medicine industry. Their application has expanded from authentication to include quality control, genetic diversity studies, and the enhancement of pharmacological knowledge. Molecular tools such as SNP analysis, metabolomics, and proteomics have enabled a more precise understanding of the chemical composition and mechanisms of action of medicinal plants, ensuring that they meet consistent quality standards and are safer for consumers. Moreover, the integration of these molecular approaches into personalized medicine has opened new frontiers for tailoring herbal treatments to an individual's genetic profile and specific health needs. The current landscape of herbal medicine benefits from the standardization and traceability afforded by molecular markers, which facilitate regulatory compliance, product safety, and consumer confidence.

Looking toward the future, the role of molecular markers in herbal medicine will continue to expand in scope and sophistication. Advancements in genomic editing, such as CRISPR-Cas9, promise to enhance the production of bioactive compounds in medicinal plants, potentially increasing their efficacy and yield while reducing dependence on synthetic chemicals. Omics technologies, including metabolomics and proteomics, will offer more comprehensive and integrative insights into plant biology, enabling the creation of more targeted, effective, and safe herbal formulations. As the demand for herbal products grows globally, molecular markers will also play a pivotal role in ensuring the sustainability of plant resources by supporting conservation efforts and the sustainable cultivation of medicinal plants.

The future of herbal medicine will likely be shaped by the ongoing integration of advanced molecular technologies, with an emphasis on personalized treatments, environmental sustainability, and global regulatory harmonization. By enhancing genetic traceability, improving bioactive compound production, and facilitating standardized practices, molecular markers will be key drivers of innovation in the herbal medicine industry. As we move forward, the convergence of molecular biology, pharmacology, and data-driven approaches will continue to revolutionize how herbal medicines are developed, verified, and utilized, ensuring their place as both reliable therapeutic agents and sustainable resources in the global healthcare landscape.

In conclusion, the journey of molecular markers in herbal medicine has transformed the way we approach plant-based therapies. From the early stages of identification to the current focus on quality control, sustainability, and personalized treatments, molecular markers have enhanced the scientific rigour and efficacy of herbal medicine. As these technologies continue to evolve, they will ensure the ongoing relevance and reliability of herbal products in modern healthcare, opening new doors for innovation, safety, and personalized care in the years to come.

REFERENCES

Ahmad, M. (2023). Plant breeding advancements with "CRISPR-Cas" genome editing technologies will assist future food security. *Front Plant Sci.* 14, 1133036. doi: 10.3389/fpls.2023.1133036

Al-Amrani, S., Al-Jabri, Z., Al-Zaabi, A., Alshekaili, J., & Al-Khabori, M. (2021). Proteomics: concepts and applications in human medicine. *World J Biol Chem.* 12(5), 57–69. doi: 10.4331/wjbc.v12.i5.57

Al-Snafi, A. E. (2016). Medicinal plants used in the treatment of cardiovascular diseases. *J Pharm Sci Res.* 8(12), 1299–1310.

Benzie, I. F. F., & Wachtel-Galor, S. (Eds.). (2011). *Herbal Medicine: Biomolecular and Clinical Aspects* (2nd ed.). CRC Press. https://doi.org/10.1201/b10787

Chen, S., Yao, H., Han, J., Liu, C., Song, J., Shi, L., ... & Leon, C. (2010). Validation of the ITS2 region as a novel DNA barcode for identifying medicinal plant species. *PLoS ONE*, 5(1), e8613.

Cuperus, J. T. (2022). Single-cell genomics in plants: current state, future directions, and hurdles to overcome. *Plant Physiol.* 188(2), 749–755. doi: 10.1093/plphys/kiab478

Fugh-Berman, A. (2000). Herb-drug interactions. *Lancet.* 355(9198), 134–138. doi: 10.1016/S0140-6736(99)06457-0. Erratum in: Lancet 2000 Mar 18;355(9208), 1020. PMID: 10675182.

Gan, W. C., & Ling, A. P. K. (2022). CRISPR/Cas9 in plant biotechnology: applications and challenges. *BioTechnologia (Pozn).* 103(1), 81–93. doi: 10.5114/bta.2022.113919

Gao, L., Xu, W., Xin, T., & Song, J. (2023). Application of third-generation sequencing to herbal genomics. *Front Plant Sci.* 14, 1124536. doi: 10.3389/fpls.2023.1124536

Ghimire, B. K., Yu, C. Y., Kim, W.-R., Moon, H.-S., Lee, J., Kim, S. H., & Chung, I. M. (2023). Assessment of benefits and risk of genetically modified plants and products: current controversies and perspective. *Sustainability.* 15(2), 1722. doi: 10.3390/su15021722

Goyal, N., Nag, A., Samarth, R. R., Upadhyay, A., & Singh, K. (2021) Identification and characterization of simple sequence repeats (SSR) markers associated with Downy Mildew resistance locus "RPV1" in grapes. *Adv Biosci Biotechnol.* 12, 371–387. doi: 10.4236/abb.2021.1211024

Hebert, P. D. N., et al. (2003). Biological identification through DNA barcodes. *Proc Royal Soc B: Biol Sci.* 270(1512), 313–321.

Hu, Z., Yang, X., Ho, P. C., Chan, S. Y., Heng, P. W., Chan, E., Duan, W., Koh, H. L., & Zhou, S. (2005). Herb-drug interactions: a literature review. *Drugs.* 65(9), 1239–1282. doi: 10.2165/00003495-200565090-00005. PMID: 15916450.

Husen, A. (2022). *Herbs, Shrubs and Trees of Potential Medicinal Benefits.* CRC Press. doi: 10.1201/9781003205067

Husen, A. (2023a). *Exploring Poisonous Plants: Medicinal Values, Toxicity Response and Other Potential Uses.* CRC Press. doi: 10.1201/b23017

Husen, A. (2023b). *Antidiabetic Medicinal Plants and Herbal Treatments.* CRC Press. doi: 10.1201/b23347

Husen, A. (2024a). *Antimalarial Medicinal Plants.* Taylor & Francis Group. doi: 10.1201/9781003378396

Husen, A. (2024b). *Medicinal Spices and Condiment Crops.* Taylor & Francis Group. doi: 10.1201/9781003387046

Husen, H., & Pant, M. (2024a). *Exploring Medicinal Orchids.* Taylor & Francis Group. doi: 10.1201/b23375

Husen, H., & Pant, M. (2024b). *Tissue Culture Techniques and Medicinal Plants: Enhancing Propagation and Production*. Taylor & Francis Group. doi: 10.1201/b23374

Hyder, Z., Hafeez Rizwani, G., Shareef, H., Azhar, I., & Zehra, M. (2024). Authentication of important medicinal herbal species through DNA-based molecular characterization. *Saudi J Biol Sci*. 31(6), 103985. doi:10.1016/j.sjbs.2024.103985

Jain, A., Jain, P., Mathur, S., & Parihar, D. K. (2024). Curcuma species DNA fingerprinting of wild and cultivated genotypes from different agroclimatic zones. *Pharmacol Res: Modern Chin Med*. 12, 100474. doi: 10.1016/j.prmcm.2024.100474

Jayesh, P., Harish, P., Ashish, P., Harshita, C., Rashmi, D., & Sarode, S. (2024). Advancements and challenges in the integration of artificial intelligence with herbal medicine research. *Int J Pharm Sci*. 2(4), 973–984. doi: 10.5281/zenodo.11060057

Kress, W. J., Wurdack, K. J., Zimmer, E. A., Weigt, L. A., & Janzen, D. H. (2005). Use of DNA barcodes to identify flowering plants. *Proc Natl Acad Sci USA*, 102, 8369–8374.

Koroma, L., Alpha, M., Kpaka, J., & Kpaka, A. (2022). Organoleptic evaluation and physiochemical characteristics of powdered plant organs of the traditional medicinal plant *Caloncoba echinata*. *Int J Innov Eng Int Sci Netw*. 2(3), 14–22.

Kumar, S., Sharma, S., & Chattopadhyay, S. K. (2013). The potential health benefit of polyisoprenylated benzophenones from Garcinia and related genera: ethnobotanical and therapeutic importance. *Fitoterapia*, 89, 86–125.

Lee, O. R., Kim, M. K., & Yang, D. C. (2012). Authentication of medicinal plants by SNP-based multiplex PCR. *Methods Mol Biol*. 862, 135–147. doi: 10.1007/978-1-61779-609-8_11. PMID: 22419494.

Li, S., Han, Q., Qiao, C., Song, J., Lung Cheng, C., & Xu, H. (2008). Chemical markers for the quality control of herbal medicines: an overview. *Chin Med*. 3, 7. doi: 10.1186/1749-8546-3-7

Li, X., Yang, Y., Henry, R. J., Rossetto, M., Wang, Y., & Chen, S. (2015). Plant DNA barcoding: from gene to genome. *Biol Rev*. 90, 157–166.

Li, X., et al. (2017). Use of SNP markers to distinguish closely related plant species in herbal medicine. *Phytochemical Analysis. Nazia Nazar, Akanksha Saxena, Anu Sebastian, Adrian Slater, Velusamy Sundaresan, Tiziana Sgamma. Integrating DNA Barcoding Within an Orthogonal Approach for Herbal Product Authentication: A Narrative Review, Phytochemical Analysis, 2024, doi: 10.1002/pca.3466

Liberty, J. T., Lin, H., Kucha, C., Sun, S., Alsalman, F. B. (2025). Innovative approaches to food traceability with DNA barcoding: beyond traditional labels and certifications. *Ecol Genet Genom*. 34, 100317. ISSN 2405-9854. doi: 10.1016/j.egg.2024.100317

Little, D. P., & Jeanson, M. L. (2013). DNA barcode authentication of saw palmetto herbal dietary supplements. *Sci Rep*. 3, 3518. doi: 10.1038/srep03518. PMID: 24343362; PMCID: PMC3865462.

Liu, M. Z., Zhang, Y. L., Zeng, M. Z., He, F. Z., Luo, Z. Y., Luo, J. Q., Wen, J. G., Chen, X. P., Zhou, H. H., & Zhang, W. (2015). Pharmacogenomics and herb-drug interactions: merge of future and tradition. *Evid Based Complement Alternat Med*. 2015, 321091. doi: 10.1155/2015/321091. PMID: 25821484; PMCID: PMC4363646.

Liu, Y., Lu, S., Liu, K. et al. (2019). Proteomics: a powerful tool to study plant responses to biotic stress. *Plant Methods*. 15, 135. doi: 10.1186/s13007-019-0515-8

Liu, Z.-W., Gao, Y.-Z., & Zhou, J. (2019). Molecular authentication of the medicinal species of ligusticum (ligustici rhizoma et Radix, "Gao-ben") by integrating non-coding internal transcribed spacer 2 (ITS2) and its secondary structure. *Front Plant Sci*. 10, 429. doi: 10.3389/fpls.2019.00429

Maddala, V. K. S., Singh, S., & Bandi, J. (2024). Soil properties influencing nutrient signaling in plant soil systems. In *Essential Minerals in Plant-Soil Systems* (pp. 373–393). Elsevier.

Mardis, E. R. (2008). Next-generation DNA sequencing methods. *Annu Rev Genomics Hum Genet*. 9, 387–402. doi: 10.1146/annurev.genom.9.081307.164359. PMID: 18576944.

Moon, B. C., Lee, Y. M., Kim, W. J. et al. (2016). Development of molecular markers for authentication of the medicinal plant species Patrinia by random amplified polymorphic DNA (RAPD) analysis and multiplex-PCR. *Hortic. Environ. Biotechnol*. 57, 182–190. doi: 10.1007/s13580-016-0064-2

Mullis, K. B., & Faloona, F. A. (1987). Specific synthesis of DNA in vitro via a polymerase-catalyzed chain reaction. *Methods Enzymol*. 155, 335–350. doi: 10.1016/0076-6879(87)55023-6. PMID: 3431465.

Newmaster, S. G., Grguric, M., Shanmughanandhan, D., Ramalingam, S., & Ragupathy, S. (2013). DNA barcoding detects contamination and substitution in North American herbal products. *BMC Med*. 11, 222. doi: 10.1186/1741-7015-11-22

Pant, M., & Husen, A. (2025). *Exploring Medicinal Bamboos.* Taylor & Francis Group. doi: 10.1201/9781003485933

Ray, S., & Satya, P. (2014). Next generation sequencing technologies for next generation plant breeding. *Front Plant Sci.* 5, 367. www.frontiersin.org/journals/plant-science/articles/10.3389/fpls.2014.00367

Salem, M. A., Perez de Souza, L., Serag, A., et al. (2020). Metabolomics in the context of plant natural products research: from sample preparation to metabolite analysis. *Metabolites.* 10(1), 37. doi: 10.3390/metabo10010037

Sao, P., Parveen, R., Khattri, A., Sharma, S., Tiwari, N., & Singh, S. (2022). Understanding the metabolomics of medicinal plants under environmental pollution: challenges and opportunities. In *Environmental Pollution and Medicinal Plants* (pp.193–205). Taylor & Francis Group.

Satam, H., Joshi, K., Mangrolia, U., et al. (2023). Next-generation sequencing technology. Current Trends and Advancements [published correction appears in Biology (Basel). 2024 Apr 24;13(5):286. doi: 10.3390/biology13050286]. *Biology.* 12(7), 997. doi: 10.3390/biology12070997.

Shinde, V. M., Dhalwal, K., Mahadik, K. R., Joshi, K. S., & Patwardhan, B. (2007). RAPD analysis for determination of components in herbal medicine. *Evid-Based Complement Altern Med*, 4, 21–23.

Smith, L. M., Sanders, J. Z., Kaiser, R. J., Hughes, P., Dodd, C., Connell, C. R., Heiner, C., Kent, S. B., & Hood, L. E. (1986). Fluorescence detection in automated DNA sequence analysis. *Nature.* 321(6071), 674–679. doi: 10.1038/321674a0. PMID: 3713851.

Tasheva, K., & Kosturkova, G. (2013). *Role of Biotechnology for Protection of Endangered Medicinal Plants.* IntechOpen. doi: 10.5772/55024

Um, J. Y., Chung, H. S., Kim, M. S., Na, H. J., Kwon, H. J., Kim, J. J., Lee, K. M., Lee, S. J., Lim, J. P., Do, K. R., Hwang, W. J., Lyu, Y. S., An, N. H., Kim, H. M. (2001). Molecular authentication of *Panax ginseng* species by RAPD analysis and PCR-RFLP. *Biol Pharm Bull.* 24(8), 872–875. doi: 10.1248/bpb.24.872. PMID: 11510476.

Uncu, A. T., Uncu, A. O., Frary, A., & Doganlar, S. (2015). Authentication of botanical origin in herbal teas by plastid noncoding DNA length polymorphisms. *J Agric Food Chem.* 63(25), 5920–5929. doi: 10.1021/acs.jafc.5b01255. PMID: 26054647.

Wang, H., Chen, Y., Wang, L., Liu, Q., Yang, S., & Wang, C. (2023). Advancing herbal medicine: enhancing product quality and safety through robust quality control practices. *Front Pharmacol.* 14, 1265178. doi: 10.3389/fphar.2023.1265178

WHO. (2008). *Traditional Medicine Strategy 2002–2005.* World Health Organization. www.who.int/publications/i/item/WHO-EDM-TRM-2002.1

Williams, J. G., et al. (1990). DNA polymorphisms amplified by arbitrary primers are useful as genetic markers. *Nucl Acids Res.* 18(22), 6531–6535. doi: 10.1093/nar/18.22.6531

Yan, S., Bhawal, R., Yin, Z. et al. (2022). Recent advances in proteomics and metabolomics in plants. *Mol Horticulture* 2, 17. doi: 10.1186/s43897-022-00038-9

Yang, F., Ding, F., Chen, H., et al. (2018). DNA barcoding for the identification and authentication of animal species in traditional medicine. *Evid Based Complement Alternat Med.* 2018, 5160254. doi: 10.1155/2018/5160254

Yang, H., Tao, Y., Zheng, Z., Li, C., Sweetingham, M., and Howieson, J. (2012). Application of next-generation sequencing for rapid marker development in molecular plant breeding: a case study on anthracnose disease resistance in Lupinus angustifolius L. *BMC Genomics.* 13, 318. doi: 10.1186/1471-2164-13-318

Yuan, H., Ma, Q., Ye, L., Piao, G. (2016). Traditional medicine and modern medicine from natural products. *Molecules.* 21(5), 559. doi: 10.3390/molecules21050559. PMID: 27136524; PMCID: PMC6273146.

Yu, J., Wu, X., Liu, C., Newmaster, S., Ragupathy, S., & John Kress, W. (2021). Progress in the use of DNA barcodes in the identification and classification of medicinal plants. *Ecotoxicol Environ Safe.* 208, 111691. ISSN 0147–6513. doi: 10.1016/j.ecoenv.2020.111691

Zaharleva, M., Santoni, S., & David, J. (2001). Use of RFLP markers to study genetic diversity and to build a core-collection of the wild wheat relative Ae. geniculata Roth (= Ae. ovata L.) [Utilisation de marqueurs RFLP pour l'étude de la diversité génétique et l'établissement d'une core collection d'Ae. geniculata Roth (= Ae. ovata L.)]. *Genet Sel Evol.* 33(Suppl 1), S269–S288. doi: 10.1186/BF03500884

Zhang, D., Yang, Y., Jiang, B., Duan, L., & Zhou, N. (2014). How to correctly identify herbal materials in market? A case study based on DNA barcodes. *Afr J Trad Complement Altern Med*, 11(6), 66–76. doi: 10.21010/ajtcam.v11i6.7

Zhang, Y., & Stommel, J. R. (2000). RAPD and AFLP tagging and mapping of Beta (B) and Beta modifier (MoB), two genes which influence ßcarotene accumulatio in fruit of tomato (*Lycopersicon esculentum* Mill.). *Theor Appl Genet*. 100, 368–375.

Zhou, J., Cui, Y., Chen, X., Li, Y., Xu, Z., Duan, B., Li, Y., Song, J., Yao, H. (2018). Complete chloroplast genomes of *Papaver rhoeas* and *Papaver orientale*: molecular structures, comparative analysis, and phylogenetic analysis. *Molecules*, 23, 437.

Index

A
Abiotic stress, 9, 116–118, 151–153
Adaptation, plant, 2, 7, 183–185
AFLP (Amplified Fragment Length Polymorphism), 1, 3, 5–7, 9, 50–52
  applications in diversity analysis, 6–7
  advantages and limitations, 6
Agrobacterium rhizogenes, 161
Alkaloids, 7, 161–163
Alleles, 4, 6, 78–80
Amplification (PCR-based markers), 1, 3, 5
Andrographis paniculata, 96, 98
Annotation, genome, 151–154
Antimicrobial compounds, 8, 183
Arabidopsis thaliana, 9, 85, 96, 145, 155
Artemisia annua, 2–3, 172–182
  absinthium, 175, 178
  afra, 175, 178
  artemisinin biosynthesis, 3, 173
  EST-SSR markers, 172–180
  vulgaris, 175
Artificial Intelligence (AI), 45, 70–71, 122, 125, 184, 195

B
Bioactive compounds, 7–8, 91–94, 183–185
  biosynthesis pathways, 7, 91–95
  genetic regulation, 78–82
Bioinformatics, 10–11, 15–16, 38–45, 69–71, 84, 92, 103, 116–119, 124, 152
Bioreactors, 101, 106, 124, 136
Biotechnology, 15–30, 91–110
  applications in medicinal plants, 15–18
  regulatory frameworks, 18–25
Breeding programmes, 2, 7, 9, 183–185
  marker-assisted selection (*see* MAS)

C
Camelina sativa, 100
Cannabis research, 166–171
  cannabinoids, 166, 168
  molecular markers in, 166–168
Catharanthus roseus, 8, 16, 104, 117, 154
Chemotyping, 116–125, 166
  vs. genotyping, 116–118
Clonal fidelity, 129–140
  assessment using markers, 129–133
Conservation genetics, 2, 7, 50–57
  in situ conservation, 8
  ex situ conservation, 8
CRISPR/Cas9, 8, 15, 33, 91, 151–160, 183
  genome editing, 151–155
Cryopreservation, 8, 16, 50, 54, 130, 136
Curcuma longa, 3, 8, 17, 41, 53, 98, 121, 133–134, 189

D
DNA barcoding, 22–24, 41, 51–52, 117–118, 183–193
DNA markers, 5–6
  applications, 6–7
  for barcoding, 22
  types, 5–6
Diversity analysis, 2, 50–55
  Dracocephalum moldavica, 50
Drug discovery, 7, 183–193

E
Echinacea purpurea, 134, 185, 188
EST-SSR markers, 172–180
  applications, 173–175
Evolutionary biology, 2–3

F
Flavonoids, 7
Functional genomics, 6–7

G
Gene editing, 151–160
  CRISPR-based tools, 151–155
Gene expression, 6, 141–148
  transcriptional markers, 141–145
Genetic diversity, 2, 7–8, 50–55
  assessment methods, 7
Genetic improvement, 33–48
  molecular *vs.* conventional markers, 33–38
Genetic markers
  definition, 1
  types, 5–6
Genotyping, 116–118
  by sequencing, 45, 62, 71, 81, 86
Genome mapping, 3, 9
Ginkgo biloba, 52, 122
Glycine max, 6
Glycyrrhiza species, 96, 98
GWAS (Genome-Wide Association Studies), 3, 8

H
Hairy root culture, 161–165
  applications in metabolite production, 162–164
Herbal formulations, 15–30
Hypericum perforatum, 98

I
Indels (insertions/deletions), 1
In vitro culture, 129–140, 141–148
  organogenesis, 141–145
ISSR markers, 4, 7

K

KASP markers, 5

L

Litchi chinensis, 52
Lupinus species, 85

M

Marker-assisted selection (MAS), 1–2, 8–9
 applications in breeding, 2
Metabolic engineering, 91, 151, 183
Metabolomics, 7, 91–110
 integration with genomics, 91–94
Metabolic engineering, 91, 151, 183
Microsatellites, *see* SSR markers
Micropropagation, 16, 129–135
Molecular markers
 advantages, 9
 characteristics, 4–5
 evolution, 3–4
 limitations, 9–10
 types, 5–6
Multiomics, 71

N

Nanobiotechnology, 17–19
Nicotiana tabacum, 85, 100
Next-generation sequencing (NGS), 1, 3, 9
 applications in genomics, 3

O

Omics approaches, 91–110
Organogenesis, 141–148

P

Panax species, 17, 41, 118, 185, 189
PCR (Polymerase Chain Reaction), 1, 3, 5
Pharmacogenomics, 123, 197
Phenotypic markers, 1–2
Phylogenetics, 3
Plant biotechnology, 15–30
Plant growth regulators, 101, 130, 136, 141, 143–144
Population genetics, 2–3
Precision breeding, 9
Proteomics, 6–7

Q

QTL (Quantitative Trait Loci), 3, 78–90
 mapping, 3, 78–82
 secondary metabolism, 78–85

R

RAPD markers, 1, 3, 5–6, 7
Rauvolfia serpentina, 2, 98, 117, 154
Reference genomes, 10
Regulatory frameworks, 18–25
Reproducibility, 4
Restriction fragment length polymorphism (RFLP), 3, 5, 34, 36, 51, 59, 96, 130, 132–133, 185
Rheum officinale, 52
Ri plasmid, 161–164

S

Salvia miltiorrhiza, 7, 84, 98–101, 123, 155, 189
Secondary metabolites, 7, 78–85, 91–110
 genetic control, 78–82
Seed banking, 8
SNP (Single Nucleotide Polymorphism), 1, 3, 4, 6
 applications, 6
SSR markers (Simple Sequence Repeats), 1, 3, 4, 6
 advantages, 6
Somaclonal variations, 50, 136
Species identification, 50–51, 117, 134, 184–192, 195
Sustainability, 183–185
Standardization, 15, 19–20, 23–25, 119–121, 156, 176, 184, 191, 194–197
Synthetic biology, 15–16, 80, 86, 92, 106, 116, 123, 161

T

Terpenoids, 7
Tissue culture, 129–140
 applications in medicinal plants, 129–132
Transcriptomics, 6–7
Transcriptional markers, 141–148

V

Variation, genetic, 1–2
 sources, 1
Vitis vinifera, 6, 86

W

Withania somnifera, 2, 7–8